MICHAEL COLLINS

Carrying the Fire

Michael Collins was born in Rome in 1930. After graduating from the U.S. Military Academy, he entered the newly independent Air Force, becoming a fighter pilot and an experimental test pilot.

He was one of the third group of astronauts named by NASA in 1963. On his first mission, Gemini 10, he set a world altitude record and became the nation's third spacewalker. His second flight was as command module pilot of the historic Apollo 11 mission to the moon in July 1969.

He is a retired major general in the U.S. Air Force Reserve and has received numerous decorations and awards, including the Presidential Medal of Freedom and the Collier Trophy. He holds honorary degrees from six colleges and universities.

Michael Collins is retired and lives in South Florida. He spends his time these days fishing and painting watercolors.

Carrying the Fire

Carrying the Fire

AN ASTRONAUT'S JOURNEYS

MICHAEL COLLINS

Farrar, Straus and Giroux   New York

Farrar, Straus and Giroux
120 Broadway, New York 10271

All photographs courtesy of NASA.

Library of Congress Control Number: 2009922238
Paperback ISBN: 978-0-374-53776-0

Designed by Kay Lee

Our books may be purchased in bulk for promotional, educational, or business
use. Please contact your local bookseller or the Macmillan Corporate and
Premium Sales Department at 1-800-221-7945, extension 5442, or by e-mail at
MacmillanSpecialMarkets@macmillan.com.

www.fsgbooks.com
www.twitter.com/fsgbooks • www.facebook.com/fsgbooks

10 9 8 7 6 5 4 3 2 1

TO PATRICIA
WITH ADMIRATION AND LOVE

I would like to thank Ferdinand E. Ruge, Master of English at St. Albans School in Washington, who taught me to write a sentence; Roger Straus III, my amiable and supportive editor; and Terry Pietroski, who ruined many of her weekends converting my chicken scratchings into neat pages of type.

FOREWORD

Charles A. Lindbergh

Perceptive, clear, and comprehensive, targeted on life's first expedition to the moon's surface, this book combines a contemplative mind and poet's eye with the essentially practical approach of a participating astronaut. Here is a fascinating autobiographical account of one of civilization's greatest accomplishments and one of man's greatest adventures. It will be read and reread as long as records last.

In these chapters, the human and technical interweave to give an extraordinary sense of NASA, its astronauts and spatial ventures. You realize the almost infinite problems involved in operating spacecraft, the intricate organization and training required to solve them, the ambitions, the frustrations, the hazards, duties, and relationships of a successful astronaut's career. It is easy to so identify with the story that you become an astronaut yourself.

Reading Michael Collins's manuscript of *Carrying the Fire* took me back in memory to my first meeting with Robert Goddard, "the father of space flight." On the porch of his home at Worcester, Massachusetts, in 1929, I listened to Goddard talk about his liquid rockets and his dreams of spatial exploration. He felt quite certain, he told me, that he would be able to build a rocket capable of carrying scientific instruments to altitudes approaching one hundred miles; and that from these altitudes, above atmospheric interference, valuable observations could be made. He thought it would even be possible to build a multistage rocket that could reach the moon; but "it might cost a million dollars," he added, with an air of dismissing the idea from consideration.

At that time, none of Goddard's rockets had reached an altitude of one mile, to say nothing of one hundred. His estimated budget for a year's experimental work was $25,000, including laboratory, launching tower, equipment, salaries, and transportation. Now I was reading about an accomplished voyage to the moon, and the United States was spending millions of dollars every day on the development of rockets.

Carrying the Fire roots into Goddard's pioneering work, and branches in modern technology's accomplishments. You must have witnessed an Apollo ascent to know the title's aptness. When a countdown at Cape Canaveral touches zero, you think the giant rocket will be consumed by bursting clouds of flame and thunder. Where I stood with astronauts three miles away from Apollo 11's pad, my chest was beaten and the ground shook as though bombs were falling nearby. Then a flame arose, left the ground behind— higher—faster—a meteor streaking through the sky. It seemed impossible for life to exist while carrying that ball of fire. Yet it did exist, as I saw later on a television screen that portrayed weightless men, countless instruments, and planet earth through a window.

The miracle of television kept me in contact with the Apollo 11 expedition until images formed by cameras a quarter million miles away showed life walking on a previously lifeless moon: pressure-suited, tank-headed beings, cleated footprints, a vehicle in the background as startling as an artist's weirdest concept of a "flying saucer."

How isolated, how lonely those two space supermen appeared! But they had each other for companionship; and through television, they were held in the thoughts of viewing millions of men and women. To be really isolated, to fully experience loneliness, you must be alone. From Armstrong's and Aldrin's spectacular movements, my mind shifted to Collins's lunar orbiting. Relatively inactive and unwatched, he had time for contemplation, time to study both the nearby surface of the moon and the distant moon-like world. Here was human awareness floating through universal reaches, attached to our earth by such tenuous bonds as radio waves and star sights. A minor functional error would leave it floating forever in the space from which, ancestrally, it came.

Only once before had I felt such extension as when I thought of Astronaut Collins. That was over the Atlantic Ocean on my nonstop flight with the *Spirit of St. Louis*. I had been without sleep for more than two days and two nights, and my awareness seemed to be abandoning my body to expand on stellar scales. There were moments when I seemed so disconnected from the world, my plane, my mind and heartbeat that they were completely unessential to my new existence.

Experiences of that flight combined with those of ensuing life have caused me to value all human accomplishments by their effect on the intangible quality we name "awareness." Scientific and technological developments have had great effect on our awareness, as the accounts of astronomers, nuclear physicists, aviators, and astronauts have shown, and as *Carrying the Fire* verifies.

After evolutionary epochs, life in the human species became aware of its awareness. Then in a comparatively short period of time, using such devices as lenses and spacecraft for assistance, man has extended his awareness outwardly through the universe and inwardly through the atom. But the assistance we receive from technological devices is countered by restrictions that they place upon us, and the same scientific knowledge that constructs our spacecraft informs us of apparently unsurmountable physical limits. We find the speed of light and the vastness of space to be incompatible with biological time. We begin to realize that a point arrives after which the distraction and destruction caused by technological enterprise reduce man's awareness. We become apprehensive of the direction in which our twentieth-century heading leads. Is it toward an affluent and spiritual utopia or a bleak dead end? Possibly we can judge best from outer areas of our technological penetrations.

As Collins takes us to the periphery of man's penetration into space, he also takes us to the periphery of human evolution. From momentary spatial limits, we look back on ourselves with new perspective. We see life compressed on a shrunken orb at the same time we feel awareness expanded to fantastically great dimensions. Can we harmonize the two? What does the future hold for man?

Again I return through a half century of time. When my interest in rockets began, I was faced with a similar question. Alone in my survey plane, in 1928, flying over the transcontinental air route between New York and Los Angeles, I had hours for contemplation. Aviation's success was certain, with faster, bigger, and more efficient aircraft coming. But what lay beyond our conquest of the air? What did the future hold? There seemed to be nothing but space. Man had used hulls to travel over water, wheels to travel over land, wings to travel through air. Was it now remotely possible that he could use rockets to travel through space?

It turned out to be possible, and it was done in a fraction of a lifetime. A superhuman accomplishment, it still seems to me, for I was born and schooled under a pre-Goddard era's apparent limitations. I felt a superhuman being in these chapters when I accompanied Collins, Armstrong, and Aldrin on their moon flight. But being human, I returned with these human astronauts to earth, and now I find myself asking: what lies for man beyond solar-system travel? What vehicle can be conceived beyond the rocket?

Of course, there is a future for us in space, just as there was in air. We can put manned stations on the moon if we want to. We probably can set foot on Mars. Such adventures continue to be tantalizing. But whether we travel in a jet transport or in a spacecraft, we still look down, or back to life on the crowded surface of the earth as both our source and our destination. Is it remotely possible that we are approaching a stage in evolution when we can discover how to separate ourselves entirely from earthly life, to abandon our physical frameworks in order to extend both inwardly and outwardly through limitless dimensions of awareness? In future universal explorations, may we have no need for vehicles or matter? Is this the adventure opening to man beyond travel through solar-system space?

Our future remains as potent as it was in Goddard's time. Past accomplishments found thoughts of greater conquests. As Goddard's dreams resulted in the spacecraft that today's astronauts are crewing, advancing man may discover that thought and reality transpose like energy and matter.

We can now determine to a large degree our physical, mental, spiritual, and environmental evolution. Limitless courses are open to us. The directions we will follow remain unknown, but books such as this by Michael Collins stimulate the mind, enhance awareness, and assist us on our way.

PREFACE TO THE
2019 EDITION

I guess another decade must find me ten years older but I don't feel that way, although some sad things have happened. Old friends, John Young and Neil Armstrong, have died, and my life has changed with the death of my wife, Patricia Mary Finnegan Collins, after fifty-seven years of marriage. I miss her dearly and daily. My beautiful and competent daughters, Kate and Ann, help me fill the void. I still live and fish near Florida's Everglades. I have hated the word *octogenarian*, but as I look at age ninety in 2020, it doesn't sound too bad.

In the world of spaceflight, nothing spectacular is being done. The shuttle era has ended, and the International Space Station so far has produced little news except for Scott Kelly's nearly year-long stay while his astronaut twin brother Mark served as a control on the ground. However, there seems to be a new optimistic spirit in the air, and plans are burgeoning, centered on building a lunar base preparatory to a Mars landing. As I have written in *Mission to Mars* and elsewhere, that planet, not the moon, has always been my favorite. I used to joke that I flew to the wrong place, and that NASA should be renamed NAMA, the National Aeronautics and Mars Administration.

But NASA it is, and today, for the first time since the days of Wernher von Braun, two names are recognizable in the world of spaceflight: Elon Musk and Jeff Bezos. They seem to be able to do things faster and cheaper than the government, and a generation that never knew Apollo is awakening to the prospect of further space exploration. Musk is a billionaire, and Bezos is the richest person on the planet. Musk is specializing in reusable rockets, but ultimately wants to colonize Mars, starting as early as 2020 with the unmanned *Blue Dragon*, and then with an expedition crew of one hundred. (In my Mars book, I thought a crew of six might be more practical.) At any rate, it's nice to know that exploration has some solid backing, and that private funds are now in the game alongside NASA's annual $20 billion or so in taxpayers' money. NASA says people on Mars is a possibility in the 2030s.

Mars. Why so long to get there? Although it's the next logical destination, it is very complicated. Different in almost all respects from Earth, Mars is nonetheless the closest thing to a sister planet, at least one close-by. But in this case, *close* is a relative term, as Mars takes a gigantic loop around the sun, as do we. Our elliptical orbits are such that we come as close as 35 million miles to each other and as far as 220 million. A rocket leaving Earth for Mars has a choice of many possible paths, but the most economical in terms of fuel is called a Hohmann transfer, and takes between six and nine months one way. Our two orbits are such, however, that a crew arriving at Mars from a Hohmann would have to wait a year or more for a favorable return alignment. Therefore, a round trip would take more than two years. Compared with Apollo's eight days, that is *huge*, and creates a host of new issues. Mission duration can turn minor annoyances into big problems. The first thing an astronaut thinks about is, "What if something goes wrong?" Well, tough—you might be a year away from the safety of Earth, so your equipment and your body better be able to handle it.

As for the environment, probably the worst problem is radiation, both solar and cosmic—high-speed particles from the far corners of the universe. As survivors of Hiroshima can attest, radiation can cause terrible damage to the human body. Even shielded from it, the crew would have to adjust to weightlessness if their craft was not producing its own gravity, probably a lot less than one G if it is rotating. Our experience with the space station shows that cardiovascular problems can ensue, and that reduced G can cause increased pressure on the eyes, from within, that can damage vision.

Once on Mars, with about a third of one Earth G, the crew should be quite comfortable (perhaps even exuberant), but must still be provided with sufficient oxygen and protected from radiation. The atmosphere is too thin to be of much help, and is quite toxic as well.

All in all, Apollo was child's play compared to a Mars landing.

One could fly directly to Mars, as outlined, or use a lunar base as a fuel-and-water depot. President Trump has opted for the latter, and so directed NASA. I have always favored a direct flight, my reasons having less to do with technical factors and more with politics and finances. Thanks to President John F. Kennedy, Apollo was a masterpiece of simplicity: "I believe this nation should commit itself to achieving the goal, before this decade is out, of landing a man on the moon and returning him safely to the earth." The *What* and *When* decided, with NASA to fill in the *How*. We had our clear marching orders, and they were very helpful in getting things done. For Mars, a prior stopover on the moon? I fear this complication would run into its own snags and delays, costs would escalate, and the schedule would dissolve into "Yes, of course we are going to Mars, but first we have to fix the lunar——." Not direct, not simple. Sometimes, in space planning, it costs less in the long run to go all-out in the first place. On the other hand, Neil Armstrong thought it was wiser to first fill in some of the gaps in our knowledge, and a lunar base could do that. Neil was a much better engineer than I, and I concede he was probably right in this case.

Perhaps the fact that I would gladly bypass the moon makes me confess that I don't think it too homey a place. The moon's jagged contours certainly stick in my memory but just barely, compared with the vision that I summon over and over again of the itsy-bitsy sphere just outside my window, motionless, cradled in black velvet. That sight—the Earth, tiny, shiny, blue of sky and water, white of clouds, with only a brown trace of land—haunts me. I can blot it out with my thumb, but back it comes, 200,000 miles away, quiet and peaceful-looking even though I know that it is neither. Surprise! If allowed only one word to describe Earth, I would say *fragile*. Give me a second choice and I'll pick *inhabited*. Now, that is a reach, because that is what I could not see. But it doesn't matter, because I did feel something—a presence—like tiny creatures crawling ant-like over Earth's surface. I couldn't help but ask, What are they? Why are they running around? How many? Where going? Is all OK with them?

On the return trip, I wasn't too busy with housekeeping chores, and Earth was the whole show, so I could let my imagination wander. As an aviator, I have always admired birds, and yes, there they were, geese, their flocks streaming by. And wolves at the tree line. Dolphins jumping—sure, plenty—and even rats scampering and hiding. But human sightings or voices, none out my window, I suppose because I had plenty of them in my ears (Mission Control), and if they were silent (rare), Neil and Buzz. No, it was the wild things that I saw. Maybe all these creatures could be drawn together. Maybe Gaia, who is their spokesperson, would talk to me. "Hey, Gaia, how's it going? ¿Qué pasa?" No answer, so I will speak for her.

Gaia has been around for a while. René Dubos (1901–82) called such a concept "a theology of the Earth." Loren Eiseley (1907–77) wrote of a living earth that could repair itself, but it took the biologist James Lovelock to choose the name Gaia as the sum total of all earthly life-forms. I guess he meant all the way, from *E. coli* and elephants to us folks. Gaia may not exist, but I do talk to her.

We lunar crews looked back at 3 billion earthlings. Almost fifty years later, the number is about 8 billion, ever increasing, with predictions of 10 billion or more by mid-century. This growth alarms some people, but not too many, I think; they have more pressing problems close to home. I tried to condense the situation in my concluding remarks in the Preface to this book's 2009 edition: "We need a new economic paradigm to produce prosperity without growth." I believe this even more firmly today. I believe Gaia is saying "Ouch!" The increased burden on our resources is being absorbed, but at a price: more CO_2 in the atmosphere, warmer temperatures, rising seas. The oceans are getting a double whammy, becoming warmer and more acidic. CO_2 may be good for our forests, but my favorite waters, surrounding the Florida Keys, are experiencing widespread coral disease, whitening and dying due to these factors. Australia's Great Barrier Reef is similarly afflicted.

Day to day, Gaia leaves me alone to navigate through the petty details of life. Living alone means all the usual bother, shopping, cooking, and so forth, plus there is an overlay, something a little special about once having been an astronaut. The fact that it was an eon ago, and for only six years, does not deter well-wishers. "What was it really like up there?" Lord, how to condense such a complex matter? Depends on the questioner's age. Usually I can get by with "cool," or lately, "awesome."

Parents frequently want guidance for their children, or the kids themselves ask, "What should I do to become an astronaut?" Times have changed so much that I am increasingly baffled when seeking a reasonable reply. When I first became interested in space exploration—thank you, Buck Rogers—there were no astronauts, no NASA. Which skills should be required for getting a job in this strange new environment? Various "experts" blew open the doors on nutty ideas. Some divers experienced "rapture of the deep" and didn't want to return to the surface. Pick a SCUBA veteran to guarantee he would willingly return to Earth. Thin air up there, pick a mountain climber. Dangerous as hell, how about a bullfighter? And so it went.

About this time, a position paper was put on President Eisenhower's desk, and he agreed with its conclusion that candidates should be selected from the tiny pool of "graduates of an accredited test pilot school." That certainly made a selection board's job a lot easier. For example, the most recent astronaut selection produced twelve "newbies" out of more than *eighteen thousand* applicants. I am guessing that the Mercury Seven, our country's first astronauts, had a hundred or so qualified competitors. The test pilot rule continued until my group of fourteen was picked in 1963.

So how to beat the odds? I am tempted to say to a kid, "Look, you are up against thousands of genius types, pick another field." But I don't. Deep down I want to say, "Read books: if you miss the astronaut thing, you are still preparing well for life." And I really want to add, "Ditch the phone, skip the movies, and avoid TV. Read newspapers, magazines, and books." That's what I do, but then I don't want to be accused of inculcating (that's a book word) my warped view in the young.

I enjoy talking with most kids, especially when there is no whirr of blades overhead. I do get annoyed when the helicopter intervenes and I am told to sign the photograph and put down "Harry, aim high." Geez, I don't know Harry. Maybe he should aim for the middle. Anyway, all I do usually is mumble platitudes about hard work and sign "Best wishes." I do, however, fairly often thank President Eisenhower. Could I be one of twelve of eighteen thousand? No way in hell.

How do I know? My qualifications sing out, loud and clear— some good things, but mostly not so much. Due to the itinerant nature of my father's job, I attended six different schools in my first eight grades. Naturally, I was exposed to some subjects twice and others not at all (i.e., music), but that didn't matter to me. I look back on it as a rare and wonderful opportunity to make so many different friends, from such varied backgrounds, locales, and cultures. *Diversity*: the mantra of today's colleges. I had plenty of that before I reached high school.

Again I was fortunate. At an age when friendships deepen and peer pressure mounts, I spent all four high school years at an excellent, maybe superb, prep school, St. Albans in Washington, D.C. Here the classes were small and the teachers experienced, dedicated, and determined to pound English, Latin, science, and math into my dull skull. I was singled out as being lazy, but today I search the literature and claim attention deficit disorder. So clearly, I can remember: The classroom door closes, the teacher starts to talk, the sun shines through the windows. My body slowly rises and exits, perches high in a nearby tree, and considers with relief the scene below and the poor wretches trapped there.

Never mind the excuses, I was a mediocre student, more interested in athletics than academics. I was captain of the wrestling team, but even that was a bit tainted, as I was also a secret smoker. Stupid.

Then on to West Point. There I graduated in the top third of my class, but that was a sorry performance, far below the numbers racked up by my West Point astronaut compatriots Borman, Aldrin, and Scott. Again I make excuses, but I was simply not very interested in most of the academics. In the sciences, thermodynamics was my favorite, if only dimly grasped. Entropy and enthalpy seemed pretty cool. In the humanities, it was Chinese history, the wild, diverse, colorful, exotic dynasties somehow winning out over Washington and Jefferson, whose lives seemed more subdued and conformist. At West Point we had an exhortation: "When the going gets tough, the tough get going." When my going toughened, I joined the Literary Guild and purchased their monthly book, usually a novel. My roommates, sweating through differential equations, thought I was crazy, but it was just my extremely mild form of rebellion.

After West Point, it was into Air Force Flying School. Academics there were a comparative breeze, a minor part of our education.

Flying was all-important, for if I washed out, I'd be a ground-pounder somewhere. Fortunately, the old T-6 prop trainer liked me, and I loved flying, and all was well. Jets and instrument-flying were more complex.

Then on to the F-86 Sabrejets, top of the line, and the 72nd Fighter Squadron, my home for four years in California and France. There I got my first taste of real danger when I had to eject at low altitude from my burning F-86. On the good side was winning one of two awards at an aerial bombing meet, competing against the best American fighter pilots stationed in Europe.

After that a turn for the worse, a stultifying stint at a maintenance officer school, then finally Valhalla, the Test Pilot School at Edwards Air Force Base, followed by a postgrad course there. I was then declared an experimental test pilot and given the choicest of assignments: the Fighter branch of Edwards Test Operations. There I joined a small elite group of stick-and-rudder maestros who had produced the Mercury astronaut Deke Slayton, my future boss at NASA and the best-qualified of the original seven, in my judgment. I was in good company, along with Jim McDivitt (Gemini 4, Apollo 9) and Joe Engle (X-15, Shuttle). That's my story, very little of which reaches the qualifications of today's NASA applicants. So I thank you, General Eisenhower.

It's a long road from picking new astronauts to sending them to Mars, but that's the single footstep that starts the journey. An astronaut's career may stretch to twenty years, so the most recent crop of twelve has only a slim chance of being on the first trip. It's more likely that the crew will come from younger men and women, perhaps even some of the MIT students whom I have visited. The Massachusetts Institute of Technology is my favorite of the technical institutions, probably because I used to go there in the 1960s to learn the inner mysteries of the Apollo guidance system, which they designed. Now they invite this old codger back to lecture, which is ridiculous, sort of like asking a shower singer to serenade the divas at the Met.

When I speak to the students, I try to beat them up a little bit. I usually start with today's technical mantra: Focus on STEM—science, technology, engineering, and math. Thus they will become educated, they are led to believe. I say no. STEM may be a beginning, but it's far from a complete education. Perhaps that is because I have worked with too many incoherent engineers. I want to change STEM to STEEM, inserting English. Others, including Buzz Aldrin, have pushed for a place for the arts—i.e., STEAM—but I say no to that also. I am a watercolor painter, but that is a frill. If I botch a painting, it's a private failing. But someone who mangles the English language in public, that's altogether different. More common is the engineer who has an important point to make, verbally or in a memo, but cannot really get it across, as it is immersed in a cloud of jargon.

But studying English can be boring. I try to use poetry as an antidote. Don't concentrate so much on the rules, but see how others have used our beautiful language. Good poems are not boring, and the best are memorable. I usually end up reciting, in a loud voice and by memory, a long passage from John Milton's *Paradise Lost*. Probably the students are wondering "Who let this guy in?" but they are kind, and even invite me back. And for that I feel very fortunate, to mingle with highly intelligent, hardworking young people.

Being an ex-astronaut is not as exciting as preparing for my next rocket ride, but after NASA, my working years were rewarding. My twenty-plus years in retirement have been fun and fulfilling in many ways. Watching my family grow, all doing well, has been paramount, but beyond that come fishing, reading, chasing the stock market, painting, and exercise, exercise, exercise. I'm not wealthy, but I have plenty to enjoy all these things in comfortable, or maybe even luxurious, surroundings . . . Lucky, lucky, lucky.

Michael Collins
August 2018

PREFACE TO THE 2009 EDITION

The years accelerate like a rising rocket, and that scares me more than the ride itself. Like most old people, I am crotchety, and—at seventy-eight—disapproving of younger customs and developments, such as the adulation of celebrities and the inflation of heroism.

Heroes abound, no doubt about it, but don't count astronauts among them. The passerby who administers mouth-to-mouth to a stricken stranger, the nurse in the emergency room who forges on while tasting spattered blood, the soldier who throws himself on a grenade to save his buddies: these people are undeniably heroes, and should be revered as such. We astronauts were good; we worked hard; we did our jobs to near perfection, but it was what we had signed on to do. It was not, in the words describing the Congressional Medal of Honor, "above and beyond the call of duty." It was not heroism.

Then there are the celebrities. What a senseless, empty concept for someone to be, as my friend the great historian Daniel

Boorstin put it, "known for his well-knownness." How many live-ins, how many trips to rehab, maybe—wow—you could even get arrested! All this can catapult an attractive youngster to the front ranks of the media, there to be consulted on the drought in the Sahel, the benefits of omega-3 fatty acids, etc., etc. (No one, not even a celebrity, talks about the national debt.) But please, don't invite me to your celebrity golf tournament.

Okay, okay, let's press on (a favorite astronaut saying). The last forty years have worked out well for me. On my tombstone should be inscribed LUCKY because that is the overriding feeling I have today. Neil Armstrong was born in 1930, Buzz Aldrin in 1930, Mike Collins in 1930. We came along at exactly the right time. I used to imagine a gigantic grandfather clock in the sky, its pendulum stuck over to the left in the "too young" position—you couldn't drink, drive, make money, or do anything really interesting. Then one night you went to bed and—voilá!—the next morning the pendulum had swung over to the right, to the "too old" side: over the hill, don't try running, getting bald. Was there nothing in the middle? For me, the middle was the 1960s: a great family, test pilot, astronaut, what more could I want? Just lucky, the right place at the right time.

In the early days of NASA, Houston was a great place to work. There were thirty of us in the astronaut office, and we were given interesting responsibilities and a lot of latitude in fulfilling them. We were generalists in that we were expected to learn the workings of the entire array of machines and the plans for using them to get to the moon. But, in addition, we were each assigned a specific specialty for which we would become the astronaut-office expert. My assignment was pressure suits and extravehicular activities, a smaller piece of the puzzle than some—guidance and navigation, for instance—but one I enjoyed and that had its peculiarities. For example, during a space walk our lives depended on how diligent the ladies in Dover, Delaware, had been in gluing together the intricate patterns of rubberized fabric that kept our fragile little pink bodies at a safe pressure. One little leak and we were dead; thank you, ladies.

Today, my life in retirement is a simple but entertaining one. I split my time between Florida and Boston, doing some watercolor painting, and fishing a lot in both places. I have a tiny paddleboat from which I gather snook in the south and striped bass in the north, getting good exercise along the way. Every once in a while I look up at the moon, but not too often: been there, done that, as I hope I have explained properly in this book.

But from my boat I keep wishing I could see Mars. I *really* wish I could see it from Washington, but not much luck there, as the space program dithers along. The days of the shuttle are about over, and NASA's next effort is centered on building new hardware to go back to the moon and then on to Mars. I fret that this long, drawn-out process will stumble somehow and get canceled, or that the time and expense devoted to the moon will put off the journey to Mars—a much more interesting place—for many decades. I regard the moon not so much as a destination, but as a direction for mankind's migration—out there, outward bound. Mars was my favorite as a child, and still is.

Fifteen years after writing *Carrying the Fire*, I wrote *Mission to Mars*. In it, I pondered what there is in us—genes, character, culture, spirit, ethos, whatever—that makes us look up into the night sky and become curious and restless. To go, to see, to touch, to smell, to learn, to understand: I think we are wanderers, and this migratory drive, this extraterrestrial imperative, will surely lead us to our neighbor Mars. The closest thing in the solar system to a sister planet, Mars is certainly an inhospitable place, but it nonetheless will be a fascinating new frontier. Imagine the wonder of living in one-third of Earth's gravity, under an antiseptic dome, with no weapons or national boundaries, recycling everything and wasting nothing; learning not only new things about a new home, but new things about our old one. The two planets have evolved quite differently, and that is important for our understanding of each.

But, after giving Mars a great deal of affectionate thought, I realized T. S. Eliot understood it better than I:

We shall not cease from exploration
And the end of all our exploring
Will be to arrive where we started
And know the place for the first time.

When I looked back at Earth from the moon, if I could use only one word to describe the tiny thing, it would have been *fragile*. A totally unexpected reaction, but *fragile* turns out, unfortunately, to be accurate in a thousand ways. World population in 1969 was more than three billion, is now more than six, and will be eight or so when the next lunar heroes and celebrities look back. I don't think this growth is healthy or sustainable, but our economic models are predicated on growth; they require it. Grow or die or maybe both. The dead zone in the Gulf of Mexico is now larger than New Jersey and still growing. The growth of death: a terrible thing to do to this planet. This one example alone makes me want to cry, and countless other catastrophes abound, some lurking in the future, many already here. We need a new economic paradigm to produce prosperity without growth.

I fervently hope some of you reading this can help reverse today's ominous trends. But as for me, I'm going fishing. Take care.

Michael Collins
October 2008

PREFACE

Despite the voluminous press coverage of recent years, and a fair number of books, especially after Apollo 11, people still don't have the vaguest idea of what it was like "up there," or what pre- and post-flight activities were necessary and how they affected the lives of those involved. I wrote this book to do that, tracing my participation in the astronaut selection process, subsequent training, the frustration of being so close yet so far away, making it on Gemini 10, being assigned to the first circumlunar flight and getting bounced off by spinal surgery, and finally orbiting the moon while Armstrong and Aldrin walked on it. Although undeniably autobiographical, I do not see it as a me-me-me kind of thing, but rather an insider's factual and simple explanation of how the machines operated, who operated them, and what it was like living in an artificial, high-pressure environment.

Being an astronaut was as exciting a job as anyone could ever have, and I hope I have conveyed that excitement. I bore easily, and I have written for people who bore easily. If I have done my job properly, the reader will be able to pick the book up at any point and find something interesting going on, because that is the way Houston and the other space places were in the sixties. There was never a dull day, and there should be no boring pages. In a way I regret not having written the book earlier, because some of the story is not as fresh in my mind as I would like. On the other hand, waiting does have its advantages, because in the intervening years I have learned more about what fascinates the non-astronaut, and I have spent more time on these points and less on some of the arcane aspects that appeal mightily to pilots but not at all to lawyers or housewives. Hopefully, I have also gained some perspective in regard to how flying in space has changed my life. But above all, I am glad that I wrote it myself. No matter how good the ghost, I am convinced that a book loses realism when an interpreter stands between the storyteller and his audience. The price I pay for going ghostless, of course, is that I cannot share the blame for awkward clauses or historical inaccuracies, but I don't mind that. A stiffer penalty has been the amount of time it has taken, virtually every weekend for the past eighteen months. My wife and children have been very understanding during this period; without their support, I could not have gone on with it, and I appreciate their forbearance.

Finally, a word about the title, and then it will be time to get on with it. Originally, I called the book *World in My Window*, quoting something I had said during the flight of Apollo 11, but the more I thought about it, the cornier it sounded. *Carrying the Fire* was born during a long, rambling telephone conversation with Roger Straus III, my editor. There is no trick to it; it is simply what I feel space flight is like, when limited to three words. Of course, Apollo was the god who carried the fiery sun across the sky in a chariot, but beyond that—how would you carry fire? Carefully, that's how, with lots of planning and at considerable risk. It is a delicate cargo, as valuable as moon rocks, and the carrier must constantly be on his toes lest it spill. I carried the fire for six years, and now I would like to tell you about it, simply and directly as a test pilot must, for the trip deserves the telling.

Michael Collins
Washington, D.C.
November 25, 1973

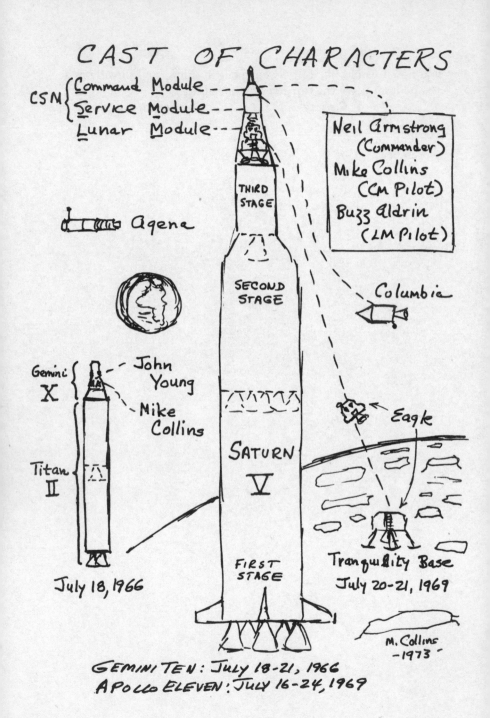

CAST OF CHARACTERS

CSM { Command Module
Service Module
Lunar Module

Neil Armstrong
(Commander)
Mike Collins
(CM Pilot)
Buzz Aldrin
(LM Pilot)

THIRD STAGE

Agena

SECOND STAGE

Columbia

Gemini X { John Young

Mike Collins

Eagle

SATURN V

Titan II

Tranquility Base
July 20-21, 1969

FIRST STAGE

July 18, 1966

M. Collins
-1973-

GEMINI TEN: JULY 18-21, 1966
APOLLO ELEVEN: JULY 16-24, 1969

10   9 _ _ _ _ _ _ _ _

8 7 6

5 4 3

2 1

There are only two ways of learning to ride a fractious horse; one is to get on him and learn by actual practice how each motion and trick may be best met; the other is to sit on a fence and watch the beast awhile, and then retire to the house and at leisure figure out the best way of over-coming his jumps and kicks. The latter system is the safer, but the former, on the whole, turns out the larger proportion of good riders. It is very much the same in learning to ride a flying machine; if you are looking for perfect safety you will do well to sit on a fence and watch the birds, but if you really wish to learn you must mount a machine and become acquainted with its tricks by actual trial.

—Wilbur Wright, 1901

I suppose Russia must test new airplanes over the Pripet Marshes, or Siberia, or wherever desolation dictates. In this country, it is Edwards Air Force Base, California—Mojave Desert country, in a vortex of the Antelope Valley wind tunnel, one hundred miles north of Los Angeles. Although I had flown over the area many times before, when I first approached Edwards on the ground I couldn't believe it. I had left the tinsel-shiny, neon-bedecked high rollers of Las Vegas a few hours before, ricocheting down the highway in an overheated 1958 Chevy station wagon, seeking Valhalla or Mecca, or at least an opportunity to fight for admission into the arcane world of high-speed flight testing. For Edwards was all these things and I had been accepted as a member of Class 60-C at the USAF Experimental Flight Test Pilot School, along with thirteen other exalted ones, mostly Americans (one Italian, one Dane, one Japanese), mostly hyperthyroid, superachieving first sons of superachievers. To this day, I am impressed by this group; I love them, they leer at me from my study wall. One of them has walked on the moon, two have circled it; two—two of the best—are dead.

But in the spring of 1960 I knew only that a nest had to be prepared for wife and infant daughter, arrangements had to be made, housing procured, forms signed, and the other necessary impedimenta gotten out of the way so that the decks would be cleared for the real action to follow.

Vaunted Edwards, the Air Force Flight Test Center, the big time at last! At least it was big, with a dry lake twenty-five miles long serving as a super runway, an earth mother for pilots in distress, for those who *must* land their aircraft immediately no matter what.

It was also dry and hot and windy and isolated, and not at all what a proper Bostonian—my wife—would expect as a nursery for her firstborn. I knew this, and winced, but I also knew that she would prevail, and neither Joshua tree nor rattlesnake nor sandstorm would dim her New England resolve, nor her ability not only to make it but to change it! After all, in a historic sense, it was a place for upstarts. Recorded history of the area spans only a few more years than does the airplane itself.

No matter how advanced the technology or sophisticated the flying machine, the lake still calls the tune, reasserting each winter the primordial dominance of nature over puny, impatient pilots. Each spring and summer, as the lake gets drier and as more high-pressure aircraft tires abuse it, surface cracks and blemishes appear, so by late autumn the lake bed appears rough and "ruined." Then come the winter rains, sparingly, but providing enough water to allow a couple of inches to accumulate on the lake bed and to be blown back and forth by the omnipresent wind. By early spring, the newly dried surface reappears, as silky as a baby's bottom, ready to take another year's traffic smoothly and safely. Of course, in recent years concrete runways adjacent to the lake bed have made the Air Force less dependent upon this annual cycle, but it is still interesting to note that the most advanced machines, such as the X-15 rocketcraft and more recently NASA's lifting body, still use the lake bed itself, and are more dependent on nature's schedule, not man's.

I had been flying F-86 Sabrejets out of George Air Force Base in nearby Victorville a few years before, so that in the spring of 1960 I was not unfamiliar with the area. I knew that Captain Joseph McConnell, our foremost Korean War jet ace, had been killed on the lake while on temporary assignment from George Air Force Base. In 1954, I had witnessed from my cockpit the fatal dive of a supersonic F-100 fighter, and followed the lifeless body of North American test pilot George Welch to earth as his undamaged parachute slowly descended. I knew about Edwards.

I also knew that despite the desolation, the one-hundred-plus heat, the perpetual howling of the wind, *this was the place*. Here the very first American jet had been tested, with a make-believe wooden propeller stuck on its nose whenever it was parked, so as not to arouse suspicion; here Captain Chuck Yeager had broken the sound barrier on October 14, 1947; here Captain Mike Collins was going onward and upward. *Ad Inexplorata*, said the motto of the Flight Test Center: "Toward the Unknown." Next to the motto of the Air Rescue Service ("That Others May Live"), this one was my favorite, and I noted with approval that it was prominently plastered on buildings and flying suits alike: a futuristic, aerodynamic shape escaping from a sandy, cactus-bedecked background into a blue-black sky. On the other hand, the insignia of the test pilot school itself gave me pause. It featured a lot of blue sky, but superimposed above all was a *slide rule*.

I signed in somewhat pensively, was assigned a neat cinder-block house, nothing fancy but white-gloves clean, and then headed on back up the highway to Las Vegas to break the good news to Pat. "A neat place, you'll love it!" At least I hoped she would, and I would, because for the first time in my Air Force career, we were due for a long and stable assignment. God knows, Pat deserved it; in less than four years of marriage we had lived in four houses, four apartments, and what seemed like forty-four motels. For that matter, I had been moving all my life at frequent and regular intervals, with never more than four years in any one spot. My father had been a career Army officer for thirty-eight years, and in the seventeen years I had lived at home, I had seen dramatic and frequent shifts in scene, from a rooftop apartment in Rome, where I was born, to a modest old colonial house in Alexandria, Virginia, to which he retired in 1945. Along the way, the family had sampled snake-infested country life in Oklahoma, bright lights in Manhattan, as viewed from nearby Governor's Island, and—most unusual of all—a couple of years' residence in Casa Blanca, which is generally recognized as the oldest dwelling in the Western Hemisphere. Built by Ponce de Leon's nephew around 1530, this imposing old fortress overlooks the harbor of San Juan, Puerto Rico. Assigned as living quarters to the commanding general of the Puerto Rican Department, as it was called in 1941, Casa Blanca was the most fascinating place I had ever seen, with seven-foot-thick outer walls, an immense ballroom, a sealed-off tunnel with a secret entrance, and a host of features not to be found in today's puny lath and plaster or dryboard construction. Even more impressive to me, as a ten-year-old, were the surrounding gardens, teeming with tropical plants and animals. I spent hours studying lizards, hermit crabs, turtles, and tiny tropical fish, and getting acquainted with such stomach-ache producers as underripe mangoes and overripe coconuts.

In Puerto Rico, I also took my first airplane ride, in a small twin-engine amphibian, the Grumman Widgeon. The pilot even let me steer a little bit, an indignity the old Widgeon endured with grace as I jerked the nose up and down unevenly, trying to heed the pilot's advice to "keep her on the horizon." My father watched all this from the rear of the plane with obvious amusement. No pilot, he preferred horses to airplanes, but as an old polo player and horse cavalryman, he did appreciate the excitement of this swift new medium, and allowed as how the Air Corps boys did have a certain juvenile appeal. In fact, he relished telling how he had, in 1911 in the Philippines, taken his first airplane ride, in a Wright machine, sitting on the wing next to Frank Lahm, who was the second military pilot to be taught by the Wrights. Frank flew the frail craft over a forest fire, and the updraft from the heated air caused a sudden lurch, which nearly dislodged Daddy (or so he said) from his makeshift perch. I was intrigued by this story, as indeed I was by Lahm himself, whom I met years later at West Point. Quiet, dignified, without pretense or affectation, this old gentleman had lived right at the cutting edge of the advances slicing through our society in the wake of the new air technology. What changes Lahm had seen in his lifetime, and not passively from an armchair, but actively from the cockpits of a series of ever more complex and fascinating machines. I was impressed, especially when I compared this solitary old eagle to the lemming-like horde of "Follow me, men, over the hill" young Army leaders I was familiar with at West Point.

As West Point graduation approached, I had to decide whether to stick with the Army or strike out in a new direction with the recently independent Air Force (to my dad it would always be the Army Air Corps). Unlike that of many young Americans, my love affair with the airplane had been neither all-consuming nor constant. In the years between the Widgeon and meeting Frank Lahm, there had been occasional passionate flings into model airplane building, but airplanes were less a part of my young life than chess, football, or girls. Also, the airplane as a career posed practical problems. One could—25 percent did—wash out of pilot training. One could be killed, practically as easily in peacetime as in war. Promotions were predicted, by those who kept book on such things, to come more slowly in the future Air Force than in the Army, because of past excesses on the part of the Air Force, which had caused a "hump" of young but senior officers, blocking the rapid advancement of those who followed. All these things, plus the entire thrust of the Army curriculum at West Point, spoke for the Army as a more sensible career choice. Against this was the wonder of what the next fifty years might bring. It had been less than fifty years since the Wrights first flew, and already we were into the jet age.

Then, too, I had a personal problem. My father's younger brother, J. Lawton Collins, was Army Chief of Staff at the time; my father had retired as a two-star general; another uncle had been a brigadier; my brother was a colonel; my cousin a major—all in the Army. With no similar entanglements in the Air Force, I felt I had a better chance to make my own way. Certainly there was no chance for nepotism, real or imagined.

So the Air Force it was, and after a pleasant month's vacation in Europe following graduation, I found myself in the front cockpit of a single-engine T-6 Texan over the flat farmland of northeastern Mississippi. It was a delightful place to be, especially after four cloying, confining years at West Point. Columbus, Mississippi, was a small, friendly town with a large girls' college, and a bachelor second lieutenant was appreciated if for no other reason than that he had access to the Officers' Club, which featured the only bar in town. But the main thing was the flying! Flying was so much fun it didn't seem right to get paid for doing that and nothing else. Fortunately it came easily to me, and I could relax and enjoy it without the constant apprehension over washing out which plagued so many of my classmates.

After six months at Columbus learning the basics, I moved on briefly to San Marcos, Texas, to learn instrument and formation flying, and then to Waco for jet indoctrination. Graduating with shiny silver wings at Waco in late summer of 1953, I was among the few chosen to go to Nellis Air Force Base, Las Vegas, Nevada, for advanced day fighter training. This was the most desirable of all assignments, since it was the sole channel into the two Fighter Wings in Korea, which, with their North American F-86 Sabrejets, were battling the MIGs so successfully. At Nellis we really learned to fly—a concentrated, aggressive course designed to weed out any-one who might be a marginal performer in Korea. It was a brutal process as well. In the eleven weeks I was there, twenty-two people were killed. In retrospect it seems preposterous to endure such casualty rates without help from the enemy, but at the time the risk appeared perfectly acceptable. We weren't sure we were going to make it through the course, but somehow we were sufficiently "psyched up" by the instructors to give it our all, despite the fact that the Korean armistice had just been signed and prospects for meeting any MIGs were growing more and more remote. We flew as well as we knew how, three and four times a day, wheeling high above the Nevada sky in fifty-minute forays, learning to shoot the guns and to develop the aerial teamwork which would keep MIGs off our tails. At night we roared into Las Vegas, driving our cars in as close a formation as we flew our Sabrejets, terrorizing the natives, gambling away our paltry salaries, snatching a couple hours of sleep before dawn, when we were expected back at the flight line, ready to hurl our little pink bodies into the blue once more. It was a hectic time, and I'm surprised to have survived. I have never felt quite so threatened since.

Because of the armistice, my destination was changed from Korea to California, and upon graduating from Nellis Air Force Base, I found myself assigned to the 21st Fighter Bomber Wing at Victorville. I had a pleasant year there, still flying Sabrejets but now concentrating on ground attack and nuclear delivery tech-niques. In mid-December 1954 our wing was transferred to France, so we picked up, part and parcel, and flew East. Christmas found us in Goose Bay, Labrador. By the New Year, we had inched along to Bluie West 1, Greenland (up the fjord at Narsarssuak). Unbe-

lievably bad weather and amply stocked bars made the going
treacherous, and we arrived at Chaumont, France, some thirty days
after departure in our supersonic jets, having averaged *four miles
per hour*.

The trip had been fascinating (I have never seen anything
from the air more beautiful than the clear azure blue of the Greenland glacier's fissured rim) and France was a new flying experience.
No more the pure clean air of the California desert, where one
could get an unimpeded view all the way from Mount Whitney to
Death Valley, from the highest point in the continental United
States to the lowest, all in an eye's blink. Now it was the Saar
Valley, leaden and flat and heavy with smoke and clouds and
greasy pollution, making it difficult for the sun to penetrate even
on the best days. No longer were we the proud eagles, flying high,
scraping our wings against the troposphere, but now we skulked
along, clinging to the protective mists of the valleys, as we practiced our new art of skimming the ground toward our imagined
targets across the Iron Curtain. Only occasionally could we escape
to the sunny shores of the Mediterranean, where once again we
practiced our trade of bombing and gun firing and flying in the
crystal-clear air. Near Tripoli, in Libya, the U.S. Air Force had
established Wheelus Air Base, a large complex where the various
fighter units from England and France could come and keep honed
the various skills which they could not practice in the crowded,
cloudy skies of continental Europe. Once a year a competitive
gathering was held among the Fighter Wings, a gunnery meet, and
in 1956 I managed to win one of the events and was presented with
a silver loving cup, which to this day I treasure above more presti-
gious honors that have since come my way.

By this time, armed with the trophy, I began to think about
how I might progress beyond my present station, not that I was
doing badly, as I had become a flight commander and now had my
own small brood to train and protect. But I was getting older and
hopefully a little wiser; at least the pilots around me seemed to be
getting younger.

For years I had been following the career of my brother-in-law, a Naval aviator, with great interest. Married to my sister Virginia, Cordie Weart had flown large patrol planes since before World War II, and had recently completed the Navy's test pilot school at Patuxent Naval Air Station, Maryland. I was fascinated by the brief glimpses of his flying which I gleaned from his letters. He was, for instance, the first Navy pilot to fly the Seadart, a small jet seaplane (a radical design never pursued). Not only that, but he seemed to have access to an endless variety of new machines, while there was only one horse in my stable, the tired old F-86. When the time came to leave Europe, in late 1957, I hoped to be able to do what Cordie had done, but I had not accumulated the fifteen hundred flying hours needed to apply to the Air Force's test pilot school at Edwards. I asked that I be assigned to a flying unit in the interim, and then be allowed to go on to Edwards as soon as I passed the fifteen-hundred-hour mark. The Air Force's reply was swift and chilling. I was to report to Chanute Air Force Base, Illinois, where I was to be a student in a nine-month aircraft maintenance officer course. After that, who knows; but it certainly didn't look good, as the Chanute training would point me away from Edwards, not toward it.

The school was dismal. The classroom work was stultifying; flying time was scarce and in obsolete equipment; classes started at six in the morning—what more can I say? In an effort to condense the misery somewhat, I started doubling up on my class load and finished the nine-month course in six months. Grand. Then I got my assignment—to be an instructor in the school! I got in an airplane and stormed back to Washington. I wandered up and down the Pentagon corridors until I finally found a personnel man who would listen to my tale of woe. "O.K.," he said, "if you don't like the school that is fine, but you are stationed at Chanute as far as we are concerned, and we are not going to send you anywhere else. Anything else you can do at Chanute is up to you." Fortunately, I had been offered a job as commander of a mobile training detachment, with headquarters at Chanute, and I now promptly accepted.

These detachments, MTDs for short, were sent out by Chanute to air bases all over the world, to flying units which were trading in old airplanes for new ones. The idea was that a couple of months before the new airplanes arrived, an MTD appeared—a training team to explain to mechanics and pilots all they needed to know to begin operating the new airplane. The composition of an MTD varied with the complexity of the airplane and a host of other factors, but I had as few as ten or as many as seventy men working for me, and it was quite an experience. For one thing, we were always on the go, and wives, girl friends, and paychecks never seemed to quite catch up with us; if they did it was even worse. We had expensive, heavy, complicated, and delicate training equipment that was easily damaged or made obsolete by factory modifications on the new airplanes. We were strangers to our hosts, yet we were dependent on them for assistance and support of all types. Our boss was a thousand miles away and harassed by a dozen other units like ours. My job description said that I was a maintenance officer, but truly a chaplain or a lawyer or a diplomat would have been better prepared for the avalanche of petty problems a traveling MTD carries with it.

Later I graduated to an FTD (a field training detachment), which was a semipermanent version of the same thing, in which the students traveled to us and the pace eased somewhat. Still, I am always amused when people say, "Oh, you've been a flier all your life, therefore you wouldn't understand that . . ." Madame, if you are wondering what advice to give your second son, your firstborn having inherited the title and the land, I heartily recommend that straight out of Oxford he proceed to Chanute Air Force Base, Rantoul, Illinois, U.S.A., and seek fame and fortune with a traveling MTD of his very own.

Thus it was that early 1960 found me at Nellis Air Force Base, Las Vegas, once again. No longer a young tiger grappling with imaginary MIGs overhead, I had my hands full on the ground setting up an FTD to teach pilots and mechanics all about the brand-new F-105, a sophisticated, all-weather, radar-equipped fighter bomber.* Challenging as this job was, my heart was not in it, as I stubbornly clung to the notion of acquiring the test pilot training which would allow me to reach the next logical plateau of my flying career.

During my MTD traveling days, I had lost no opportunity to fly whatever airplanes I could, sweet-talking strange operations officers, making myself available nights and weekends or whenever, and taking trips that no one else wanted. In this way, I had finally built up my flying time past the magic fifteen-hundred-hour mark, and I immediately fired off an application to test pilot school, and now waited at Nellis for my rejection slip. When I flew I usually managed a southwesterly detour to peek at the great lake bed at Edwards, so near, yet so far away. Although I had some nice recommendations, there was nothing in my background to make me believe I would be accepted. There must have been thousands like me. Therefore, when the fateful letter finally arrived, assigning me to "Class 60-C, USAF Experimental Flight Test Pilot School, commencing 29 August 60, course duration 32 weeks," I could not have been more pleased had I been offered a flight to the moon. Our crew of three (Pat and year-old Kate) moved the two hundred miles from Nellis to Edwards.

"Through these portals pass the world's finest pilots" is a

---

* It might be interesting to digress a moment and discuss the Republic F-105 and some of its predecessors. The F-84 was Republic's first production jet, and it was affectionately referred to as the "Hog"—apparently because it was not as lean, trim, or swift as some of its contemporaries. A swept-wing version was promptly dubbed the "Super-Hog," and when the F-105 appeared, what else could it be but the "Ultra-Hog"? This name gave way, however, to the shorter and more descriptive title which it enjoys to this day, the "Thud." Never known to spring lightly from the ground, the F-84 has been described as the "world's fastest tricycle," and pilots have proposed various changes to make takeoffs a little less exciting. My favorite is the sandbox modification, wherein the pilot—seeing the end of the runway approaching at an alarming speed—pulls a handle and releases a trickle of sand in front of the nose wheel. The F-84, sensing this, thinks it has passed out into the sandy overrun and promptly takes off! The F-105, after a shaky start in squadron service, has turned out to be a wonderful machine, versatile and sturdy.

fairly common boast and can be found over the operations door of a variety of flying units. In the case of the test pilot school, however, I really believed it. It remained only to view these paragons, who had been handpicked from all over the Air Force and who would be my friends and friendly competitors for the next nine months.

First was fellow student Frank Borman, greeting us all at the door like a politician with a close election coming up. Fresh from a teaching assignment (thermodynamics at the military academy) and armed with a masters from Cal Tech, Frank was a tough competitor who flew well and who raced through the academic curriculum with slide rule practically smoking.

Then there was Greg Neubeck, who had three thousand hours in the T-33 jet trainer alone—half again as much flying time as I had all together, yet he was a couple of years younger. How had I managed to take so long to get so little done—no advanced degree, a piddling two thousand hours' flying time, thirty years old, and nothing special in my record to offset these deficiencies?

There was little time for feeling sorry for myself, though, as Tom Stafford, the chief of the Performance Section, was determined to get us off on the right foot. Doing the work of three himself, old Mumbles soon had us immersed in the intricacies of measuring the various performance parameters of new aircraft. The general pattern was: first, by classroom lectures, to learn the theoretical aspects of a particular type of test, then to make one or more flights trying out the new technique, and finally to analyze the reams of test data acquired during the flights. This last part, the reduction of thousands of bits of information ("data points") into an intelligible report, complete with charts and graphs, was a tiresome, time-consuming process. It began in flight, when information was handwritten on a kneeboard, or instruments were photographed by a special camera, or an oscillograph traced on graph paper the record of thirty or more specific measurements. After each flight, the developed film and oscillograph paper were delivered to us; and weekends and nights would find us hunched over a desk calculator or peering at a film projector, trying to reduce this overwhelming amount of information into a terse report which would gain Stafford's approval, so that we could move on to the next test series and begin the miserable process all over again. Ah, the glamour of a test pilot's life!

As the months of schoolwork wore on, I was getting myopic and crotchety, but I knew that all was well down at the other end of the flight line where the *real* test pilots were. They, the chosen few assigned to Test Operations, were zooming through the skies in gleaming F-104 Starfighters, while this student was chugging along in a T-28 powered by a washing-machine engine. They were wearing silver pressure suits; my greasy flying suit was threadbare, especially in the seat. If I could only hang on a little longer, get creditable grades, write another half dozen dazzling reports full of never-before-revealed tidbits of aeronautical lore, then perhaps I too would graduate, be sent to Test Ops (Fighter Branch, of course), and maybe even be given a clean flying suit.

In the meantime, life was not 100 percent drudgery. We amused ourselves by swapping stories about our earlier flying exploits, and Harley Johnson—our class character—would talk by the hour about his motorcycle racing career. We were all impressed except Jack Tyson, who owned a 1918 Bentley and who was impressed only by 1918 Bentley owners. We also had some memorable parties, usually on Friday night after a typical tedious week of academics, quizzes, a little flying, and a lot of data reduction. Then we would gather and drink and ignore our wives and discuss our future (after graduation) and complain about the few good jobs available and sometimes even sing, ditties whose words don't seem worth repeating but which generally described the tribulations of poor pilots in a non-comprehending world.

We also began to develop considerable confidence in our own abilities, and grudgingly had to admit that there was some merit to the test pilot school's approach, that the long hours were paying off. We learned to do things with airplanes which normally are not considered necessary or even possible. We learned to control the airspeed, for example, to the nearest knot; we learned to observe, remember, and record every last movement of a bucking, heaving, spinning plane; we learned to organize our tasks so that not one minute of precious flight test time was wasted. We set high standards and met them and were proud of it. We lived in a sloppy world, but we were precise, very precise. Our great fear was that we would not be able to use this precision, that we would be sent off somewhere to the bush leagues, that we would be assigned to fly endless circles in the sky while an engineer in the back of the plane twiddled with the dials of some new electronic box. For every bona-fide test pilot job in the Air Force, there are ten which masquerade under the name. The majority of our class would fill these quasi-test-pilot positions, and we knew it; and we waited, and we worried, for the Air Force personnel system is not noted for its sensitivity or perspicacity. Finally the word arrived. Jim Irwin, who a dozen years later would walk and ride on the moon as one of the Apollo 15 crew, was one of the disappointed ones. So was Harley Johnson, and so in fact were most of the class. Borman and Neubeck were to stay on as instructors in the school, an honor, but the only available slot in Fighter Ops went, by what alchemy I will never know, to Mike Collins.

From that point on, it was all downhill. I can't recall our graduation speaker or what he said, a sad fact I have ruminated on more than once when I have found myself sweltering in a borrowed gown on a rickety platform about to deliver a graduation harangue of my own. In truth, the only graduation speaker to make any lasting impression on me was Roscoe Turner, who in 1953 had come to the graduation of our primary pilot school class at Columbus, Mississippi. The most colorful racing pilot from the Golden Age of Aviation between the world wars, Roscoe had had us sitting goggle-eyed as he matter-of-factly described that wild world of aviation which we all knew was gone forever.

Roscoe had flown with a waxed mustache and a pet lion named Gilmore; we flew with a rule book, a slide rule, and a computer. To get to the moon would require a very big computer, a whole basement full of computers, in fact. But in the spring of 1961 at Edwards, I was somewhere between Gilmore and moon, in the realm of Starfighters and Super Sabres and Delta Darts, sophisticated machines which bridged the gap between the colorful past I knew I had missed and the complex future I did not know was coming. I was content. I was off to Fighter Ops. I even went shopping for a Model A Ford.

2

Plans, coordinates, and conducts flight test programs on experimental and production type aircraft to evaluate and report the flight characteristics, performance, stability and functional utility as a military weapon system. Evaluates and reports the suitability, serviceability and functioning of installed components and equipment. Conducts test support flights as required in the accomplishment of military and contractor flight test programs. Represents the Air Force Flight Test Center as directed in conferences and meetings which concern the flight test missions.

—Duties of a Test Pilot, from Air Force Regulations, 1962

In addition to the above, test pilots have to have a touch of schizophrenia: they must think about a new airplane, not only in terms of how difficult it is for *them* to master, but how it will be for the average pilot who has not had their specialized training and experience. If something is not quite right, what should be done? On the one hand, perhaps nothing; on the other, maybe it had better go back to the factory for expensive and time-consuming modifications. Perhaps a gentle warning to the customer might suffice.

With each new plane comes a book, a thick book, the Pilot's Handbook, his bible, and it is chock-full of terse admonitions describing various sins of omission and commission. These are categorized as to their seriousness, and put in little boxes to make them more eye-catching.

 Operating procedures, practices, etc., which will result in personal injury or loss of life if not carefully followed

 Operating procedures, practices, etc., which if not strictly observed will result in damage to or destruction of equipment

Note An operating procedure, condition, etc., which it is considered essential to highlight

When to ignore, when to warn, when to fix? This is the most important judgment the test pilot must make. In general, he finds himself arguing the case for the pilot at the expense of the project manager's precious schedule. "Harry, I don't care if we've gotten this far without an accident or not. I'm telling you, the canopy jettison handle sticks out too far, and someone is going to snag it with his flying-suit sleeve if he gets in a hurry reaching for the TACAN channel selector. Be careful? Hell yes, we're careful, that's why it hasn't happened so far, but how about some poor second lieutenant in bad weather, at night for the first time, when he's clanked up anyway? A CAUTION? So what should it say? Beware the handle, or don't get clanked up in bad weather at night? Come on, Harry. Besides, we are getting so many CAUTIONS that the handbook is getting to be the size of the Bronx phone book. This one needs fixing."

And so it goes; not bad training for someone who might inadvertently jettison a canopy or hatch in lunar orbit. In fact, to me the amazing, the incredible, the fantastic (pardon the slip into astronaut parlance) fact is that so few crew mistakes and mishaps have occurred during Gemini and Apollo flights. The opportunities for error are almost limitless, and only superlative design permits virtually error-free operation of a machine as complex as an Apollo command module. As will be discussed later, the astronauts were hired early enough to participate in the design phases of Gemini and Apollo, and in my view this was one of the wisest decisions NASA made.

A test pilot, more than any other type of aviator, must be objective. It is all right for a squadron pilot to fall in love with his airplane; it is all he has to fly, and he might just as well enjoy it because it has already been designed; it exists in its present form and no one is going to change it now. He can afford to put on his blinders, compile a long list of prejudices in its favor, and develop an almost religious fervor when discussing it. He can look with scorn or pity on the unwashed from the neighboring squadron who are not so privileged.

The test pilot cannot fall into this trap. Just because he has spent years flying Convair products doesn't mean that Lockheed's system is not just as good. He must learn that the Convair's delicate feel is great for an all-weather interceptor at high altitude, but that Lockheed's heavier stick forces are much more practical when working close to the ground. He must carefully analyze the possible uses to which a new airplane might be put and judge it accordingly. Lindbergh's *Spirit of St. Louis*, for instance, had a longitudinal instability which generally would have been considered objectionable. However, Lindbergh didn't mind, because the constant attention he had to pay to the elevator control helped to keep him awake during the long lonely hours over the Atlantic.

As in most other professions, a test pilot does not instantaneously find himself at the top of the heap; rather there is an apprentice-journeyman system—at least there was at Edwards—to guide and advise and develop the neophyte. The "new boy" at Fighter Test, for instance, finds he has inherited the most undesirable jobs, paramount among which is that called the "barrier." The barrier is a device for slowing down airplanes in a hurry, hopefully without damaging them. It is a last-ditch affair, designed to grab a speeding jet at the very end of the runway and prevent its crashing into trees and houses or whatever lies beyond. The barrier usually consists of a single heavy cable stretched across the runway which snags the airplane by popping up in front of its main landing gear, or by grabbing a tail hook, which the pilot lowers when he sees that he is overshooting the runway. Once the speeding jet is securely attached to the cable, the jet's tremendous kinetic energy must somehow be absorbed within a few seconds in order to bring the beast to a super-quick stop. Various energy-absorbing schemes have been used, such as dragging heavy chains across the ground, towing a scoop through a closed conduit filled with water, or using hydraulically actuated aircraft brakes. Each new system must be tested and verified for a wide variety of weights, speeds, and aircraft types. All this is done at Edwards, and when the telephone rings in Fighter Ops, the old heads have the uncanny ability to know when it is the barrier-test engineers calling for a pilot. Disappearing into the woodwork, mumbling excuses about writing test reports, they vanish—leaving the shiny new boy to be offered up as a sacrifice to the voracious maw of the barrier god.

Arriving at the appointed spot, a deserted runway ending at the lake bed's edge, the new test pilot is quickly strapped into the oldest, most dilapidated jet he has ever seen, a model long since abandoned by even the South American air forces. Peering down through cobwebs and birds' nests, he is shown the one and only shiny new gauge in the cockpit—the airspeed indicator, all-important in delivering a precise reading of the amount of kinetic energy (one half the mass times the velocity squared) the engineer demands for a given test run. The engineer has consulted his slide rule, charts, computer, and astrologer, and screams up to the cockpit over the whine of the engine, which the sweating pilot has finally managed to start with laconic advice from a disgusted mechanic who obviously feels personally insulted by having been sent this imbecile. "Say, er . . . er . . . is it Collins? O.K., Collins, we need eighty-two knots on this one, no faster please, Collins." They need the name to put on the accident form.

Lining up at a respectful distance from the barrier, Collins examines the adversary as a *novicero* might scrutinize his first fat Pamplona bull. Only in this case, he is the bull, he must do the charging, he must steer this juggernaut toward the artful dodger at the end of the runway. He feels eyes on him, but there is no applause from the crowd as he revs up the tired engine, releases brakes, and steers toward the center of the barrier. Eighty-two knots, eh, he'll show them . . . seventy, seventy-five, oops . . . eighty-five . . . oops . . . ninety . . . slow her down . . . eighty-two on the button . . . damn . . . eighty-seven. Thump! Out of the corner of his eye he notices that the thump was the barrier cable passing under the airplane's wheels. He jerks the throttle to idle and braces himself for the jolting deceleration to follow. Nothing happens. By this time, the end of the runway has come and gone, and he is trailing a large cloud of dust as he zips out onto the lake bed, finally slows down enough to turn around, and heads back. Someone, mistaking the dust for smoke, has called the Fire Department, and an escort of bright red trucks with flashing red lights follows him back to his parking spot, where a welcoming committee is gathered. The mechanic examines his battered old machine with studied amazement as if he had never seen one like it before. The test engineer is mesmerized by his own calculations. "Eighty-two knots, Collins. Did you have eighty-two when you passed over the cable?" Collins admits to an extra five, the heinous number is recorded for posterity with much grumbling and clucking, and the engineer departs in a pickup truck. The Fire Department leaves en masse (coffee break?), and the pilot dejectedly finds his way back to Operations. His boss barely looks up from the acey-deucy board. "Forgot to drop the tail hook, Collins; test is rescheduled for the morning." Collins never forgot the tail hook again.

Like the Biblical seven lean years and seven fat years, the pendulum at Edwards seems to swing back and forth; there are never the right number of test pilots or test programs, but always too many or too few. The Air Force should be able to keep a stable of prototype aircraft flying, winnowing and pruning, and finally selecting only the best for production, but given the McNamaras of this world, the system is not allowed to work that way. McNamara decreed that the F-111 would be a great success long before the test program began; in fact, he decreed that it would be everyplane for everypurpose, sort of like building a car to drive Daddy to work, or to handle Mom's groceries, and to mix concrete on weekends, except in May when it would be busy practicing for the Indianapolis 500. Fortunately, the Navy had the smarts and the guts to decline McNamara's kind offer, even if he was the Boss; but in 1961 the Air Force Flight Test Center was gearing up for the great F-111, earlier programs were finishing up and closing down, and work was hard to find. The seven lean years were beginning.

Also, for the first time, Edwards had a competitor—NASA, with its burgeoning space budget and far-out plans for landing a man on the moon before the end of the decade. On top of that, the rumor was out that they were hiring astronauts again. They had, of course, already picked seven in 1959—the *Original* Seven, who were all military test pilots. These men had also been exposed to greater public scrutiny than any group of pilots, engineers, scientists, freaks, or what-have-you in recorded history. And the reaction had been uniformly good. All of them came through as Gordon Goodguy, steely resolve mixed with robust muscular good humor, waiting crinkly-eyed for whatever ghastly hazards might be in store for them "up there." There were no disturbed psyches in the bunch either, for the wide publicity included lurid accounts of their sessions with psychiatrists, their ability to stand icy dark desolation or searing heat in a bake oven. There had also been extensive security and morals checks, "background investigations" as they are called, which had gotten all skeletons out of their closets, although I recall Al Shepard's lopsided, malevolent grin a few years later. "I still have a few secrets." They were *la crème de la crème*, and the nation loved them.

Except for the old heads at Edwards. Some of them were disgruntled because they had not been selected (too old, too tall, no degree, or simply rejected), but others had actively avoided the selection like the plague. In retrospect, I can't say whether their reasons were obtuse or perceptive, but they do seem to me to be a fascinating indication of the time and place. Man, they were here to *fly*, not to be locked up in a can and shot around the world like ammunition. They were master craftsmen, artists, they were Jonathan Livingston Seagulls—they flew, in smooth control, in command. They flew day after day, in various machines to prove various things, but they *flew*. In Project Mercury, on the other hand, one rode; granted there had been only one Mercury flight so far, Al Shepard's fifteen-minute up-and-downer, but he was a passenger, man, a talking monkey.

At Edwards the old heads clucked with disapproval: Why all the fuss? Not one quarter of the skill, finesse, or flying technique of an X-15 flight was required here, yet the public went wild. *Life* magazine showered them with money to tell their "personal stories" to the eager nation; but had X-15 pilot Bob White been offered a similar deal, the Air Force would have summarily turned it down. Besides, who were all these people anyway? Slayton was the only one who had ever been assigned to Fighter Ops (Cooper had been in *Engineering*, for God's sake), and old Deke must have taken leave of his senses to forsake a fighter test assignment to get shot off somewhere in a tin can.

Not having accumulated sufficient credentials to be so snotty, the new boys watched and wondered. I certainly had had no childhood dream of flying to the moon or anywhere else, but the idea was damned appealing, it had been endorsed by the President,* and it beat the barrier four ways to Sunday. For me, the

---

* On May 25, 1961, President Kennedy, in a special message to Congress on "urgent national needs," had included this oft-quoted, yet still stirring passage: "I believe this nation should commit itself to achieving the goal, before this decade is out, of landing a man on the moon and returning him safely to the earth. No single project in this period will be more impressive to mankind, or more important for the long-range exploration of space; and none will be so difficult or expensive to accomplish."

clincher came on February 20, 1962, with John Glenn's magnificent three-orbit flight aboard *Friendship* 7. Imagine being able to circle the globe once each ninety minutes, high above all clouds and turbulence! Kennedy's moon loomed a bit closer now, and the old heads at Edwards no longer seemed to be above all competition. So despite the fact that I was picking up some interesting piecework, flying odd jobs in the F-100, F-102, and F-104 jet fighters, I was more than ready when, in April 1962, NASA announced that it would select a new group of astronauts to supplement the Immortal Seven.

NASA's terse announcement was encouraging, in that the agency was sticking with test pilots. This narrowed the field considerably, as there just weren't that many certified test pilots around, even though NASA was opening the selection to civilian as well as military pilots. A degree in one of the biological sciences or engineering was required, and candidates could not be more than six feet tall or over thirty-five years old, as of the day of selection. Five to ten were to be picked, with June 1 the cutoff date for applications. My application was in before the ink was dry on the announcement.

The Air Force decided not to pass all applications on to NASA, but to conduct some preliminary screening of its own. At the end of May, I was summoned to Washington for an interview by Air Force personnel specialists, and those of us who passed that were called back to Washington in the latter part of June for a couple of days of indoctrination in how to make a favorable impression on NASA during their forthcoming tests and interviews. We promptly dubbed this "charm school," and weren't quite sure whether to be offended, to scoff at its validity, or to take its lessons to heart. Perhaps the Air Force had not been pleased that, of the Original Seven, only three were Air Force officers.

At any rate, like would-be radio announcers, we read selected passages aloud, and these were critiqued at great length by a man who held a Ph.D. in education. We made impromptu speeches. We were told how to talk, dress, stand, and sit (wearing knee-length socks, of course, so no gross expanse of hairy leg intruded upon whatever Ph.D.'s in education think about), how to answer questions (not too long, not too short), how to drink at parties (long drink, make it last). We even got pep talks from General Curtis LeMay, Air Force Chief of Staff, and General Bernard Schriever, Research and Development Commander, who were almost as good as Roscoe Turner. But, in my opinion, the apogee of the course was surely reached the day we learned how to hold our hands on our hips (doesn't anyone learn anything at home any more?). Thumbs forward, ladies! Thumbs to the rear, gentlemen! To reverse would not only be gauche (perhaps even worse) but, of course, would result in instant detection and rejection by NASA, the old meanies. I have studied this matter exhaustively in later years, and have found burly construction workers with thumbs forward and mincing fairies with thumbs aft, but then they obviously are not charm-school grads.

Armed with the kind of self-confidence this information engenders, I was ready for the next hurdle, which was a real one— passing a five-day physical exam conducted by the Air Force School of Aerospace Medicine at Brooks Air Force Base, Texas. I had been there the year before, when our test pilot school graduating class had, with a bit of prodding, "volunteered" to help the medics gather some base-line data on healthy patients by undergoing a wide variety of tests. So in early July 1962, on my second trip to Brooks, I knew what to expect. Monday morning you arrived, fasting, to be greeted by a pleasant lab technician, part Dracula and part leech, who takes what seems to be a quart of blood. You then breakfast on a large beakerful of glucose, sickeningly sweet, and are punctured with more needles at regular intervals during the remainder of the morning, which is interspersed with various other tests. By noon the medics have measured, by the level of sugar in your blood, any tendency toward diabetes, and you have taken their measure as well; from now on it is an adversary proceeding. Inconvenience is piled on top of uncertainty on top of indignity, as you are poked, prodded, pummeled, and pierced. No orifice is inviolate, no privacy respected. You are a secondhand car being inspected prior to a coast-to-coast trip. Cold water is poured into one of your ears, causing your eyeballs to gyrate wildly as conflicting messages are relayed to your brain from one warm and one cold semicircular canal. Your body is taped with electrocardiogram sensors and you are ordered onto a treadmill, which maintains its inexorable pace up an imaginary mountain road. As the tilt becomes steeper, the heart rate increases, until it finally reaches 180 beats per minute. Then you are allowed to rest, and your heart's recovery time is measured. You are made comfortable, strapped onto a flat table which is suddenly jerked upright, measuring your cardiovascular system's response to a sudden shift in the gravity vector. You blow into a bag, as rapidly and deeply as possible, and your pulmonary efficiency is measured. The pressure inside your eyeball is ascertained by an ingenious metal cup placed against it, since increased pressure can be the harbinger of glaucoma. Your fanny is violated by the "steel eel," a painful and undignified process by which one foot of lower bowel can be examined for cancer or other disease processes. Your eyes and ears are tested, with an unbelievable attention to detail, by some of the foremost specialists in the world.

Then the shrinks take over where their more stable compatriots leave off. Thrust and parry. What are inkblots supposed to be, anyway? Is one crotch in ten pictures too many? How can I describe the blank, pure white piece of paper this year? Last year I said it was nineteen polar bears fornicating in a snowbank, and the interviewer's face tightened in obvious displeasure over my lack of reverence for his precious cards. Hostile, they said I was. But not this year. This year I want to fly to the moon, badly I want it, and I will describe that white card in any way that will please them. Perhaps I see a great white moon in it, or a picture of Mother and Dad, with Dad a little larger than Mother. Second-guessing the shrinks is not easy.

By week's end I am weary, but fairly confident that no obviously disqualifying defect has emerged. I do have a list of minor ailments and abnormalities; years later I learn that it is normal to be abnormal in a half dozen or so measurements, but at the time I was not privy to anyone's problems except my own. I just assumed everyone else was Jack Armstrong; I knew that each litter must have a runt, and perhaps in this one it was me. I didn't care as long as I passed.

Brooks was the first opportunity I had to see the non-Air Force competition, the civilian, Navy, and Marine pilots who wanted to fly to NASA's moon as much as I. They were an impressive lot, gregarious and outgoing, smiling even with needles in their arms and thermometers in their mouths, obviously able to cope with all this and anything else an unkind fate might throw their way. I felt my competitive hackles rising. Smiling I was not, but I felt that I could handle a flying machine as well as anyone, and I just hoped I would be judged on that basis and not by a wild list of ancillary attributes dreamed up by charm-school chaperons or psychiatric soothsayers.

Back at Edwards in one piece, I barely had time to brief Pat on the physical wreck she had married before I had to scurry off to Houston for more tests and interviews. By now our group had been sifted, winnowed, and pruned to the point that, out of nearly three hundred candidates who had originally seemed to meet all of NASA's criteria, only thirty-two (thirteen Navy, six civilians, four Marines, and only nine Air Force—but they were all charming) were left. The interviewing was, therefore, getting to the serious stage, as one out of three or four would be selected. I wasn't very confident as I left Edwards, mainly because, with barely one year's experience in flight testing, I knew my credentials were marginal.

Ah, Houston! Even today the name has a magical, lyrical ring. I can't imagine calling from the moon with, "Say, Minneapolis, we have a problem up here" or "Salt Lake City, this is Astonisher 19, do you read?" No, Houston it had to be, and any blemishes the place had were certainly invisible as far as I was concerned. I checked into the Rice Hotel as if entering the pearly gates. NASA

had instructed me to use a fictitious name, which added a tantalizing overlay of mystery to my already exciting mission.

The next morning was a return to reality, as another series of
tests began. NASA seemed interested in measuring our powers of
observation, and took an approach I had never seen before—or
since. We were shown two movies, one a grand tour of our solar
system in which we flew by each planet in turn, and the second a
series of underwater shots of reef life, teeming with flora and fauna.
Then we took out paper and pencil and described, as best we
could, each planet visited and each fish glimpsed. We also took
another battery of psychological tests ("Are you a snob or a
slob?"), and were interviewed by a panel of NASA physicians, who
had before them the results of our Brooks physical exams. They
were a chatty lot, and I learned a bit more about my ailments (all
minor) and a few tidbits about some of my compatriots. Apparently the Brooks shrinks had forgiven me the polar bears, and even
liked my test results.*

Then came the main course—the technical interview, a meat-
and-potatoes session conducted by people who wanted to know
what we had done and what we knew, and who didn't really care
whether we held our hands on our hips while answering. Deke
Slayton was there, and Al Shepard, and Warren North, who was in
charge of astronaut training and who was also an ex-test pilot. John
Glenn drifted in and out of the room, asking an easy question or
two and smiling at the answers. The other three were tough,
friendly, but no-nonsense, and their technical questions required
answers of substance and precision. In some cases I felt I gave
satisfactory answers; in others I clearly had not. I had not, for

---

* A Beetle Bailey cartoon says it all, in describing what bothers psychiatrists about
astronauts. We eavesdrop on a conversation between Dr. Bonkus and Captain Scab-
bard, who is Private Zero's boss.
    DR. BONKUS: I'm worried about that boy.
    THE CAPTAIN: Private Zero? Why?
    DR. BONKUS: He has no hang-ups, no worries, no phobias . . .
        no problems, no neuroses, no fears.
    THE CAPTAIN: What *has* he got?
    DR. BONKUS: I don't know—that's what worries me.

example, known what reliability data had been accumulated on the Atlas booster to date, and my guess was far off the mark, on the optimistic side. Today most people take it for granted that these rockets will not fail, but in 1962 fewer than nine out of ten Atlas launches were successful. When my half hour was over, I left the impassive trio poring over their papers, preparing for the next victim.

That night we perspiring hopefuls had cocktails and dinner with our inquisitors, as well as some of the other key people from NASA's fledgling Manned Spacecraft Center. Scott Carpenter and Gus Grissom were there, so of the Original Seven, I had now met all but Gordo Cooper and Wally Schirra. They were an impressive bunch, photogenic, personable, with quiet assurance their common denominator. They somehow stood out in a crowd; if they had been chosen by a public-relations man, he could not have made a better choice. Yet I guessed that PR had had little to do with it. The emphasis in their selection was reported to have been on breadth of test pilot experience and on physical conditioning. As NASA gained experience, these two factors decreased somewhat in importance and were replaced by an emphasis on youth, education, and scientific specialties. But in the early days, the environment of space was expected to produce a variety of surprises, and who better to cope with them than a tough old bird who had been around the block once or twice. I think it was good reasoning.

The director of the Manned Spacecraft Center, Bob Gilruth, was also at the dinner: a delightful, bald-headed teddy bear of a man, reedy-voiced, twinkly-eyed—certainly not impressive, but capable of putting together an impressive team. The more people I met, the better I liked the whole operation. After dinner, someone suggested that we repair to a local strip joint which was featuring "Amateur Night." Aha! Was this the final test of the day? I weighed the upside potential versus the downside risk, in good stockbroker fashion, and decided not to go—admittedly a decision tinged with a bit of paranoia. Thus, I returned to Edwards forever in the dark as to how the amateur lovelies had made out, and with not much more information about that other group of thirty-two amateurs.

Back at Edwards, the really tough part—the waiting—began. Of the nine Air Force finalists, six were stationed at Edwards, and we warily kept tabs on each other, on the lookout for any abnormal behavior that might indicate someone had heard good news from NASA.

On August 19, 1962, I wrote my father:

> I suspect that there will be close to ten people selected and I suspect that the breakdown will not unduly favor either the Air Force or the Navy (they say, of course, that everyone will be selected on merit regardless of his branch of the service). I strongly suspect that at least one civilian will be included, for propaganda purposes, if nothing else, and Neil Armstrong will be on the list unless his physical discloses some major problem. I say this because he has by far the best background of the six civilians under consideration, and he is already employed by NASA.

Today, the military chauvinism expressed in this letter surprises me; otherwise it was about on target. In mid-September, when the word arrived, Neil's name was on the list, with one other civilian (Elliot See), four Air Force, three Navy, and no Marine. Three of the Air Force selectees were from Edwards—Frank Borman, Jim McDivitt, and Tom Stafford. The fourth, Ed White, a classmate of mine from West Point days, was involved with all-weather testing at Wright-Patterson Air Force Base in Dayton, Ohio. The three Navy troops were Pete Conrad, Jim Lovell, and John Young. In my opinion, this group of nine was the best NASA ever picked, better than the seven that preceded it, or the fourteen, five, nineteen, eleven, and seven that followed.

My own failure was, of course, quite a blow, even though I had never really expected to make it. I wrote my father that I had guessed four out of nine correctly (Armstrong, Borman, Lovell, McDivitt), but I found it impossible to conduct a really satisfying postmortem. The logic which would explain Ed White (athletic,

gregarious, more in the mold of the Original Seven) did not explain Stafford (the schoolteacher), and so on. Nor did Gilruth's
letter to me provide much usable information.

> The impression that you made on our Selection Committee was
> favorable. Overall, however, we did not feel that your qualifications
> met the special requirements of the astronaut program as well as
> those of some of the other outstanding candidates.

The *special requirements*, eh? Well, if I didn't have them, could I
get them? Or was the door irrevocably closed behind me? Being a
perpetual optimist, I focused on two points: first, there were strong
rumors it wouldn't be more than a year or so before the burgeoning
space program required selection of a third NASA astronaut group;
and second, Conrad and Lovell were rejects from the Original
Seven's screening process. Therefore, since I felt my basic problem
was lack of experience, hopefully I could grow into a successful
candidacy next time around.

In the meantime, my job at Fighter Test, which had seemed
the best in the world, no longer quite had a larger-than-life quality.
I knew people in Houston and I envied them.

With NASA growing by leaps and bounds, the Air Force
became increasingly edgy about the role it might be assigned in the
exploration and exploitation of this new medium grandly called
Space. Space appeared to be the great wave of the future, and
certainly military uses would be found for it. The Russians, with
their lead, had probably already learned to put BOMBS up there, and
we couldn't get them down. We must learn to neutralize them, or
at least inspect them, and besides, we should have our men—if not
bombs—up there too, where they could perform the classic first
step of any military action—reconnaissance. No doubt about it,
space was the new high ground, and the Air Force had best prepare
to occupy it.

Many things were needed to do this—money, political support, technology, people. And many different kinds of people, including qualified crew members. NASA seemed to be having success training test pilots to be astronauts (was there a difference?), and the Air Force decided to do the same. The USAF Experimental Flight Test Pilot School was promptly renamed the USAF Aerospace Research Pilot School, and the course was extended and the curriculum broadened to include the fundamentals of space flight. In this way, the Air Force hoped to build a cadre of officers trained to fly any new vehicles which, in those confusing days, might come along. To ease the transition from test pilot school to ARPS (rhymes with carps), as the new school was called, a couple of "postgraduate" classes were conducted. In these, selected graduates of the test pilot courses were brought back for six months of additional instruction, while the staff struggled with the formidable task of overhauling the curriculum to provide the basic skills needed in this new space age.

When I was offered a chance at the third postgrad class, I eagerly filled out the application. NASA and its astronaut program were in my blood now, despite my recent failure, and anything which would qualify me for the next NASA selection was worth a try. So I took a six-month leave from Fighter Ops, and in October 1962 reported to ARPS down at the other end (the "wrong" end) of the flight line.

Only two years ago, I recalled, I had gone through all this—books, classes, flight-test data, weekends, dog work—and while it had been fun and had tapped the competitive juices the first time through, frankly I was not looking forward to it again. I was just trying to get some of that "experience" NASA had thought I was lacking. Or maybe they were just being polite and I would never ever qualify to be an astronaut, but it was worth a bloody go anyway, and here I was. Here also were a strange assortment of classmates, a mixture of the old and the new, mostly the old. Some, like Charlie Bassett, were graduates of the most recent test pilot class and had been held over, but most were more like Ed Givens, who had graduated from the basic course four years before and who had held a wide variety of assignments since then. Each of the two, incidentally, had received the outstanding student award in his class, so this was not a group to be taken lightly. It also included Jim Roman, whose undergraduate degree was in mechanical engineering, following which he had gone to medical school and then into Air Force pilot training. An engineer-physician-pilot! Here was Greg Neubeck again, still full of the same steely determination he had shown in our test pilot class; Joe Engle, who went on to fly the X-15 before being picked by NASA a couple of years later; Doug Benefield, one of the few multi-engine types, with vast experience in the test-flying world and with the delicate touch of a fighter pilot; and so it went, a small class of ten with a wide variety of flying experience.

In fact, we students had about as much experience of one type or another as did the staff, who were struggling to design new "space" courses to teach us, to keep a chapter or two ahead in a textbook which had not yet been written. Consequently, we had an easy time of it; we were treated as equals, and the cutthroat competition of the test pilot course was replaced, temporarily at least, with a no-grades gentlemanly informality.

We also had some interesting flying to do, centering on the F-104. Built by Lockheed in the mid-fifties, the Starfighter was the first Mach 2* fighter in our inventory. It had a long fuselage with a sharply pointed nose, a single huge engine, and tiny short wings. The wings measured only seven and a half feet from root to tip, and were so razor sharp on the leading edge that special protectors were installed whenever the plane was on the ground. The plane was designed ("optimized") for all-out speed, and it did go like hell, preferably in a straight line. When slowed down, it was simply too heavy for the amount of wing area (too high a "wing loading"), and if the engine quit, you'd better be over the Edwards dry lake, because it dropped like a stone.

* Mach number (named after the German Ernst Mach) is the speed of an airplane relative to the speed of sound in the air it is passing through. Mach 2 equals twice the local speed of sound. It is used instead of mph because it is a better indicator to the pilot of aerodynamic peculiarities.

The F-104 was our toy at ARPS, and we flew it well. We flew it high and fast, and low and slow, and we pretended it was a spacecraft and tried to do spacecraft-like things with it. We dressed up in pressure suits, climbed to thirty-five thousand feet, got it going as fast as was legal (Mach 2), and then pulled back on the stick and zoomed upward as high as it would go. Trading kinetic energy for potential, up we would go, ever more slowly, until we floated over the top of a lazy arc in a not-so-bad simulation of the weightlessness of space. Using this technique, we would stagger up to around ninety thousand feet (I think Jim Roman held the record in our class, to the chagrin of us old test pilots). At ninety thousand feet, the sky overhead is so dark blue as to be almost the pure black of space; the weather, clouds, indeed practically all the atmosphere is below you; the horizon becomes a more sharply defined line, and you can detect the curvature of the earth. So near yet so far away! Grab a quick glimpse, then back to mundane chores in the cockpit, for the zoom up into this very thin air has not been without casualties. First, the afterburner portion of the engine blew out routinely at around sixty thousand feet. Next, the main-engine temperature would begin to climb dangerously, so that if the engine did not flame out of its own accord, it had to be shut down, usually in the vicinity of seventy-five thousand feet. Third, as soon as the engine quit, cabin pressurization was lost, and the only thing between the pilot and a blood-bubbling death was the pressure suit, inside which a safe if somewhat reduced pressure was maintained. Since pressure-suit gloves had been known to pop off, we were especially careful to check the wrist locks prior to these zoom flights, and then we would tape the locks down in place with adhesive. Space-age baling wire! A mixture of sophistication and ingenuity, or crudity and stupidity? I would have plenty of time to think that one over, and other crew equipment design problems like it, in the years to come. In the meantime, one more glance at the glorious view, then back to this real-world illustration of what goes up must come down—and hopefully over Edwards with the engine working and in time for lunch, because my appetite had truly been whetted.

We spent time indoors too, some of it bone-crushing boring, some fascinating. We derived all the equations necessary to prove to the most hardened skeptic that a spacecraft *could* orbit the earth; the higher the slower, the lower the faster. Right? Right—if at one hundred miles, one orbit takes an hour and a half; if we go up to 22,300 miles, one orbit takes twenty-four hours. So what? Well, the earth also takes twenty-four hours to go around once, so that the 22,300-mile satellite will stay over the same spot on the ground, i.e. be synchronous with the earth's rotation, which is fiendishly clever if you want to use it as a communications-relay satellite. Interesting applications of boring equations.

Some of this was merely an extrapolation of what we already knew, pushed up into a higher, faster realm. On the other hand, some of the space rules were the very antithesis of airplane rules. Consider the rendezvous problem, with one vehicle behind another and the laggard wishing to catch up. In an airplane, the answer is quick and simple: use increased engine thrust to increase speed and approach from the rear; then reduce relative velocity to zero to fly in formation with the target. Try this in a spacecraft and it simply will not work: adding thrust toward the target puts one into a higher orbit, and high orbits are slower, remember, so instead of catching up, one falls behind. No, against all instincts, one must apply thrust away from the direction of the target, drop down into a lower, faster orbit, and then transfer back up into the original orbit at precisely the right point in the catch-up trajectory. To complicate matters, the speeds and distances involved in the orbital situation made the pilot's eyeball useless a great deal of the time, and gimmicks such as radar and computers have to be carried on board and operated in conjunction with a whole bevy of geniuses on the ground, who in turn are armed with their own, more powerful, radars and computers.

When our minds became saturated with the equations and other mathematical gibberish pertaining to all this—"orbital mechanics," it was called—the ARPS staff would have the good sense to whisk us away somewhere for a breather. Once we visited a couple of aerospace contractors' factories, where not much was being built but where advanced design groups were delighted to explain to us their far-out plans for zinging off to Mars using atomic power, or for building a huge machine that would fly through the atmosphere, scooping up and liquefying the propellants required to enable it to zoom up into space.

We also spent a couple of weeks with the Air Force space doctors in San Antonio (some of the same ones who had administered the NASA physical a few months before), where a large and competent staff had been assembled over the years to consider problems running the gamut from ear blockage in diving jets to the cardiovascular effects of weightlessness. These were sharp, inquisitive, highly skilled professionals in the burgeoning field of aerospace medicine—and they welcomed us with open arms. We were invited into the inner sanctum. None of this cold doctor-patient relationship, we were behind the scenes; they kept no secrets from us, and I wondered why.

First, they gave us a basic course in physiology, a very swift and deft survey of the human body and its components (just what the hell does a spleen do, anyway?). Then they supplemented these classroom lectures with a look at their laboratories, where various experiments were in progress, mostly involving anesthetized dogs. Some of these were beauties, purebred shepherds and setters that had strayed or been abandoned or stolen. I confess I spent more time empathizing with the victims than I did learning the intricacies of the medical principles involved; perhaps I felt like a victim-to-be. Although I would never have the vivisectionists marching in my behalf, I began to feel a little like a volunteer dog, ready to add his bit to the storehouse of medical knowledge waiting to be tapped by flights into weightlessness. I began to feel like one of those data points I used to fuss over in test pilot school; there, at least, I used to produce data points; here I *was* one. Could this be why the medics were all of a sudden so friendly? Did they want helpful, understanding, cooperating data points, rather than suspicious, hostile ones? Or were they, as frustrated space buffs, simply welcoming comrades in the long struggle toward manned space flight, where medical problems were justifiably on the list along with their engineering counterparts? At any rate, speculation at this point was useless, and as a group we relaxed and enjoyed the bizarre change of viewpoint.

And parts of it really were bizarre, such as the day we were all herded downtown to visit the city morgue. An autopsy was in progress on an elderly lady who had died of peritonitis, resulting from a punctured bowel. Waiting their turns on a sideboard were a couple of premature-infant corpses. I'm sure a morgue is never like a florist shop, but the stench of peritonitis is especially bad, and our class was about evenly divided between those who had to leave immediately, those who stayed, barely, and those who took an active interest in the proceedings. As a member of the middle third, I tried to figure out just what the bleeding hell we were doing there? Did those who left fail the course? Was this another manifestation of the medics' trust in us? Was this supposed to make us at home with death on earth, in case we had to cope with death in space? Or were we really supposed to learn something here, from the awful obscene jumbled pile of poor old lady parts in front of us? Peritonitis is no way to go, baby, that's all I learned.

Back at Edwards, in more familiar surroundings again, our class entered the home stretch and began to focus on the military student's usual worry: what is to become of him? What assignment will the great impersonal Pentagon personnel machine belch out this time? The first time through test pilot school, I had been offered an instructor's billet and had turned it down, gambling successfully that I would end up in Fighter Ops. This time, having come from Fighter Ops, I was presumed to be going back there, and no other options were offered. Yet, I still had to have a piece of paper from the Pentagon, and it could say Timbuktu, for God's sake. What had I been thinking of, to volunteer for this strange school that derived equations and dissected old ladies? I deserved Timbuktu, or at least Wright-Patterson, where they "test flew" planeloads of wonderful new black boxes ad nauseum.

Fortunately, my fears were soon allayed and I was given my walking papers back to Fighter Ops. At almost the same time (May 1963), the persistent rumors about another NASA selection were confirmed, and I eagerly scanned the announcement, which was officially released June 5, 1963. The requirements were: "Be a United States citizen . . . have been born after June 30, 1929 . . . be six feet or less in height . . . have earned a degree in engineering or physical science . . . have acquired 1,000 hours jet time, or have attained experimental flight test status through the Armed Forces, NASA, or the aircraft industry . . . be recommended by present organization." Compared to the 1962 criteria, the maximum age had been reduced one year, and test pilot certification was changed from mandatory to preferred. NASA clearly wanted to broaden its base, and later emphasized that it had conducted a talent search by soliciting recommendations from "the military services, various reserve organizations, industrial aerospace firms, and other organizations such as the Society of Experimental Test Pilots, the Airline Pilots Association, and the Federal Aviation Agency." For me, there were good and bad aspects to the announcement. Fortunately, I still was under the age limit (by sixteen months) and I met all the other requirements; on the other hand, broadening the field to non-test pilots was clearly going to open Pandora's box, and I had no idea what kind of supermen were apt to fly out.

The selection process itself followed the pattern established the previous year: by the mid-July deadline, a NASA selection committee had screened 271 apparently qualified applicants and reduced their number to thirty-four. The thirty-four of us went off to Brooks for physicals, most for the first time, a few for the second, and me for thirds. By now I was greeted there like an old friend; even the "steel eel" seemed almost friendly, and that means it's time to start worrying. As elder statesman, I sat on a bench in the corridor and directed traffic to the various clinics with a savoir faire that must have been maddening to my fellow victims. I was even allowed to skip the psychiatric evaluation, the theory being that I couldn't have gone completely buggy in one short year. I appreciated that, but was mildly disappointed that I couldn't tell the blank card story one more time. "You see, once upon a time there was this id, er, I mean kid, who wanted to fly to the moon. Naturally the first person he turned to for advice was his psychiatrist, who . . ."

I also found out that the previous year's thirty-two finalists and their assorted ills had been cataloged in a 276-page document which, for me at least, made fascinating reading. Witness the following:

> Only seven candidates denied that they had ever been involved in legal action. Minor infractions appeared to be the rule, the vast majority of which are speeding offenses. In the vast majority of instances the legal offenses were indeed minor and rarely of more serious magnitude. Offenses listed include:

Failure to yield right of way	1
Failure to stop at stop sign	2
Speeding	16
Illegal parking	6
Two traffic violations	7
Three or more traffic violations	5
Driving without a license	1
Moving traffic violations	3
Other minor traffic violations	1
Disturbing the peace	1
Wreckless driving	1
Driving while intoxicated causing an accident	1
Illegally shooting birds in a field	1

I really had to admire that "wreckless" driving, a status which, like virginity, once lost can never be regained. I also felt, I hope, the proper degree of revulsion over such aberrant behavior as illegally shooting birds in a field. Where would it end?

I also discovered that the mean IQ of the 1962 group had been 132.1 on the Wechsler Adult Intelligence Scale. My own score in the Miller Analogies Test, which measures verbal abilities, had been the highest in the group, whereas my mathematical reasoning and engineering analogies scores were much lower. I was reminded of Pete Conrad's battle cry: "If you can't be good, be colorful!" —except that Pete is both very good *and* very colorful.

I also did my stint on the treadmill for the third time, my cardiovascular conditioning apparently getting better with each passing year. If the data could be plotted on a curve, it would show that at age thirty-one I was ready for a wheelchair, while by the time I reach sixty-two I will be ready to try out for the Olympics. Actually, there is an interesting story behind this, involving exercise and smoking, but I want to save it for a later chapter in which I hope to debunk the whole myth of space flight vis-à-vis physical conditioning and to describe a sensible adult exercise program.* I also got my "lean body mass" (as opposed to "body fat") measured again, by an esoteric process involving injecting radioisotopes into the bloodstream and measuring their diffusion. I had been told the previous year, by this measurement, that my weight of 165 pounds was exactly right, but this year I was advised to take off a few pounds, because I weighed 163! I didn't argue the point, this being but one measurement out of a jillion I had by now accumulated, but instead I gratefully toted my bloated carcass and even more oversized medical folder back to Edwards.

In 1962 it had been impossible to explain to my non-believer of a wife why I wanted to transfer our happy home from Edwards to Houston, trading the high risks of test flying for the absolutely freakish world of space flight. However, after living through the selection process, seeing my enthusiasm for it, and becoming

---

* All this from a veteran member of Athletics Anonymous, that stellar organization which, when you feel the urge to exercise, will send someone over to drink with you until you return to your senses.

gradually accustomed to the idea of living in Houston (out of the Edwards wind), Pat was by now at least amenable to the idea and was keenly following the nerve-racking daily developments of this contest. It was a traumatic time for all of us. Pat had had two babies in rapid succession: Ann Stewart, born in late 1961, and Michael Lawton, who was only five months old.* My father, ever more deaf but enjoying robust good health and satisfying golf scores, died very suddenly of a heart attack on June 30. Granted he was eighty, but it seemed that for a man who had walked gracefully on his hands at sixty, he had been prematurely and capriciously taken from us. Furthermore, I knew that he would have enjoyed beyond measure a vicarious participation in the space program, for he was always ready to visit a new place or to entertain a new notion. The exploration of the moon would have surely been on his list, with perhaps a twinge of regret that he couldn't have made the trip himself. Now he would never know whether his son would have such a chance, and I felt a great sadness. At least he had known about his grandson, and seemed especially to enjoy our naming him Michael Lawton. Frank Lahm died the same week he did, and it seemed that a little of me had died too.

At any rate, when I left Edwards on September 2, 1963, to go back to Houston for interviews, I carried with me a sense of finality, a fatalist's foreboding that a moment of truth was coming; a test which, failing, would keep us at Edwards for a relatively conventional life, or, passing, would raise us to a new level of consciousness and achievement. In retrospect, this analysis seems more than a little overblown, but that's how I felt at the time, and I was really "psyched up" for the inevitable interview. Oddly enough, the interview itself seemed a lot easier this time, despite the importance I placed on it. For one thing, I was not among strangers any longer, and for another, the ARPS training had

---

* Ann received my mother's maiden name, and in a dazzling display of family planning was born on Halloween, which is also *my* birthday. Michael, given the name of my father's mother's family, was born February 23, 1963. These two Edwardsites joined Bostonian Kate, who was four at this time.

provided me with a good deal of information I had not had the year before. Also, I had a better idea of what they might ask, and had studied for their questions. Even Deke Slayton and Warren North seemed to have mellowed a bit. Houston, on the other hand, seemed unchanged—still a magic place. The Houston ship channel, for instance, was described as "too thick to drink and too thin to plow," but for me it had the allure of a virgin trout stream as I cruised along its banks looking for potential places to live. There was no "amateur night" this time—everything seemed more serious, especially my attempts to convince NASA that it couldn't fly to the moon without me.

Unconvinced myself, I eased on back to Edwards and quietly resumed my life in Fighter Ops, pretending all was well. Back to the barrier, or whatever flying was available; I was injured now, vulnerable as I had never been before, waiting for this last-chance word from NASA, a word which could not have been terribly important to Gilruth, or whoever would send it, but which had the capacity to destroy my self-satisfied equilibrium, to banish me to the bush leagues forever. The weeks became a month and more. Finally, on October 14, I left Edwards for a short trip to Atlanta, Georgia. I was to stop at Randolph AFB, Texas, to pick up a T-38 jet trainer and fly it on to Atlanta, where I would act as an instructor and check out a Lockheed pilot in the small trainer. As I was filling out the necessary forms prior to takeoff at Randolph, I was called to the phone. Houston calling, said the operator, and after an interminable series of clicks and squawks, on came the gravelly voice of Deke Slayton. Never one to use a paragraph when a phrase would do, Deke allowed as how he was ready to hire me, if I still wanted to come to work for NASA. *Still wanted to!* Here I'd been holding my breath for two years, and this guy, sounding as excited as a hostess offering a second lump of sugar, wanted to know if I had changed my mind! I don't have any idea what I said in return, but it must have been intelligible to Slayton, for he grunted, told me to be in Houston on October 18, and hung up. Somehow I got my T-38 to Atlanta and my Lockheed pilot checked out. Then, instead of heading back to Edwards, I detoured by Houston, arriving this time with the self-conscious smugness of the new club member. The occasion was the public announcement of our selection, at a press conference, but it was mainly a picture-taking and get-acquainted session, a very pleasant day that had been a long time coming.

We were the third group, or the Fourteen, making a grand total of thirty astronauts. We were sometimes erroneously called the "Apollo astronauts," as opposed to the Seven (the "Mercury astronauts") and the Nine (the "Gemini astronauts"). We were seven Air Force, four Navy, two civilian, and one Marine. We were younger (average age thirty-one) and better educated (5.6 years of college) than the two previous groups, but not as experienced. "College specialties are key items in selection . . ." said the Washington *Evening Star*, pointing to the fact that only eight of fourteen held test pilot certificates, and singling out Buzz Aldrin's Ph.D. thesis on manned orbital rendezvous. In retrospect, we were in the same tradition as the previous two groups, despite the press's natural tendency to highlight differences. It was not until 1965, when NASA selected five scientist astronauts, that a new breed appeared. Some of our group of fourteen thought we were a bit different, however, especially those who had not been through previous selections, and I was horrified to hear Rusty Schweickart say that our group would provide the first lunar landers. I expected them to be from the Seven and the Nine, with no such early chance for us upstarts. As it turned out, we were both right, Neil Armstrong being a Nine and Buzz Aldrin a Fourteen.

I returned to Edwards, and a week or so later, a Bob Gilruth letter made it official. This year, instead of harping on the impression I had made on the selection committee, he struck out in a new direction. "Your background and proven abilities assure me that you are fully aware of the responsibilities you are assuming by accepting this appointment. I am certain that your forthcoming association with the Manned Spacecraft Center will be of great benefit to our country's space program." I was glad he was certain, because I surely was not, but at least the runt of the litter was going to get a chance to find out. I had come a long way from Casa Blanca, but not because of any remarkable foresight on my part. I had had no master plan. The choices had simply been there and I had taken them, one step at a time. A free education at West Point or an expensive civilian one? Army career or Air Force? Pilot or ground officer? Fighters or transports? Test pilot training or more of the same? Fighter Ops or Wright-Patterson? Edwards or Houston? With one part shrewd logic and nine parts blind luck, I had qualified for Houston, and was now off to a different life, perhaps even to the moon.

To honor the occasion, Fighter Ops hosted a going-away party for Charlie Bassett and me. The two-man Gemini was in vogue then, and damned if they didn't give us one—a garbage can painted flat black, with a window for each of us—and damned if it didn't look like a Gemini. After that, the only thing left was to sell the Model A and the '58 Chevy, load Pat, Kate, Ann, and Mike onto an airliner, and take off for a new life in Houston.

3

A gentleman of 32 who could calculate an eclipse, survey an estate, tie an artery, plan an edifice, try a cause, break a horse, dance a minuet and play the violin.

—James Parton, *Life of Jefferson*

Like a pretty girl glimpsed fleetingly on a crowded street, Houston had seemed exquisite, perfect, absolutely without flaws or blemishes. It seemed that way especially at night, when viewed from the air, the city lights shimmering cheerily below filmy clouds covering the flat black coastline. But now it was time to discover the blemishes, to examine the warts at close quarters, to find out what this NASA was all about. The first thing I found out was that NASA in Houston had some reassuring similarities to Edwards. At least there was flying of sorts. The elaborate complex of buildings known as the Manned Spacecraft Center (MSC) had not yet been completed, so NASA had rented and borrowed temporary office space wherever it could be found. I finally located my new office in a renovated World War II barracks at Ellington Air Force Base, which was only a couple of miles from the MSC site and which is also where NASA keeps its fleet of astronaut-training airplanes. At the time, we had a half dozen old T-33 jet trainers, essentially identical to the Lockheed Shooting Star, which came out in the closing days of World War II, and a couple of F-102 jet fighters, which were fairly modern. These planes were used to maintain our "proficiency"—that elusive quantity and quality which says that one is at home in the air (at least in a familiar airplane), that one is not going to be airsick, or become distracted by a strange new environment, or panic when some emergency arises. How real, or valuable, this "proficiency" is I do not know, but I do know that there is a world of difference between flying a simulator and flying a real machine. I think the key to it is that one can only go so far in pretending that life and limb are at stake. In the middle of the fanciest simulation, one can nip out to drink a cup of coffee or answer the telephone, but if a fog bank rolls in off the Gulf of Mexico and causes Ellington to go below landing minimums as you begin your approach, then the adrenalin pumps, the abdominal muscles tighten up, the pulse rate increases, and life or death decisions get made, well or poorly, but they get made. Even in a battered old T-33.

The fleet also served as transportation. As we soon learned, the geographic distribution of the Gemini and Apollo contractors was well balanced. In the Los Angeles area was North American Rockwell, which made the Apollo command module and parts of the giant Saturn V. In northern California, Lockheed built the Agena, the Gemini's target vehicle and auxiliary propulsion engine. St. Louis was home to McDonnell, which built the Gemini spacecraft, while its Titan II booster was assembled in Baltimore. NASA's center at Huntsville, Alabama, was responsible for the Apollo boosters, the Saturn IB and the Saturn V. The lunar module came from Grumman's Bethpage, Long Island, factory, and the Apollo guidance system was worked out at M.I.T. in Boston. We lived in Houston and launched from the Cape. These

were just the main places involved, and the list should be supplemented with dozens of major subcontractors sprinkled around in
out-of-the-way places like Grand Rapids, Michigan. At any given
time, at least half a dozen different places would be actively seeking, or demanding, an astronaut's participation in some design
review, meeting, simulation, or PR visit. Our small fleet of airplanes was, therefore, a tremendous help in allowing us to go places
with a flexibility airline schedules do not permit.

Later we got more, and better, airplanes as we traded in our T-
33s and F-102s for Northrop T-38s, the Air Force's most modern
jet trainer. A twin-seater capable of supersonic flight (Mach 1.2
straight and level, Mach 1.5 in a dive), the T-38 is a small sleek
craft which cruises high and fast and can cover a thousand miles or
so—not bad with a paltry two-hour fuel supply. Like the F-104, it
is not especially good at low speeds, and in fact, the T-38 jiggles,
shakes, and buffets its way around the landing pattern to the point
where it is difficult to recognize an impending stall, since the traditional stall warning of airframe buffet cannot be used. The Navy
apparently found this low-speed buffet condition unsatisfactory
and has not purchased any T-38s as basic trainers, but the Air
Force uses them very successfully in its pilot-training program and
has reduced its accident rate substantially with them. The T-38 has
one other shortcoming: its small jet engines are quite fragile and
are easily damaged by chunks of ice, which build up on the lip of
the engine intakes, and then crack off and get sucked into the
engine. The resulting damage rarely causes the airplane to crash,
but it does require expensive major repairs; hence flying in areas of
suspected icing is prohibited. This restriction was particularly annoying for the Houston operation, since we were expected to be
places on time. It frequently cost the government a lot of money to
delay a test if we were late, and on any winter day some part of the
United States may be covered with icy clouds.

Despite these shortcomings, the T-38 was a welcome change
from the antiquated T-33, and being able to fly it is a dividend

which an astronaut manages to mention in the first five minutes when encountering old buddies who are leading up to, "What do you guys do down there in Houston, anyway?" Partly this is because flying a sleek new speedy jet on your own schedule is a status symbol of sorts, but more basically it is that this type of flying is exciting, demanding, even exhilarating at times; a wonderful outlet for pent-up desk frustration, a third dimension rarely available in our two-dimensional world. Just consider the roll control of the T-38 for a moment. If the pilot moves his hand an inch or two sideways, he instantaneously commands three thousand pounds per square inch of hydraulic pressure to actuate cylinders which deflect large ailerons into the speeding slipstream. With the aileron on one wing up, and the other wing down, the aircraft has a tremendous rolling moment applied to it and reacts by corkscrewing through the air more swiftly than the eye or hand can accurately follow. In less than one second, what is upright becomes inverted and then upright again; the clouds and sky are up, down, then up again; the earth is down, up, down. What power, to command the position of the earth! What glee, in being able to do it smoothly and precisely, with the horizon perfectly aligned before and after the roll, with an even and uniform motion during the one swift second the hand is at work.

So the T-38 brings with it great satisfaction, far beyond the superficiality of cocktail party surprise. "Sure, I left Houston after work, refueled at El Paso, and landed in L.A. before sunset. Would have been here sooner except I had to shut down one engine just past Phoenix." See her eyes widen. But as it brings satisfaction, it also brings sadness, for on rare occasions the T-38 smashes into the ground with a good friend or two in it, and then the trip to Arlington Cemetery wipes out all the small talk, and all the joy of an aileron roll, for a long time to come.

First it was Ted Freeman, who had the rare misfortune to strike a large snow goose while approaching Ellington. With both engines stopped, and possibly blinded by the bloody impact, Ted delayed his ejection procedure a second too long and plunged into the ground before his parachute had fully deployed. Then it was a double tragedy. Charlie Bassett and Elliot See, circling under a low cloud cover at St. Louis, trying to get to work at McDonnell, smashed into the very building they were seeking and ricocheted out into a parking lot, where at least they died without taking anyone else with them. Finally it was C. C. Williams's turn, coming home to Houston from Cape Kennedy. As he turned westward over Tallahassee, his T-38 unexplainedly went out of control, rolling, rolling, rolling; then diving vertically at supersonic speed into a not-so-shallow sandy grave.

Four highly experienced, expensively trained astronauts dead in airplane crashes! Was it worth it? As I write this, nearly eight years after Ted Freeman's death, I could fall back on statistics and say that if the T-38 took one good man from us every two years, then it gave more than it got, in terms of that elusive feeling of confidence, of "proficiency," which pilots need to cope with difficulties of the third dimension, be they two hundred feet, or two hundred thousand miles, above us. I could say that had these four not perished, four or more would have died during space flights. Or I could say that they at least were flying, not driving a Volkswagen as was Ed Givens when he was killed. Yet I cannot really believe these arguments; I only know that the T-38s must keep flying, just as the salmon keep swimming, and hopefully no more will die— salmon or men—until they have completed their journey.

But in the spring of 1964, in my new office at Ellington, I was not concerned about T-38s or death or any unhappy thoughts. I was the smallest fish in the biggest pond I had ever seen, and I was pleased as hell. Yea, verily this was the big time; the seven lean years were over, and the fattest years were coming up. They were to be doubly fat: Deke Slayton had suggested to us (he never promised anything) that we might expect a couple of flights apiece, and in the meantime things on the ground weren't half bad either, with a "personal story" contract with *Life* magazine and Field Enterprises. This contract, which brought us $16,000 a year for the first two years and then trailed off abruptly and sadly, was a source of great debate, in the press and on the home front. Pat felt that to take money under any circumstances for our participation in the space program was wrong, since the taxpayers had financed it and we should not gain personally from a public venture. On the other hand, I felt strongly that the contract was legitimate and that we were entitled to it. The argument was a complex one, and continues to this day, at least in certain elements of the press and astronaut worlds.

It is no secret that James Webb, NASA's voluble, aggressive, extraordinarily competent administrator, was dead set against it, and had so ruled. It was a closed issue until John Glenn caught President Kennedy's ear, and gained his approval, and of course that was it. I don't know exactly what John told the President, but the conventional argument ran something like this: these men had suddenly been thrust into the national spotlight and would be caught in its glare for years to come. The leaders of the country had a responsibility to see that the astronauts, representing a national program, were able to carry off this necessary and sometimes onerous public portion of their assignment, and among other things this meant money. Money for clothes, money to hire baby-sitters, money to bring wife and children to the right places, money to provide a proper home, money which simply was not there under the salary scale of junior military officers. The contract would also prevent infighting among the astronauts, since the money would be split evenly regardless of crew assignments, and therefore greed would not be allowed to rend this intimate and friendly team. Also, consider for a moment the possibilities in holding *Life* magazine a virtual captive: certainly after being invited into the home, and hearing wife emote and children prate, the stinkers weren't going to turn on their host (even a paid host) and write nasty articles. No, the contract almost guaranteed peaches and cream, full-color spreads glittering with harmless inanities. Against this, of course, must be weighed the vitriol of the rest of the press, which was left out in the cold,* but at least the stories coming from inside hearth and home would be friendly. In short, the contract seemed to be a practical and tasteful solution to the problem, and Glenn apparently had no trouble convincing a receptive President. Astronauts were *in* that year.

Just what did the contract say? Stripped of its legal verbiage, its only point was that *Life* and Field owned the "personal stories" of each astronaut and his family. Now what is a personal story, and why should it not be in the public domain? In the case of the astronauts themselves, it is certainly hard to define "personal." Test pilots are taught to perceive, to remember, to record every impression in flight—so that later, on the ground, they can report, as fully and as precisely as possible, exactly what happened. No one disputed this point, so that *what* happened during a space flight was discussed publicly at the post-flight press conference in as much detail as the press could stomach. But, of course, that was not sufficient. What they really wanted to know was: beyond all that technical crap, what did the crew *feel*? How did it feel to ride a rocket, what thoughts were racing through your mind as you plummeted toward the sea with the parachutes not yet open? How scared were you, anyway? This is what *Life* paid to find out, and what the others pried to find out without paying, and in truth, neither unearthed very much. *Life*'s little extra certainly wasn't worth the money. I suppose this was mainly because, as technical people, as test pilots whose bread and butter was the cold, dispassionate analysis of complicated facts, we were frankly embarrassed by the shifting focus. It didn't seem right somehow for the press to have this morbid, unhealthy, persistent, prodding, probing pre-

---

* Not surprisingly, the "pencil" press felt the bite of unfair competition more strongly than did the other media, and not surprisingly *The New York Times* howled as loud and long as anyone. Therefore, it came as a mild shock when, years later, long after this contract was terminated, *The New York Times* wrote a special contract with the crew members of Apollo 15 for a series of their personal by-lined articles. Naturally the *Times* did not escape unscathed, but the murmur of editorial disapproval seemed muted to me, perhaps because people were sick of the whole issue by then, or perhaps because for the first time these stories became "fit to print."

occupation with the frills, when the silly bastards didn't even understand how the machines operated or what they had accomplished. It was like describing what Christiaan Barnard wore while performing the first heart transplant. Furthermore, we weren't trained to emote, we were trained to repress emotions, lest they interfere with our very complicated, delicate, and one-chance-only duties. If they wanted an emotional press conference, for Christ's sake, they should have put together an Apollo crew of a philosopher, a priest, and a poet—not three test pilots. Of course, they wouldn't get them back to have the press conference, in all likelihood, because this trio would probably emote all the way back into the atmosphere and forget to push in the circuit breaker which enabled the parachutes to open.

At any rate, it was tough sledding indeed for *Life* or anyone else to bring a tear to the eye of these mechanical men, but at home—ah, that was different! Flip the armored beetle over and inspect the soft underbelly. What did little Sarah Jean think about Daddy's impending departure, temporary or permanent? How did Mom feel when Dad was up? (An astronaut wife I know, when asked this by a female reporter, blinked a time or two and then deadpanned, "Honey, how do you feel when *your* husband is up?" End of interview.) But no doubt about it, home was where the "personal stories" were, and this is what the contract was all about. I feel that it was perfectly proper to extract compensation for this invasion of privacy. I know that in politics all the family is fair game to some extent, and I recall that when I came to Washington to become Assistant Secretary of State for Public Affairs, several of the "Witches of Washington" were more than a little put out because my home was not open to them for an "in-depth family profile." However, I have never been a politician, and as an astronaut I felt that the stresses on my family deriving from my weird occupation were sufficient without exposing them to a constant stream of well-meaning but insensitive reporters who would exacerbate the doubts and fears which most certainly were there, as they would be in the minds of any normal, sentient wife and offspring.

Even more fundamental than that, I felt that I was paid to do my job to the best of my ability, and I put in long hours toward that end; but when I came home and that front door clicked behind me, that was it. My family life was my business and mine alone, and if I chose to open that door, it was perfectly proper to expect extra compensation for it.

One more comment about the contract: not only did it bring in some extra dough, split evenly among the thirty, thirty-five, fifty-four, or sixty-five families involved, but also it served as a beautiful excuse for turning off any non-contract interviews. In fact, once the word got around, the regulars would not even bother bugging wife and kids, because they already knew the answer would be "Sorry, no personal interviews, we are under contract." This was not true during the chaotic times when a husband was up, but that is a later story. Of course, the men still had professional, or non-personal, or whatever interviews, and these were scheduled on Fridays. If possible, we would schedule ourselves to be away from Houston on Fridays, but if we weren't, we grunted, hot under the TV lights, and sat walleyed while passing back and forth the same old cold potatoes. "Frightened? Yes, of course, to some extent, but you must realize that on this next flight we won't have time, what with the addition of eleven new medical experiments, which are . . ." *Sic transit media.*

Back in the mainstream of Houston life, far from the prying eyes of the reporters, we plodded and played our way toward the moon. I am afraid it was more plod than play, in the case of the Collins family at least, as neither Pat nor I were too interested in the various social opportunities available in town. In the first place, "in town" was twenty-eight miles away, as we lived neither in Houston nor in Galveston, but midway between the two, on the alluvial mud flats near the shore of what some master of irony had named Clear Lake. The terrain near this mud puddle was as flat as Edwards's dry lake, yet was not offered as evidence of an aberrant quirk of nature, as was Edwards; here we were supposed to believe that this was proper and just, that people would find nothing strange about living voluntarily in an area where nothing rises higher than a telephone pole on the Gulf freeway. Once converted to this pool-table perspective, one learned that twenty-eight miles was really only a one-cushion shot away, and if that one cushion was on the sofa of a very oil-rich hostess, so much the better.

Like professional tennis players and bullfighters, astronauts are the kind of freaks that hostesses adore. Usually their veneer of good manners is thick enough to last the evening, especially in the hands of a skilled hostess who knows when to break up a group before latent hostilities surface. Unfortunately, most parties include one middle-aged man, slightly inebriated, who wants to challenge the tennis player to a match at dawn, draw a little blood from the bullfighter, or expose the astronaut as some kind of witless super-monkey. Fortunately, the Manned Spacecraft Center had been built in Texas instead of New York, and the genuine warm hospitality of the informal Texans went a long way toward compensating for these occasional encounters.

Houston was also a great sports town, with the Oilers the premier team of the American Football League and the Astros doing whatever baseball teams do. Later the domed stadium came along, the first one in the country, and then watching the Oilers really came into its own. It was no longer just a sporting event but an all-purpose happening, centering on a social center called a skybox.

These skyboxes were high up, wedged under the roof, as far away from the playing field as one could get and still be inside the stadium. While this might be undesirable territory in a conventional open stadium, here the skyboxes were much sought after, and rich Texans queued up for a chance to lease one. The average skybox was larger than the name implies, as they were more suite than box, generally including two dozen seats in front and a room in the rear which could be outfitted with a bar and buffet of ample, or even luxurious, proportions. Each proud occupant tried to think up a decorative or culinary motif to set his skybox apart from all the others, and at half time one made the rounds, sampling *la spécialité de la maison* at each stop. From the bar rooms of the boxes, the playing field was out of sight, but not to worry, because, of course, the progress of the game could be followed on closed-circuit TV. Get the picture? Usually it was well into the third quarter by the time one lurched back to his seat, with Bloody Marys, popcorn, lobster Newburg, and Pearl beer all bubbling in his belly. I never could tell whether the late-afternoon cries of the crowd were due more to partisan fervor or to intestinal distress.

But the earth turns on its axis, and Monday follows Sunday, and then it was a different ball game. In the astronaut office, Monday morning was the time for a general group meeting, at which travel plans for the coming week were discussed and problems of general concern were aired. I was always fascinated by these gatherings, usually presided over by Al Shepard, which could be raucous, or hostile, or humorous, or argumentative, or informative—but which were never dull. "Eagles never flock," says a current advertisement; put thirty of them in a room and they hop and fidget, squawking mightily, flapping an occasional wing, pecking each other, unified only as they leave, flying out in a tight formation designed to amaze all the starlings and turkeys.

Who were these thirty? What were they like, what did they have in common? Were they all as alike as the papers indicated? Perhaps I should begin with the facts and follow with my own speculations and prejudices. I can, for instance, state their dimensions without fear of contradiction, as I have on my desk a chart covering sixteen (the Seven plus the Nine), with thirty-two anthropometric measurements listed for each man. From Pete Conrad's 138 pounds to Wally Schirra's 190, this storehouse of vital statistics reveals medial malleous height, biacromial diameter, and my all-time favorite, the distance from crown to rump. Sort of says it all, doesn't it? As I scan the chart, I'm not impressed. They seem a scurvy lot, with calf circumference a paltry 15 inches on the average, and they huddle together in protective statistical clumps, with Glenn, Slayton, Lovell, McDivitt, Stafford, and White all virtually the same height (5 feet 10½ inches), for whatever that is worth. Gus Grissom is shortest (5 feet 6.3 inches), John Young widest (19.9 inches at the shoulder), and Ed White has the longest reach (34.4 inches). Male chauvinism being what it is, nowhere can we find the beauty contest's convention of bust, waist, hips, but a little arithmetic puts it in the vicinity of 38-34-38. Ugh.

It is more difficult to measure the mentality of this group. I suppose the 132.1 mean IQ of the thirty-two finalists in the third group would be too low an average, because most of these were not selected. I recall that when, in 1966, I sat on a selection board which picked the Nineteen, the average seemed to be pushing 140, and there were two whose score exceeded 150 (supposedly the "genius" level). I doubt that there are, or were, any true geniuses in the group, but there weren't any dummies either, especially in regard to their engineering or mathematical reasoning. Criticism of their mental abilities, therefore, must center on some other area. Perhaps they were deficient in aptitude for the arts and perhaps they were not able to verbalize particularly well. I think it is fair to say that they had all the flaws of specialization which can be found in almost any segment of today's compartmented society. They might have been narrow, but within the limits of their responsibility they were good, they were true experts, and they were a pleasure to be around. The fact that all wanted badly to fly in space, to fly to the moon, to make the next flight—all this pent-up desire, all this competition—made for a lot of pressure, long hours while in Houston, and frequent trips out of town. This frenzy of activity to convince Slayton and Shepard of our true worth gave us all a common characteristic, but our individual differences more than offset any tendency toward bland conformity and made for a very diverse group.

At the risk of alienating friends, and with an awareness that it is unfair, if not impossible, to reduce any one human's unique qualities to a sentence or two, here comes my attempt to do just that. I present them in the order in which they come to mind.

Scott Carpenter A nice guy, but kind of out of it. Left the program early when it became obvious that his one Mercury flight was all he was going to get. Got hooked on underwater exploration, later got into the wasp-breeding business (yes, wasp-breeding).

Wally Schirra Oh ho ho! Could make a good living playing Santa Claus in a department store. This affability is backed up by a larger-than-life ego, but you have to admit that he is the only one to fly on all three in the series—Mercury, Gemini, and Apollo. His Apollo flight was especially gutsy, coming after a fatal fire, but then the spacecraft wouldn't *dare* blow up with Wally on board.

Deke Slayton The super straight shooter, honest, no-nonsense—grounded by the medics in an absurd auto-da-fé involving irregular heartbeats. Should have flown to the moon and back many times over by now, but has not gotten past his Houston desk, where he presides over all the astronauts and a lot of the engineers—and the program is better for it. The best boss I ever had, with the possible exception of William P. Rogers.

John Glenn The only one I don't really know, as he was leaving as I was entering. One thing for sure, though, he's the best PR man in the bunch.

Gordo Cooper Kind of went downhill. Flew well on Mercury, not a bad job on Gemini 5, but Apollo seemed too much.

Al Shepard "Big Al," and big in many ways. Shrewdest of the bunch, the only one to get rich in the program, he ran the astronaut office as merely one part of his far-flung empire. No teddy bear, Al can put down friend or foe alike with searing stare and caustic comment.

Frank Borman Aggressive, capable, makes decisions faster than anyone I have ever met—with an amazingly good batting average, which would be even better if he slowed down a bit. Attracted to money and power, in the long run Frank will probably be the most successful of the group, not counting Neil, who will, of course, occupy a special place in history.

Jim McDivitt One of the best. Smart, pleasant, gregarious, hard-working, religious. Thought by some to run a little scared, his thoroughness was legendary.

Pete Conrad Funny, noisy, colorful, cool, competent; snazzy dresser, race-car driver. One of the few who lives up to the image. Should play Pete Conrad in a Pete Conrad movie.

John Young Mysterious. The epitome of the non-hero, with a country boy's "aw shucks—t'ain't nothing" demeanor, which masks a delightful wit and a keen engineer's mind.

Neil Armstrong Makes decisions slowly and well. As Borman gulps decisions, Armstrong savors them—rolling them around on his tongue like a fine wine and swallowing at the very last moment. (He had twenty seconds of fuel remaining when he landed on the moon.) Neil is a classy guy, and I can't offhand think of a better choice to be first man on the moon.

Jim Lovell Like his good friend Pete Conrad (who inflicted the horrendous nickname of "Shaky" on him), he stands out in a crowd. A smooth operator, Shaky would do better in the PR world than in the engineering or technical end of things.

Tom Stafford Fantastic memory and eye for technical facts and figures; does less well with people. Politically ambitious, Oklahoman Tom projects the image of a schoolteacher, rather than the professional pilot he is, or the romantic entrepreneur he would like to be.

Donn Eisele Who? Lost in Wally Schirra's shadow on Apollo 7, Donn in 1972 became Peace Corps director in Thailand.

Mike Collins O.K. if you're looking for a handball game, but otherwise nothing special. Lazy (in this group of overachievers, at least), frequently ineffectual, detached, waits for happenings instead of causing them. Balances this with generally good judgment and a broader point of view than most.

Buzz Aldrin Heavy, man, heavy. Would make a champion chess player; always thinks several moves ahead. If you don't understand what Buzz is talking about today, you will tomorrow or the next day. Fame has not worn well on Buzz. I think he resents not being first on the moon more than he appreciates being second.

Rusty Schweickart A blithe spirit, eager, inquisitive mind, quick with a cutting retort, not appreciated by the "old heads." Mildly non-conformist, with a wide range of interests, contrasting sharply with the blinders-on preoccupation shown by many astronauts.

Dave Scott A Jack Armstrong, all-American boy, the last one you would expect to get involved in a shady stamps-for-sale deal, Dave should instead be remembered for his three stellar performances aboard Gemini 8, Apollo 9, and Apollo 15. One of the best.

Gene Cernan Relaxed, jovial, a pleasant companion. After Scott, the second in our group of fourteen to make three flights, two of them to the moon.

Dick Gordon Lots of balance, lots of common sense—one of the easiest to get along with. Likes to party, but never at the expense of getting next day's job done. If the New Orleans Saints don't start doing better, I'll be surprised (he's their VP now).

Al Bean Pleasant, persistent, relentless pursuit of required information—give him an office boy's desk and within a week he will know what the president of the company does. Very pleasant fellow to be around, especially if you like spaghetti, which is all he eats on a trip.

Bill Anders Intense, energetic, dedicated, no drink, no smoke, no nonsense—used to be inflexible and a bit immature until he became executive secretary of the National Aeronautics and Space Council in Washington, a job that would teach anyone humility and flexibility. Bill is now one of the Atomic Energy commissioners.

Walt Cunningham Outspoken, blunt, small chip on shoulder; strange mixture of Marine fighter pilot and Rand Corporation research scientist; a complex man alternating between genuine warmth and outright hostility.

I have drawn the line at the living, and exclude from the list of thirty Grissom, White, Chaffee (Apollo launch pad fire, Cape

Kennedy, January 27, 1967); and Freeman, See, Bassett, and Williams (T-38 accidents). *De mortuis nil nisi bonum.* Naturally all these men had their good and bad points, but with no opportunity for rebuttal on their part, I think it is better to omit description rather than present an intentionally unbalanced view. With regard to the later astronauts, there are just too many for me to describe individually. Some, like Ed Givens, were very close friends, while others I hardly knew at all.

If the thirty astronauts were a hard-working and effective group in 1964, so was the parent organization, NASA's Houston facility, the Manned Spacecraft Center. At Edwards I had learned never to get between an office building and its parking lot at 4:30, because the civil servants would assuredly trample you to death in their frantic dash for home. At MSC, people were more job-oriented, the clock was not so vital, and there was a pervasive excitement, a feeling of important work to be done, even with supper cold on the table. I have seen such dedication since, in the higher reaches of the State Department and in other Washington offices, but in 1964 in Houston it was a new experience for me, and it must have been for others as well, because the enthusiasm seemed to permeate all ranks, not just those who were highly paid and expected to act accordingly. Part of it was the bizarre nature of the work, and part of it was the timetable. Had not President Kennedy said *before* the end of the decade?

Even more important, the goal was clearly and starkly defined, as it rarely is in government. People knew that each day was one day closer to putting man on the moon, and balanced their progress against the time remaining. In 1964, of course, there was a lot of time remaining, but there was also a tremendous amount of work to be done, an enormous reservoir of doubt that had to be drained off before the task could even be considered a reasonable possibility for the sixties. Today the televised rovings across lunar hill and dale seem utterly routine, but in 1965 a friend of mine bet me $100 that no man would reach the moon (much less return to earth) before June 1972. My friend was not a layman, but an experienced test pilot and engineer who was employed at MSC and who was later selected as an astronaut. I didn't give him odds on the bet, either.

In 1964 in Houston, a lot of answers had still to be provided before any rational person could assess the chances of success. Eminent scientists, like Tommy Gold of Cornell, fretted over a possible dust layer on the surface of the moon whose thickness might just exceed the height of the lunar module (LM). Others postulated a charge of static electricity on the LM which would cause whatever dust was there to become attracted, and adhere, to the lander, obscuring the astronauts' view out of the windows. Lunar soil was thought by some to contain pure metallic elements which, when introduced on dirty boots into the pure oxygen in the LM cabin, would spontaneously burst into flame. Meteorites were known to exist in space, and with no atmosphere to protect it, the moon would clearly be an inviting target for them. Who knew how many there might be, or at what rate they might be impacting its surface? One has only to look at the moon's craters with naked eye to realize meteorites were not just a theoretical menace, but how real a threat to man they would be no one knew. At least the moon was well past the earth's Van Allen belts, which promised a healthy dose of radiation to those who passed and a lethal dose to those who stayed. Or how about a solar flare, that sporadic belch of energy spewing unpredictably from the sun?

These were the main problems of the environment, but even more worrisome were some of the unknowns involving transporting men and machines across a quarter of a million miles of hostile vacuum. The round trip would take about eight days, which, of course, meant that problems must not demand solutions in less than half that time. Including human problems, and no one knew what they might be. By 1964, the United States had completed its Mercury program and Russia its Vostok flights. Gordon Cooper had orbited the earth for thirty-four hours aboard *Faith* 7 in May 1963, followed the month after by Valery Bykovsky aboard Vostok V for eighty-one orbits, or slightly over five days. Gordo hadn't had any physical problems, but the Russian literature was a little disturbing in that several of their cosmonauts had reported feelings of malaise and nausea, which were considered to be outgrowths of the weightless state. We intended to take a detailed look into the matter by flying a Gemini, first for eight days, then for fourteen, but in the meantime we didn't know what to make of the Russian reports. Our own medics were not particularly helpful, either, as a number of them had been prophets of doom and gloom from the very beginning. It had been like pulling teeth to get them to admit that man had escaped unscathed from each successive foray into weightlessness. At the very beginning they had said that even a few seconds of weightlessness would impair bodily functions: one would not, for example, be able to swallow properly, especially fluids, and those nutrients which did reach the stomach would not be properly assimilated. Even worse, the heart and lungs would become confused at best, or incapacitated at worst, and the efficient human machine would quickly grind to a halt. How quickly remained to be seen, but as the data poured in, the medical threshold was not removed, but simply pushed a little bit farther backstage, to reappear in full force at each successive press conference.

The data from airplanes was, of course, severely limited, as even the most advanced could maintain weightlessness, or "zero G," for only twenty-five seconds as they raced through the sky in a gigantic parabola. Test subjects frantically drank, and swallowed, and voided, and carried on every possible test during these crucial few seconds, but alas, no contraindications to space flight appeared, so the medics grudgingly moved the "health barrier" back to the next threshold. When Al Shepard seemed to be his normal, healthy, obnoxious self after fifteen minutes of weightlessness aboard *Freedom* 7, the medics moved the decimal point over one place and said, "Well, yes, man can endure for minutes, and perhaps even for hours, but for days—horrors!"

So little by little, as the Mercury and Vostok results proved them wrong again, the medics stubbornly retreated another yard or two, raising hell with the mission planners and the equipment designers. In regard to Gemini, for example, they had a theory that if an astronaut spent a couple of days in space and then was subjected to the sudden stress of re-entry into the earth's atmosphere, his autonomic nervous system would forget to perform all its vital functions, and among other snafus, it would allow blood to pool in his lower extremities and cause him to pass out because of insufficient blood to the brain. Even worse, once passed out, the astronauts would remain strapped in their seats, with heads up and feet down, so that the blood would never return to their upper body and they would die!

Living and working with these people was like having an aunt who lives in a haunted house, or a close friend who sincerely believes in astrology and can't stop talking about it, especially delighting in reading you your horoscope on bad days. None of it is to be believed, but it's pretty goddamned difficult to ignore. What makes it even worse is that sooner or later the medics are going to be right, their negativism is going to be vindicated, and some bad medical thing is going to happen to someone in space. The next big chance was the Skylab program, but everyone made it up there for nearly three months, so how about an eighteen-month round trip to Mars? Dr. Charles Berry, the biggest blabbermouth in the space program, has already said that an all-male crew might become too testy for this task, so the medical community is again abuzz with speculation over the psychological, if not the physical, barriers to space flight. The truth of the matter is that the space program would be precisely where it is today had medical participation in it been zero, or perhaps it would be even a little bit ahead, because we could have done without all the impedimenta and medical claptrap, such as blood-pressure cuffs, exercise ergometers, and urine-output measuring devices. All they did was add weight, and complexity, and rob time and energy from tasks of greater value.*

* There have been, however, a number of unexpected benefits to sick people on earth as a result of the development of space medical equipment and monitoring techniques, so the time and money spent have not been entirely lost.

But if the medical problems existed mainly in the minds of the doctors, the engineering problems existed in everyone else's mind, including the astronauts'. How much fuel did it require for a successful rendezvous and docking around the moon, when no one had ever tried it in earth orbit? Half of the fuel we planned to carry, or twice as much? What would the temperature be inside the spacecraft during the constant sunlight on the way to the moon? With the sunny side baking and the shady side freezing, what equilibrium conditions would there be inside, where the softies lived? The electronics might endure variations of a couple of hundred degrees, but the human body demands ridiculously narrow limits of temperature and pressure. Could we guarantee those limits? How about humidity? With some equipment colder than the rest, would not beads of moisture condense, as on the outside of an iced-tea pitcher on a humid summer day, and if so, would not this moisture cause arcing and shorting of delicate electronics?

And how about the guidance system on board the spacecraft? Could it really find its way to that magic spot in the sky where the moon would be three days after launch, and then navigate back to earth again? The allowable limits were spectacularly small. On the return trip, the atmospheric "re-entry corridor," or zone of survivability, or whatever you wanted to call it, was only forty miles thick, and hitting a forty-mile target from 230,000 miles is like trying to split a human hair with a razor blade thrown from a distance of twenty feet. Granted, the primary responsibility for keeping the razor blade aimed precisely toward the absolute center of the hair would be the job of powerful ground tracking radars, coupled to gigantic computer complexes. But suppose the spacecraft experienced a communications failure and was unable to receive steering instructions from the ground. Would the astronauts be able to take over this delicate and vital assignment? The key to it would involve measuring the angle between a selected star and the moon's or earth's horizon, but how accurately could this be done? Assumptions were being made and equipment was being designed, but suppose mistakes were being built into the system?

Suppose the sextant design was too crude for the precision required, or suppose the instrument became warped slightly by virtue of the fact that it was built on the ground (one G) but had to perform in space (zero G)? For that matter, was the earth's horizon that clearly defined when seen from above the atmosphere? What would you see, the terrain horizon, or the atmospheric horizon, or some line in between? Perhaps it would shimmer, or move or vary with the angle of the sun's rays, or perhaps a hundred other unknown factors might creep into the equation. Suppose all was known to a fare-thee-well, but then some absent-minded technician at Cape Kennedy loaded the wrong number for the earth's diameter into the spacecraft computer, or, more likely, suppose the crew procedures incident to taking such a measurement were so involved and complicated that the astronauts made mistakes. Suppose each individual system aboard the spacecraft was a design masterpiece in itself, but when all the systems were added together they required twenty-six hours a day of crew attention? Suppose—suppose—suppose.

If I appear to be asking too many questions and not providing any answers, I am merely trying to convey the flavor of MSC in 1964, when we had a mandate to fly to the moon but few hard facts with which to work. It was primarily a question-asking operation at this stage, defining those things which needed answers, and trying to decide whether they could be answered on the ground in test chambers, or by astronauts flying simulators, or whatever; or whether they had to be answered with flight test data—which meant inclusion in the Gemini series or in one of the early Apollo test flights.

No one knew how many of these flights would be required, but clearly we test pilots had to modify our thinking to some extent. We were accustomed to a very conservative, stair-step approach to testing, in which as few unknowns as possible were tackled on any one test flight. Take *speed*, for instance. It might take several dozen test flights to coax a modern fighter plane up to its maximum speed, and extensive ground analysis of flight test data would be performed between each flight to make sure it was safe to proceed. Each step on the ascending staircase would be subjected to individual, unhurried scrutiny, a luxury which we discovered was simply not available to us when we switched from airplanes to rockets. No test pilot in his right mind would, for instance, go faster than the speed of sound on his first flight in a new airplane: aerodynamics are quite different on each side of the sound barrier, and the handling qualities of an airplane are sorely tested in the highly sensitive region near Mach 1. But consider the Gemini spacecraft, mounted on top of a Titan II rocket. When it goes, it goes! All the way to orbit, banging through the speed of sound, from step A to step Z with no intermediate stops, with no opportunity to reconsider or redesign along the way. In the case of the Gemini-Titan, it was possible to run two unmanned test flights before putting Gus Grissom and John Young aboard the third one; in the case of the Saturn V, again two unmanned flights preceded the Borman-Lovell-Anders flight of Apollo 8. Of course, it is true that the Air Force had a lot of experience with the Titan before the Gemini was mated to it, and it is true that NASA had conducted an involved unmanned test series with the earlier Saturns, the I and IB, before the V came along. But the Saturn V was truly a new beast, far heavier and more powerful than any flying machine the world had ever seen, and it was deemed sufficiently trustworthy to send men to the moon on only its third flight.

On December 21, 1968, I was sitting in the Mission Control Center in Houston with my eyes glued on a huge wall display which showed a moving dot (Apollo 8) streaking across the screen, straddling a line representing the ideal path in the sky which Apollo 8 must follow to set a safe course for the moon. My job was a simple one. If the dot deviated from course by an agreed amount, I was to radio Borman to shut down the third-stage rocket and return to earth, if he could. Too bad I'm not more of a betting man, because I'm certain that in 1964 Borman would have wagered a year's pay that no such preposterous plan would be followed for the third flight of the Saturn V. Had I added that it would be only the second manned flight of the Apollo command and service modules, I'm sure I could have gotten at least three-to-one odds. But if 1964 was for astronauts a year of technological doubt, a year for questioning scientists, engineers, and flight planners, it was also a year of great spiritual exuberance. We had not been with the space program long enough to become jaded by constant close contact with its marvels. Nineteen sixty-four was a great vintage year for Burgundy wine, as well, and it was a pleasure to know that the longer the wine waited in its bottle, the better it became—up to a point. When the time to drink came, years later, in some cases it would be harsh and bitter, but in others delicious and rewarding past expectation. Nineteen sixty-four was a year of promise, a year without manned space flights, but a year in which knowledge was being harvested and bottled for future use.

Wise or not, we were on our way, and I remember 1964 as a vintner might: "Young, still tasting of tannin, rough around the edges, but solid, with body, a spicy bouquet, and a strength suggesting greatness in five years' time."

4

We are all ignorant. We are just ignorant about different things.

—Will Rogers

But could I continue to be ignorant about *so many* different things? Where to begin, what to study, how to take the first step on the proverbial thousand-mile journey? Thoughtfully, NASA had provided a sugar teat for the fourteen of us, a tidy little course of classroom instruction intended to bridge the gap between aeronautics and astronautics, to minimize the technological shock we might otherwise experience. Although I have a basic dislike for schools, I was pleased about this one, for several reasons. First, I wanted to know what—out of the huge grab bag of possibly pertinent disciplines—NASA really considered of paramount importance. Second, the school was being offered by the same agency that had just completed the highly successful Mercury series, so that certainly it would be a practical course, organized by pragmatic doers rather than abstruse theoreticians. Finally, this would be the *last* school; it had to be, for after NASA, what was there? Surely this was the Supreme Court, and this young attorney was being given a chance to argue his case. If the case involved flying to the moon, so much the better.

The 240-hour course was divided into fairly predictable categories, for the most part.

Astronomy—15 hours
Aerodynamics—8 hours
Rocket propulsion—12 hours
Communications—8 hours
Medical—12 hours
Meteorology—5 hours
Physics of the upper atmosphere—12 hours
Guidance and navigation—34 hours
Flight mechanics*—40 hours
Digital computers—36 hours
Geology—58 hours

The emphasis on computers seemed a little heavy, but then I did not realize the extent to which the Gemini and Apollo spacecraft would be flown by the astronaut using the computer. But everything else on the list seemed in balance. That is, until the last item: Geology—58 hours? Apparently, we were not only going to fly to the moon, a couple of us anyway, but we were expected to act as prospectors once we got there. Further, these fifty-eight hours were designated as Training Series I, and were to be followed by Training Series II, III, IV, V, and VI! The geology classes were also different in that they were not just for the fourteen of us, but for *all* astronauts, so we fledglings got our first close look at our senior partners. In a ramshackle World War II "temporary" frame building on the Ellington Air Force Base flight line, we assembled

---

* I should hasten to add that flight mechanics is a misnomer; it does not involve machinery in flight, or fixing it, but rather is the mathematics and physics of satellite orbits and other trajectories.

a couple of times a week for learned lectures by various members of a joint team of NASA and U.S. Geological Survey scientists, and for laboratory work.

It was a bizarre scene. Boxes of rock samples were on each desk, each sample with a number neatly painted on it, and each astronaut was given ample opportunity to paw through the boxes, to come to know and greet his favorites by number, if not by mineral content. Remember the tired old story about the convicts who amused themselves by reciting, not jokes, but the numbers they had assigned to each memorized joke in their collection? Years later when Dave Scott discovered his famous "Genesis Rock" on Apollo 15 and launched into a lengthy description of it, he need only have said, "Number 408." We would have understood.

But in 1964 the numbers and descriptions came slowly indeed. We had to learn a whole new vocabulary and a whole new way of looking at rocks. "Gray and lumpy" would not suffice; it was now "hypidiomorphic granular, porphyritic, with medium-grained gray phenochrists." "Soft and crumbly?" Sorry, no sale; how soft was it, as measured by the Mohs hardness scale,* which we had all been told to memorize? We even got out our microscopes to study crystal structure and our old chemistry books to review elements and their valences. We all know that H_2O is water and that there is none on the moon, but not even your friendly jeweler would tell you that turquoise is really $CuAL_6(OH)8(PO_4)4 \cdot 4H_2O$. Who would suspect that $(Fe^{2+},Mg)Ti_2O_5$ would be discovered at Tranquility Base in 1969 and that this new mineral would be called "armalcolite," a name derived from the initial letters of Armstrong, Aldrin, and Collins. Certainly no one in that 1964 Ellington classroom could foresee exactly how this strange new subject of geology would be used. Did the key to the greatest imaginable adventure, the journey of the century, lie somewhere in this numbered rock pile? Would the first lunar astronaut be the best identifier of

---

* In case anyone has forgotten, the 1 through 10 Mohs scale stretches from the softest to the hardest mineral: 1. talc, 2. gypsum, 3. calcite, 4. fluorite, 5. apatite, 6. orthoclase, 7. quartz, 8. topaz, 9. corundum, 10. diamond.

terrestrial rock samples? Perhaps, like a spelling bee, we should all stand and be quizzed, sitting down with each mistake, until only one was left standing, and he—the rock champion—would be decreed first man on the moon. Bullshit!

The grumbling began among the Mercury veterans, and was quickly picked up by the second group, the Nine. Did the geologists really think we were going to carry microscopes to the moon, or scratch rocks *in situ* to determine their hardness? What matter their chemical composition, as long as we could bring them back to earth for analysis? That's what we were concerned about, getting their precious rocks home! Never mind all this theoretical jazz; one guy would just scoop up whatever was around the lunar module, obviously grabbing any rocks which appeared dissimilar, and then get back inside and hightail it out of there. Had anyone suggested that astronauts would *leave the LM unattended*, mount a jeep-like vehicle, and drive off for a couple of miles, I think Slayton would have marked him down as a weirdo indeed.

Our group of fourteen, being new, did more listening than talking, but as we got to know each other better, a variety of viewpoints surfaced. My own thinking was especially shortsighted in that it stopped abruptly at the first lunar landing. I thought that scores of preparatory flights would be required, and that the first landing would signal success and the end of the program, rather than the beginning of lunar exploration, with subsequent long-duration traverses complete with lunar rovers and scientific instrument packages. On the other hand, I believed (and still do) that lunar bases would one day be established and that man would fly to Mars. But these efforts would not be Apollo; President Kennedy had said, ". . . to land a man on the moon and return him safely to earth." Were we not planning to do twice as much as he had asked, just by landing two men one time? This was Apollo to me, and it would be difficult enough. I also thought that, although the fourteen would have excellent opportunities to fly a time or two, the actual lunar landing crew would be composed of the three very best and most experienced men, and they would be found among the Seven and the Nine. Therefore, I was not particularly attentive when Walt Cunningham explained that he, as a scientist, should be the first because of his serendipity, as he called it, which might cause him to notice something on the lunar surface the rest of us non-scientific types might overlook. I had no idea that crews would be assembled long before anyone knew which flight would make the first landing, and that Armstrong, Collins, and Aldrin would come together not because they were the three best, but because of a complex matrix of decisions involving Gemini as much as Apollo, and even involving my own health.

Meanwhile, we fourteen worked hard at mastering the geology course despite our misgivings as to its utility; after all, we were full-time students now anyway, and geology was simply one course among many. We wanted to do well at all of them, for assuredly a day of reckoning would come when good students would be re-warded by flying before poor students; and the name of the game was to get airborne and get some experience, so as to have a better crack at the lunar landing(s?). Unfortunately, many of the other courses seemed to be as far off the mark as the geology classroom sessions. My critique for a typical one follows:

> This course was designed for someone who is going to either (a) build a better computer, or (b) repair and replace components of the existing computer. It was not a course for the pilot, who needs to know how to operate the computer and how to detect malfunctions. This comment applies to astronaut academics in general: give us the pilot's viewpoint, not that of the repairman or the circuit designer.
>
> July 17, 1964

On the good side, the courses were not graded but were informal give-and-take sessions which allowed us to get to know each other, and which provided time on the side to prowl around the scattered NASA offices at Ellington and at the new site of the Manned Spacecraft Center, which was nearing completion a few miles to the south of Ellington. As we got to know the people and facilities better, we felt more at home, and we even began to sound like old-timers, making ever more facile use of the specialized technical jargon which marks the NASA employee.

NASA-ese is no worse than Air Force-ese or State Department-ese; I suppose each has its place, although none of them seems a desirable substitute for English. Examine the sentence "Jones and Smith don't get along well" translated into: (1) Air Force-ese: "It is considered that effective utilization of the potential allegiance between Jones and Smith is not being harmoniously exploited"; (2) NASA-ese: "The interface between Jones and Smith has gone divergent"; (3) State Department-ese: "Messrs. Jones and Smith's abrasiveness vis-à-vis each other is counterproductive to their bilateral relationship; each considers the other a *bête noir*." NASA-ese is not longwinded—it is just different, and in addition there are some no-nos in its vocabulary whose use marks one as a gauche outsider. For example, just as the terms "joy stick" and "tailspin" cause professional pilots to shudder, so does "capsule" make astronauts wince. Capsules are swallowed—one flies a spacecraft. It is particularly galling when people like Jim Fletcher, the NASA administrator, use the term. Another thing one must learn is the proper use of units: spacecraft velocity, for example, is measured in feet per second, not miles per hour (or furlongs per fortnight, which is probably as sensible a unit as feet per second). One must learn the language if one is to study a foreign culture, and the language of space flight is as complex and precise as the maneuvers it describes, a fact that dismays NASA public-affairs officials at flight time when they have to translate the steady stream of "delta vees" or "go to poos" coming from the "capsule" into language the public can understand. Perhaps public support might have been stronger had the public really understood what the astronauts were talking about; on the other hand, I suppose some mystique is desirable in any enterprise.

At any rate, we were learning the lingo and the basic fundamentals of space flight well enough to be allowed to escape the nest, to get out into the real world as representatives of NASA, as bona-fide astronauts. We were given the chance to order calling cards, or business cards, with our names and titles (Major, U.S. Air Force, NASA ASTRONAUT) on them, printed alongside a Technicolor NASA insignia. This honor I declined, not because I wished to hide the fact I had become a NASA ASTRONAUT, but because I had a secret fear that people would burst out laughing if I dropped this miniature totem on their desk or through their mail slot, or even if I transferred it from one moist palm directly to another. At least my old friends, who had given me a Gemini garbage can, would have laughed, and they were more important to me than those who might have been impressed. As our formal school in Houston drew to a close (with the exception of the apparently interminable geology classes), we were asked by Shepard to list all our compatriots in order of merit, from one to thirteen, the best being the one we would most like to accompany us on a space flight. After a good deal of thought, I picked Dave Scott to top the list, turned in my list and class critiques to Al, and was rewarded with a mock diploma, complimenting me on finishing "Basic Grubby Training" and giving me permission to "leave town without a den mother."

Leave town we did, singly and in groups. In fact, my income tax records show I made twenty-seven trips away from Houston in 1964, most of them lasting two or three days. Six of them were geology field trips, and the remainder covered a variety of other responsibilities which I will describe later. Our first geology trip was to the Grand Canyon and was one of the most interesting, partly because it was the first trip and partly because of the natural beauty and awesome grandeur of the place. Of course, the idea was to give us an opportunity to augment our classroom knowledge with field experience, to see not just a half-pound chunk of 801 on a desk top, but miles of it clearly delineated in a horizontal bed, deposited during the Devonian period on top of Cambrian period rock, and in turn covered by Permian limestone much later (a mere 200 million years ago). About a dozen different rock formations have been exposed as the Colorado River cuts ever deeper into the Arizona desert, crashing along nearly a mile below the rim, and carrying with it nearly a million tons of sediment a day. As we descended the south rim along the famous Kaibab Trail, we examined and recorded ("hypidiomorphic granular") each succeeding layer of rock, passing from the upstart limestones and sandstones through the older shales and finally, near the bottom, discovering very old rock, the tortured and baked Vishnu schist, over two billion years old. The flaw in all this was that, with the exception of the schist (a metamorphic rock) and a lone outcrop of granite (igneous), all the rocks we saw were sedimentary types, having been water-deposited, and many contained fossil plants and animals, including shark teeth. Of course in 1964 no one knew what rock types awaited us on the moon, but no reputable scientist believed we would find sedimentary rock there, especially with any signs of fossilized flora or fauna. In fact, we used to joke about carrying a few fossils to the moon to mix in with the lunar rock samples, and I am surprised that no lunar crew has announced the discovery of a fossil or two on the radio, just to shake up the geologists in Mission Control.

Nevertheless, it was a beautiful trip down in the Grand Canyon, and knowing that we would have many other trips to volcanoes, impact craters, ash flows, and other lunaresque places, we relaxed and chipped rocks with our hammers, enjoyed the magnificent scenery, and wondered at the changing world we were entering. At the top of the rim, for example, the trees were nearly all scrubby pines, but below the cliffs grew the magnificent Douglas firs typical of southern Canada. The combination of high altitude and cool shelter somehow made it possible for these strangers to thrive, at least in a narrow band. Farther down, the plants gradually changed from Canadian to Mexican, with a true Sonoran desert zone at the very bottom, complete with lizards and yucca trees. In between we found a fascinating temperate zone, and even saw a family of mountain goats. I confess that on this trip—indeed on all geology trips—I found the flora and fauna much more interesting than the rocks. Some in our group, like Roger Chaffee, were becoming damned good geologists, perhaps because they truly enjoyed it or perhaps simply because they were quick studies. However, I never quite did get into the spirit of the thing, and spent nearly as much time engaging in rock-throwing contests with Gene Cernan as I did filling my field notebook with maps of outcrops and such.

Our trip down into the canyon took nearly the whole day, so we spent the night in a charming inn at the bottom, and the next morning those of us who wanted to rented burros for an expedited ascent. I chose to ride, but picked an animal which stopped walking whenever I stopped kicking, so I got as much exercise as if I had been afoot. I also had plenty of time to contemplate the rapid pace at which I was speeding toward the moon. From supersonic jets at Edwards, I had progressed all the way to kicking a burro up out of the Grand Canyon. Just as the jets had their tricks, so did this creature; it seemed I was perpetually more involved with the transportation system than with the use to which it could be put. A scientist I would never be, at least not a geologist, but perhaps a transportation technologist deserved a role somewhere. I certainly was willing to share my burro with a deserving scientist or two, provided they did their share of kicking and I got to steer.

The Grand Canyon trip was followed by a long series of forays to other places of geologic interest. Some were truly unique, such as Meteor Crater, Arizona; and the hole left in the desert by a nuclear explosion at the Mercury, Nevada, test site. Others were more representative of a class of geologic feature (the caldera at Valles Crater, New Mexico; the cinder cones at Sunset Crater, Arizona), or a type of material (lava near Bend, Oregon; basalt and ash flow in the Marathon Basin of Texas). In other cases we were more interested in mapping geological structures, such as at Philmont Ranch, New Mexico. The heavy emphasis on the western United States lies in the fact that this region tends to be less densely covered with obfuscating vegetation, and more nearly resembles the moon, with its naked maria and its many craters, be they of impact or volcanic origin. On the more practical side, this area was also close to Houston and was well known to our instructors. Later on, more esoteric spots would be visited, such as Mexico, Alaska, and Iceland. I missed those trips, in favor of Gemini crew duties, but then I am getting ahead of my story.

The geology trips tended to be the most relaxed, but as the focus shifted away from geology and into engineering problem solving, PR visits, and so forth, we found that our trips were nearly always overscheduled and we had to hustle to keep up. The locals always wanted to get as much mileage as possible out of us, and I was continually amazed to find that the title "astronaut" made us instant experts in so many fields. We found naïve people who sincerely believed that if an astronaut said it, it must be so, and shrewd people who enticed us into making pronouncements that could be used to reinforce their own parochial viewpoints.

But before we got too heavily involved in this "real world," the world of equipment design and mission planning, we had a couple of other training courses to complete as part of our basic education. These were the survival courses, designed to teach us what to do if the spacecraft came down unexpectedly in an uninhabited part of the world. Fortunately, the geometry of the situation (launching toward the east from Cape Kennedy at 28 degrees north latitude) dictated that the Gemini flights would trace a sinusoidal path back and forth across the equator. In the case of Apollo, the fact that the moon's orbit is only 5 degrees out of the earth's equatorial plane caused the recovery areas to be near the equator. (I only know these things because I stayed awake during flight mechanics.) In short, we didn't have to worry about the very cold regions, thank God, and had only to plan for jungles, deserts, and oceans. In the case of an unscheduled ocean landing, one had no choice really but to stay put, using the meager supply of drinking water sparingly and hoping the weather was calm and sunny enough to permit successful use of a solar still, a simple plastic bag device which used the sun's energy to evaporate sea water, condensing pure drinking water from it. But the jungle and the desert offered more complicated choices, a greater array of hazards perhaps, but also many more escape options, if one was well enough informed.

So we were off to Panama, the fourteen of us plus Pete Conrad (our den mother), to the Air Force's tropical survival school, where we spent a couple of days in the classroom followed by a couple of days living off the land. The classroom work was most illuminating, and this city boy sat frozen into attentiveness as we discussed a few dos and don'ts. Our bible was Air Force Manual 64–5, entitled *Survival*, and it makes jolly good bedtime reading. In fact, I have kept two copies of this little gold mine to this day, for one never knows.

The manual opens on a cheery note: "Anything that creeps, crawls, swims, or flies is a possible source of food." Then it gets a bit too specific for my taste. "People eat grasshoppers, hairless caterpillars, wood-boring beetle larvae and pupae, ant eggs, and termites." Not me, babe! Oh yeah? Read on. "You have probably eaten insects as contaminants in flour, corn meal, rice, beans, fruits, and greens of your daily food, and in stores in general." No wonder the supermarket has been less crowded lately.

How about something for the person of more conventional taste? "Look on the ground for hedgehogs, porcupines, pangolins, mice, wild pigs, deer, and wild cattle; in the trees for bats, squirrels, rats, and monkeys. Dangerous beasts—tigers, rhinoceroses, elephants—are rarely seen and best left alone." I'll say! Besides, to hell with all that, I'll just be a vegetarian. How about a recommendation along that line? "The taro grows 2–3 feet high and has yellowish-green jack-in-the-pulpit flowers. Cook the large heart-shaped leaves well, preferably with lime juice, before eating, otherwise they will irritate your mouth and throat." Or mushrooms, maybe? "Poisonous fungi cannot be detected by unpleasant taste or disagreeable odor." I had seaweed one time in a great Japanese restaurant, so how about that? ". . . Some have too much lime carbonate or are too horny to be eaten. Others are covered with slime." Isn't there anything a little closer to home? "Sweet potato vines are easy to recognize; they look like morning glory vines." Now what the hell do morning glory vines look like? Just as I am ready to throw in the towel, I get a little encouragement. "Poisonous plants will be met with in the tropics, but in no greater proportion to the non-poisonous kinds than in the United States." Bully! At least there is some solid practical advice scattered about ("Don't eat toads") with with I can wholeheartedly agree.

Suppose I give up the idea of eating entirely, and just sit there and wait to be rescued? Be careful where you sit—". . . will sting you if you touch them, and their sting is like that of a wasp. Avoid many-legged insects." You bet I will, but will they avoid me? "Scorpions are real pests, for they like to hide in clothing, bedding, or shoes . . ." Would they bite a fellow Scorpio? Snakes? Oh, no problem there. "Poisonous snakes are less abundant than most people think." Than most people think, for heaven's sake! They don't have the vaguest idea of what I'm thinking, which is, if snakes are so goddamned non-abundant, then what is? "The . . . crocodile . . . is very dangerous. It is abundant."

That does it. I'll need help, and quickly. "Call or clap your hands to attract attention. Don't be afraid to be an object of amusement to the natives. Be ready to entertain with songs, games, or any tricks of cards, coins, or string which you may know." They gotta be kidding! "Rock salt, twist tobacco, and silver (not paper) money should be used discreetly in trade . . . Someone may understand a few words of English. If not, use sign language; natives are accustomed to it because they communicate a lot by signs themselves. State your business simply and frankly." You bet I will. Get my ass out of here! While waiting for transportation: "Leave the native women alone at all times . . . always be friendly, firm, patient, and honest. Be generous but not lavish. Be moderate." Moderation in all things, well, almost all things. "Don't worry about lack of bowel movement; this will take care of itself in a few days." On my diet, I don't see how it possibly could.

But no matter, put all fears away, leave civilization cheerfully and confidently behind, and put your trust in Air Force Manual 64–5. After all: "You are probably safer from sudden death in the jungle than in most big cities." Armed with this dubious reassurance, and a dull machete, I stepped briskly off a helicopter and plunged straightway into the green wall of awaiting jungle. I had to know for myself; there were some things I simply had to confirm: "The sea cucumber can and does shoot out his stomach when excited."

We were divided into teams of two for this three-day outing, and my partner was Bill Anders, who turned out to be a jewel, in two very important ways. First, he was an outdoorsman, an avid fisherman who liked nothing better than to keep walking until he found a stream no one else had fished, and he was an experienced camper, wise in the ways of providing substitutes for the creature comforts we city dwellers had come to expect. Second, and no less important, he was a finicky eater, but more about that later.

Our first task was to hike a couple of miles through the jungle to our assigned camping area. I got my first surprise of the day when, looking on the ground as instructed by Air Force Manual 64–5, I not only did not spy any "hedgehogs, porcupines, pangolins, mice, wild pigs . . ." I didn't see anything moving at all. Furthermore, the trees were devoid of the promised "bats, squirrels, rats, and monkeys." I suppose we were making such a din and clatter as we proceeded that any but the most aged and feeble creatures were staying miles ahead of us. Or maybe it was that this particular part of the jungle was simply *empty*. Was that possible? I don't know, but it was certainly quiet, and as my stomach got emptier and emptier, my visions of plump little creatures roasting merrily on a spit became not only much more attractive but obviously much less realistic. "Anders, what the hell are we going to eat?" "Collins, you couldn't be hungry already, we just got here."

By the time we got to our camp and got our meager possessions organized, night was falling, and we became trapped, helpless victims of the jungle until dawn. I settled down in my makeshift hammock, stomach gurgling its disappointment, and again considered the matter of the moon. Did it seem any more reachable from Panama than from the Grand Canyon? Were we really drawing closer in our quest, or did not each new experience merely result in the discovery of other unknowns that needed to be sampled? Was not the moon staying forever one step ahead of us, just as described in one of Zeno's famous paradoxes: you see a turtle in the distance moving slowly ahead and you race to catch up with it. However, when you reach place A, where the turtle was when you first saw it, the turtle will have moved on to B. By the time you reach B, the turtle will have gone on to C. Therefore, no matter how many times the process is repeated, the turtle will always be ahead of you, and you can never quite catch up. Turtle, moon, or even dinner—none seemed reachable. To add insult to injury, Anders had somehow managed to attract a huge swarm of mosquitoes, which he generously shared with me. Swatting and scratching, nursing a terrible headache, I finally drifted off into fitful sleep, a hungry, disillusioned, mosquito-bitten, would-be philosopher, with a very bleak future.

The next day dawned bright and dry, at least, and we busily set about looking for food. We spent the morning trying unsuccessfully to catch minnows in a small stream and foraging futilely for some edible plants. Finally we were visited by the school director and his staff, who cheerfully informed us that really the only things worth eating in these here parts were the palm trees, that is, certain palm trees. It turns out that those little whitish disks you find in your heart-of-palm salad are part of a largish cylinder of edible stalk buried inside the trunk of certain varieties of palm tree. The trick is to be an expert palm-tree identifier, because it takes a couple of hours to chop down a palm tree with just a small machete, and to whack through the tough fibrous exterior of the upper trunk to expose what could be either a delicious tender heart or more of the same inedible woody pulp, depending on the tree. To me, if you've seen one palm tree, you've seen them all. With great hopes but little science, Bill and I finally selected a likely candidate and amateurishly went into our act, whacking away like mad until it finally toppled over. Out of the severed trunk boiled thousands of ants, but before they drove us away, we could clearly make out the fact that the area of the heart was discolored and decayed. With great trepidation we picked a second victim; this time we hit the jackpot, and a couple of hours later we were able to make off with our prize, a heart perhaps two feet long and five inches in diameter, more than enough to keep the two of us in salad for a day or so.

We were still having no luck in the protein department, so again our softhearted instructors came to the rescue. They had found iguanas! All of us astronaut twosomes were called together to share this good news, and we buzzed excitedly around a couple of unfortunate victims, who regarded us unblinkingly from the spot where they had been unceremoniously dumped. They could not run away, because their front legs and rear legs had been tied behind them. Slowly the shocking nature of these gruesome bonds became evident: the big lizards were tied with their own tendons, which had been stretched by jerking them loose from the extremities to which they were still connected. The exposed tendons were then used as cords and were neatly knotted. Although cruel beyond belief according to our SPCA standards, this method of restraint is apparently the accepted one in many tropical countries, and the mute, impassive creatures may spend several days in what must be agony between the time of their capture in the jungle and their eventual sale and slaughter in the city markets, where their chicken-like flesh is considered a delicacy. Although threatening in appearance, prehistoric and dragon-like, iguanas are actually sluggish, harmless fellows, and deserve better treatment. Therefore it seemed an act of charity to slaughter them quickly, which we did, dividing up the pieces evenly between teams.

Bill and I returned to camp with ours, and within minutes had a cheery fire going, with water boiling in a tin can, into which we plunked savory chunks of iguana. I remember that the last piece in, a front leg, kept floating to the surface with small supplicating hand extended, despite my best efforts to poke it back down. Bill watched all of this in uncomfortable silence, and finally muttered something about his share of iguana being available if I wanted it, as he really wasn't hungry. What a partner, what a buddy, not ever hungry! I thanked him profusely and sat cross-legged by the fire (which kept the mosquitoes away), munching on crunchy heart-of-palm, happily gulping great chunks of iguana, burping contentedly, and thinking that the jungle seemed not such a bad place after all.

The next morning was rainy and I was glad that it was our final day in the jungle, because once wet it is very difficult, under the dense canopy of trees, to dry out again. We also met some Choco Indians, by arrangement, and were most impressed by their chieftain, Antonio, a man who appeared much younger than his years (forty?) because of his superb physical condition and unlined face. He also had a great impassive dignity, and seemed surprised at nothing, not even the news that we were practicing in his jungle so as to be able to fly to the moon. Perhaps he did not believe it. Years later, in Washington, I met Antonio again, at the Smithsonian, in the hall containing the original Wright Brothers airplane and, almost directly below it, the Apollo 11 command module. I patiently explained to Antonio about Apollo 11, about flying to the moon, and I learned from his interpreter that he was not particularly interested in the Apollo 11 craft. He didn't disbelieve my story, he simply could not relate it to the ugly triangular chunk of heat-shield-covered machinery. He liked the idea of flight to the moon, and obviously knew a lot about it—that it was a satellite of earth, and so forth—but the machinery he was interested in was the Wright Flier. This he could understand, the mannequin lying prone on the lower wing, and the crude engine powering a pair of wooden propellers. Here was useful technology.

Little expecting that I would ever see Antonio again, I took leave of his jungle, paddling pleasurably down a river in a Gemini life raft, and then meeting a larger boat which whisked us back to civilization. Somewhere along the way I picked up a couple of hundred companions, chiggers, evenly distributed from the waist down. I cannot adequately explain to the unchiggered what they are missing. My dictionary says simply that chiggers are "the parasitic larva of certain mites." It doesn't say they are also abominable little red creatures who burrow into your skin, where they ultimately die. Uneasy in their terminal tunnels, they either jump about or dig deeper, or secrete some irritant, or something; at any rate they itch like crazy. Friends are always pleased to offer remedies. "Rub them with Scotch and sand. They'll get drunk and stone each other to death." They merely itch worse when they (or you) have a hangover. "Have you tried an ice pick?" The most popular notion (false) is that they can be suffocated, and I have heard doctors recommend covering each spot with clear nail polish. Why clear rather than blush pink (my natural color) I cannot say, but I can say with authority that it doesn't work either, nor does the iodine-like chigger medicine sold in pharmacies. The only thing to do is wait ten days for the truculent little bastards to die or depart, leaving behind a cratered field of battle not easily forgotten by the landowner.

Back in civilization once more, we spent our last night in Panama hosting a bachelor party for C. C. Williams, the last of the whooping cranes, the first and only—at that time—single astronaut. Perhaps because we were overjoyed to be back at the bar, or perhaps because we all liked C.C. so much and relished his company, or perhaps because we were trying to deaden the nerve conduit between chiggers and brain, I know not the reason, but we did in fact have a real wallbanger of a party. Charlie Bassett was master of ceremonies, and he was a regular William Jennings Bryan, delivering a far-ranging tour de force describing not only the beauty of C.C.'s bride to be, C.C.'s many virtues, the condition of the Panamanian jungles, NASA's approach to flying to the moon, the economic condition of the country, the moral fiber of the younger generation . . . I think, although my own recollection of it is far from clear, we had to remove him physically from the podium and deposit him, still in good voice, back in his room. The next morning, bleary-eyed and bedraggled, we piled aboard one of NASA's small turboprop transports and headed back for the "mainland," whiling away the hours by playing poker, napping, or merely sitting and scratching.

Compared with the jungle, our desert-survival training was clearly an anticlimax. It's not that the desert is not a fascinating place; it certainly is, teeming with life despite the desolate face it shows the amateur observer. But the survivor's task in the desert is much more simply defined: conserve water. Water, above all else, decides who will live or die, and no amount of determination or cunning can overcome that stark reality. The human body needs a certain minimum intake of water each day or it ceases to function. It can even be spelled out in tabular form:

DAYS OF EXPECTED SURVIVAL IN THE DESERT UNDER TWO CONDITIONS

Condition	Max. daily shade temp °F	Available water per man, U.S. quarts					
		0	1	2	4	10	20
No walking at all	120	2	2	2	2.5	3	4.5
	110	3	3	3.5	4	5	7
	100	5	5.5	6	7	9.5	13.5
	90	7	8	9	10.5	15	23
	80	9	10	11	13	19	29
	70	10	11	12	14	20.5	32
	60	10	11	12	14	21	32
	50	10	11	12	14.5	21	32
Walking at night	120	1	2	2	2.5	3	
until exhausted and	110	2	2	2.5	3	3.5	
resting thereafter	100	3	3.5	3.5	4.5	5.5	
	90	5	5.5	5.5	6.5	8	
	80	7	7.5	8	9.5	11.5	
	70	7.5	8	9	10.5	13.5	
	60	8	8.5	9	11	14	
	50	8	8.5	9	11	14	

You can see that without any water in a 100-degree desert you will live five days if you stay put or three if you start walking. Nothing you can do will change these numbers for the better, only for the worse if you make mistakes, such as walking by day instead of by night. If you have a little water, so much the better, as the table shows, but again physiology rules, and contrary to popular opinion, no amount of water rationing or discipline will help. In fact, men have been found dead on the desert with canteens full.

So when we were dumped out in the Nevada desert near Reno one blazing August morning, our job was simply to learn how best to conserve our body fluids and how best to signal for help. All this had been explained to us in the classroom, and now Charlie Bassett and I were to spend a couple of days practicing. First, we had to make clothing, all important in slowing down the evaporation of perspiration. You never see an Arab in Bermuda shorts and a T-shirt, do you? The burnoose and flowing robes are not only modest but functional as well. Fortunately, all downed airmen have parachutes, and the yards of nylon make wonderful tents, bedrolls, robes, etc. Once Charlie and I had fashioned proper clothing from ripped parachute panels, we were ready to find or build a shelter. Not finding one, we decided to burrow into a hillside, picking a spot as protected from the sun as possible. The trick here is to get up off the ground a few inches if possible, as it can be 30 degrees cooler a foot above the ground than it is on the surface, and to keep the sides of your shelter or tent or whatever open, to allow free circulation of air. Having done this, one can only relax, conserving energy and fluids, whiling away the hours by fiddling with a survival radio and signal mirror. The slower the motion, the better, so Charlie and I lay there inertly, chatting idly and reading paperback books while waiting to be "rescued." Although the temperature of the sunlit sand was 148 degrees that day, we were fairly comfortable in our hideaway. Then we motored to Reno, with a monumental thirst assuaged by a variety of fluids dispensed at great expense by surly folk apparently more interested in spinning wheels and flying dice than in a pair of dehydrated desert rats.

Fortunately, no spacecraft has yet come down in an area where jungle or desert survival training has been needed. Neil Armstrong and Dave Scott brought Gemini 8 down suddenly and unexpectedly in the Pacific Ocean rather than the Atlantic, but they were picked up by a destroyer stationed there for that contingency purpose. Scott Carpenter's *Aurora 7* overshot its landing area by 250 miles, but a helicopter quickly reached him. Neither crew was in the water over three hours. Thus, as in the case of so much astronaut training, the information has never really been needed, but it nonetheless was prudent to try to prepare people for as many variations upon the expected theme as possible. I think this has been one of NASA's great strengths in the conduct of the Gemini, and, even more so, the Apollo series. Large groups of capable men have sat for years around conference tables with engineering schematic drawings and asked themselves, "What happens if?" In some few cases the answer is simply "Tough," there is no out, no solution, no alternate course. But these are few and far between. In most cases, systems have been designed with a built-in redundancy or redesigned whenever safety reviews have revealed weaknesses. Despite careful design, however, the prudent manager (or crew member) still expects trouble, asks "What happens if," clearly thinks through the possible choices of action, and selects the best—in the quiet, unhurried calm of the conference room, rather than waiting for the ominous voice on the radio: "Say, Houston, we have a problem up here . . ." When the voice comes, Mission Control has a reference library to which it may turn, a compendium of training sessions, simulator runs, data from previous flights, in addition to thick books full of recommended procedures for each failure which has been considered a reasonable possibility.

The most famous failure during a manned space flight was the oxygen tank rupture during Apollo 13. This is an example of the system at its worst and at its best. The possibility of a service module oxygen supply tank leak had been considered years before, but it was not deemed catastrophic because there was a second tank just like it, and they emptied evenly, so you always had some oxygen left; analysis showed you could always make it back to earth from the moon if you hurried. You had only to make sure to work your valves so that none of the oxygen from the remaining good tank vented overboard through the leak. O.K. so far. But when the tank on Apollo 13 blew, it did so with such destructive force that it ruptured the line leading from the *other* tank, a possibility the designers had never considered, and it rapidly became apparent that the two tanks were not truly independent of each other, as they were designed to be. It became equally apparent that the command module, with Lovell, Swigert, and Haise inside, was rapidly running out of oxygen, which meant they would shortly be unable to generate electricity, produce drinking water, or breathe. Death for all three, because of a faulty plumbing design? It very nearly was, except that fortunately Apollo 13 was on its way *to* the moon, instead of returning from it, and the lunar module was still attached, with all its supplies of oxygen, water, and electricity untouched. Could they be used, could the LM be pressed into this unfamiliar role? Yes, it not only was theoretically possible but it had been considered before and documented in detail. The appropriate volume was snatched from the shelf, and Mission Control and the crew were soon working on the life-saving procedures which allowed them to limp home. Needless to say, the oxygen-supply system was redesigned before the next flight. That the vast majority of the emergency library has remained untouched over the years is no indictment of it. It was one of NASA's wisest investments.

Whether the investment in our jungle and desert training was worth it is a moot point, because I doubt that what we learned there would have made the difference between life and death in any situation I can visualize. However, strictly from a personal point of view, there is a good deal of satisfaction to be derived from these "feel so good when they stop" situations. One cannot really appreciate being chigger-free without first sampling chiggers. Living in temperature-controlled, water-tight enclosures, with food and drink instantly available, causes us to forget—or worse yet, never to know—the pleasure of simply feeling warm and dry and full. I don't want to exaggerate the hardships experienced in these two short outings, but I do know from previous Air Force survival training that it doesn't take long to make discomfort a lifelong recollection. In 1955, for instance, I spent five days and five nights in the Bavarian Alps, sleeping by day and walking ten miles or so each night over rugged terrain. The grocery list shared by two men for this period was as follows:

1. One chunk of veal the size of your fist, with which to make jerky
2. Two small trout, to smoke
3. One live bunny rabbit, on a string, with pink eyes and long ears, to do with whatever you wished. Don't ask
4. One immense head of cabbage
5. Four or five large beets
6. About a half-dozen medium-sized potatoes

It took a long time after this jaunt for me to sit down to a full meal without absolute delight at my good fortune, yet I am embarrassed to mention something like this when I consider people who have been POWs for years, or those who have rarely known a full stomach in all their lives. Suffice it to say that there is some merit in educational programs, such as Outward Bound, which try to introduce jaded schoolchildren, and perhaps less jaded adults, to the new dimension of living outdoors, by their own wits, within their own self-contained resources, if for no other reason than to return them to their cities more appreciative of what the cities have to offer.

Back in the city of Houston, the fourteen of us were bringing ourselves "up to speed" by a combination of formal classroom instruction and informal mingling with the old-timers, not exclusively the astronaut old-timers, but the engineers, managers, and flight controllers as well. All this romping through Grand Canyon, jungle, and desert should not obscure the fact that we could have flown handily to the moon without such trips, but we could never have made it without the solid foundation of technology that was available in Houston. We were smart enough to know this, and spent long hours—longer than our wives appreciated—in an attempt to absorb a sufficient base of information upon which to build when we were turned loose on our own.

NASA is (or at least used to be) flooded with requests for astronauts to make public appearances, and there had to be some system of deciding which engagements to honor and which to politely decline. As Deke Slayton delicately put it in a memo: ". . . Nothing engenders hatred more rapidly than having an astronaut appear at the Podunk Center Elks Club on his own after having officially turned down the local congressman's request for a Chamber of Commerce or Rotary Club dinner. The message is: No astronaut has authority to accept speaking engagements or public appearances . . . without approval from higher authority. When in doubt, call . . ." As requests came in, they were routed to NASA headquarters in Washington, where the public-affairs officials (I assume) weighed them on the basis of political clout, persistence, prestige of audience, etc., and then grouped them into neat packages. These packages were organized, insofar as possible, within certain limits of geography and time, so that one astronaut could fulfill all of them in one week. Hence the affectionate title of "week in the barrel" was given to this duty, the genesis of the term deriving from one of Al Shepard's obscene jokes. When people asked about the strange name, we always said it had something to do with shooting fish in a barrel, and that is close enough.

Naturally the requesting group always wanted John Glenn or one of the other "flown" astronauts, but when I was first involved, there were only six veterans as opposed to twenty-four rookies, so chances were they got someone (Mike Who?) they had never heard of, and Mike Who, in turn, got to visit a bunch of places he had never heard of. I think the Russians have a better system, which is that one only becomes a cosmonaut *after* making a space flight and the novices are kept strictly under wraps; their names—I am told—never appear publicly. I have an even better suggestion, which is to have stand-ins for each astronaut who would faithfully go into training as a flight for his principal approached. While the real crew was being whirred around in centrifuges, their PR stand-ins could be practicing limp handshakes. They could learn to sign autographs speedily and legibly, with flourishes, and they could eat creamed chicken and peas three times a day. Best of all, they could memorize speeches about what it was like up there, and get their delivery down pat. "I simply can't tell you how *terribly* pleased I am to be heah today" for the big Eastern cities or "Well, folks, I'm tellin' yew" for the Midwestern towns. Then, as soon as the flight splashed down, the real crew could go on vacation and their stand-ins would immediately spring into action, for a week—a year—a decade in the barrel, whatever the traffic would bear.

Unfortunately, my helpful suggestions along these lines were never implemented, and so after nearly a year at Houston, we fourteen were deemed sufficiently seasoned to be sent out as NASA representatives, and our names were added to the list of potential barrel victims. The only people who were not eligible were those assigned to a crew in training for a specific flight, and that was probably the largest single incentive for us neophytes. Why do you want to go to the moon, Collins? Because it's either that or a week in the barrel, that's why!

Once in the barrel, there are some helpful rules to follow, but not many. Rule 1 involves travel arrangements: get there as late as possible; leave as early as possible. The reason for this is that no matter what the printed schedule, they are going to keep you jumping the entire time you are there. "On our way over to the reception, we'll just stop by the hospital, it won't take a minute." You can't say no to that. I've staggered back to my guest cottage at midnight, bone-weary after fifteen hours on my toes, and been greeted by the genial local host and thirty of his cronies, who are just getting warmed up. "Surprise party! We thought after today you'd like to relax and meet a few of these folks . . ." The next day, at the next whistle stop, they can't understand why you are not more exuberant, happy to be an astronaut, and delighted to spread the NASA gospel. Rule 2 is to stop worrying about what you say, it really doesn't matter. This rule applies especially when talking to the press out in the hustings. I always had a secret fear, especially in the early days, that the shallowness of my technical knowledge would be exposed. "Tell me, Astronaut Collins, what will be the eccentric anomaly of that orbit, to three decimal places?" in sonorous tones. "I'm sorry, sir, I don't know," in a small piping reply. This reverse Mittyism is ridiculous—all the guy wants is some little thing that will make both of you look good, and if you're misquoted, who reads the West Wissioming *Daily Astonisher*, anyway? Rule 3 is don't blow your cool. I used to get angry at being shoved, or asked to sign fourteen autographs by one lady who pushed in front of a nice kid who had been waiting patiently for one, or at seeing the schedule slip further and further behind. Don't bother. Detach yourself. Repeat a favorite Omar Khayyam quatrain over and over. I suggest "The moving finger writes . . ." Pretend you are not down on the ballroom floor, surrounded, but are up on a chandelier, swinging gently, looking down at the bizarre scene below with detached amusement. I can't say I have always applied my third and final rule with great success, but if you can manage it, it certainly does make the time go faster, and it has helped me to preserve energy and sanity on a number of trying occasions.

I guess the truth of the matter is, I simply don't enjoy PR work and there is no point pretending I do. I find this very difficult to explain to people, especially in declining speaking engagements, a task I do today on a regular basis without help from NASA. It is simply one of those cases where the whole does not equal the sum of its parts. Each individual request may be eminently reasonable, but when added together they produce a life style that is clearly unworkable, at least for me. Yet it gives me no pleasure to decline; rather there is an overlay of guilt, for a couple of reasons. First, it is clear that the idea of putting tax dollars into space has enjoyed uneven public support, and the charge is frequently made that NASA has done a poor job of "selling the program." Now I don't agree with that point of view, but who knows?—it may be true. I am certainly a poor judge of what the public could be, or should be, sold. Perhaps I should be doing my PR bit, advocating continued exploration, for I certainly believe in that. A second source of guilt derives simply from being so negative so much of the time, but it's been years since my first week in the barrel and I want out. Perhaps it would help me gain some perspective if I relived one incident in that unforgettable week.

You see, there was this guy who had ten thousand (count 'em) Boy Scouts lined up. They were converging on this Midwestern suburb from a three-state area, and naturally their gathering would not be complete without an astronaut, especially a flown one. NASA headquarters was impressed by the numbers, and of course Boy Scouts are the premier group of non-voters around, so a stop in Shaker Heights, Ohio, was officially sanctioned and added to the appropriate week in the barrel. Little matter that it was *my* week, and not only was I unflown, I was unbadged, being probably the only astronaut who had never been a Boy Scout, and I wasn't about to join on my thirty-fourth birthday, which that day happened to be. So I was a bit wary as I drew up to the appointed place at the appointed time. It was a shopping center, a nice one, with a covered mall containing a high circular dais, upon which I dutifully assumed my assigned seat. Milling around it were not ten thousand, not ten hundred, but perhaps ten Boy Scouts, whose ranks were swelled for short periods by curious camp followers, shoppers out for a Saturday-morning bargain who wanted to know what this impediment in their path was all about. There were helpful signs, which said things like ASTRONAUT APPEARANCE and HE WILL SPEAK. I recognized, with a growing knotting of the abdominal muscles, that this meant me. Before I could speak, however, came *The Introduction*, which must have been longer than the speech that followed. The situation reminded me of a William Jennings Bryan story. When the famous orator came to one small farming community, he was introduced by the local politician, who, seeing the grand and unusual spectacle of a large attentive crowd, put his all into his delivery; finally, after much bombast, Bryan was allowed to proceed. When it was all over, two old farmers assessed the situation. "He were really sumpin', weren't he?" "Yes, sir, and that feller what came after weren't no slouch either." While my introducer was droning away, interrupted only by periodic squeals from the PA system, I felt a persistent pulling at my pant leg. I didn't want to kick a Boy Scout in the head, but neither did I wish to insult the speaker by turning my back on him just as he was building to a crescendo, so I studiously ignored the tugging. It was no use—the tugger simply spoke up, "Psst, buddy!" "PSST!" in a deep voice. Finally I wheeled around. "Hey, buddy, aren't any of the *real* astronauts coming?"

5

This foolish idea of shooting at the moon is an example of the absurd length to which vicious specialisation will carry scientists working in thought-tight compartments.

—Professor Bickerton, Speech delivered to the British Association for the Advancement of Science in 1926

Little by little the fourteen of us were edging toward accreditation as "real" astronauts. The next step was to put us to work, making us contributing members of the team rather than simply students in the classroom and in the field. Both Gemini and Apollo were in the formative stages, and it was early enough to change designs, based on what the potential crew members had to say. Since most of us had been involved, as test pilots, in seeing new machines grow from drawing board to flight line, hopefully we would be able to transfer our knowledge to this new realm, to look with jaundiced eye at what the theoreticians had put together, to call upon the strong points of some similar programs and to avoid the pitfalls of others. This at least was the theory, and I think it worked well.

The first step in the process was one of "vicious specialization," in which areas of responsibility were delineated and the technological pie cut up into slices. Each astronaut was then given a slice of his own to devour. We had been individually polled, by Al Shepard, as to what specific responsibilities we thought we were best qualified for, and I had indicated a desire to work on the development of the pressure suit, or space suit, as the public calls it. I made this choice somewhat hesitantly. Clearly this was not the largest piece of the pie: it was not in the thick of things, such as cockpit design, or guidance and navigation, and therefore it might cause me to be overlooked as an early candidate for a flight. On the other hand, designing a light, flexible, practical garment was a fascinating challenge, combining as it did a lot of rigorous engineering with a touch of anatomy and anthropology, and more than a little black magic. Furthermore, I felt—grudgingly or gladly, I'm not sure—that my own education and innate abilities were inferior to those of some other Fourteens in the highly mathematical areas. For example, to really understand guidance and navigation, one must dig into its attendant mathematical base, and this jungle is full of ominous creatures with names like *vector* and *tensor*.*

Therefore, I was pleased when Al issued a memo announcing our assignments:

Buzz Aldrin—mission planning
Bill Anders—ECS, radiation and thermal
Charlie Bassett—training and simulators, operations handbooks
Al Bean—recovery systems
Gene Cernan—spacecraft propulsion and Agena
Roger Chaffee—communications, DSIF
Mike Collins—pressure suits and EVA
Walt Cunningham—electrical and sequential, non-flight experiments

---

* An abridged dictionary says a tensor is "a generalization of the concept of vector that consists of a set of components usually having a double row of indices that are functions of the coordinate system and have invariant properties under transformation of the coordinate system." See what I mean?

Donn Eisele—attitude and translation controls
Ted Freeman—boosters
Dick Gordon—cockpit integration
Rusty Schweickart—future programs and in-flight experiments
Dave Scott—guidance and navigation (G&N)
C. C. Williams—range operations and crew safety

One moon, neatly carved up into fourteen pieces. Why these fourteen, and what do they mean translated into English? And where were the old heads? Last question first: Glenn had left, Carpenter was still there but was working with the Navy on underwater projects. The Gemini 3 and 4 crews had been picked and were working full-time on their respective flights: Grissom and Young to fly 3, Schirra and Stafford back-up; McDivitt and White on 4, Borman and Lovell back-up. Cooper and Conrad, pending their assignment as Gemini 5 prime crew, were looking after eight of us new kids, plus covering the progress of the Apollo command and lunar modules. Armstrong and See, who would back up Cooper and Conrad on Gemini 5, were assigned to oversee the other six of us, as well as being responsible for overall astronaut operations and training. Al Shepard was our boss, as chief of the astronaut office, and he in turn reported, as did a couple of hundred other people, to Deke Slayton, who was assistant director for Flight Crew Operations of the Manned Spacecraft Center. And that makes thirty.

Translations and amplifications follow:

Buzz Aldrin—mission planning This involved going to an interminable series of meetings at MSC, attempting to define how many flights would be required to get to the moon and how they should be organized for maximum efficiency. The rendezvous problem was probably the biggest single unknown from the crew point of view, and there were almost endless variations in technique for bringing two vehicles together in earth or lunar orbit. Should the interceptor approach from above or below, in dazzling sunlight or in inky black, at a great overtaking speed or very slowly? These were not philosophical questions, but practical ones which by proper analysis could be converted into gallons of fuel remaining, or mathematical formulae describing probability of success. Since one of Gemini's main tasks was to investigate rendezvous for later application to Apollo, Buzz's task was a broad one which required an intimate knowledge of both programs. In addition to rendezvous, mission planners were concerned with almost every facet of space flight, with heavy emphasis on scientific and medical experiments, and how they should be distributed flight by flight. Generally speaking, experimenters wanted things done as they might be done in a laboratory, with little regard for competitive demands on the crew's time, and with little understanding of the constraints imposed by the environment in which they would be operating. It was the mission planners' job to whittle the experiments down to size and cram them on board the most logical flight. It was a sensible use of Buzz's education, which included a doctoral dissertation on orbital rendezvous.

Bill Anders—ECS, radiation and thermal ECS stands for environmental control system, alias the plumber's delight. A baffling array of pipes, tanks, valves, connectors, switches, filters, heaters, fans, sensors, etc., whose details almost obscured its central purpose, that of providing an environment of 100 percent oxygen at a pressure of five pounds per square inch (to prevent body fluids from boiling in the vacuum of space). Temperature and humidity limits also had to be controlled, so that the crew could not only breathe but live comfortably for the fourteen days or so they would be up. The system also had to cope with the carbon dioxide which humans exhale, and arrange for potable water to be stored or produced. While the ECS determined the environment within, and was a hardware design problem, radiation and thermal refer to the outside environment and are less concerned with hardware than with limitations on its use. The Apollo command and service module (CSM), for instance, has severe thermal constraints. By that I mean, if you hold the CSM steady in any one position in space for a long time, the up-sun side gets too hot and the down-sun side too cold. Fluids within start boiling and freezing, and all kinds of bad things happen. These constraints, if unreasonable, must be solved by hardware redesign; if reasonable, they must be clearly defined and fed to the mission planners so that "work arounds," i.e., ways of avoiding the problem, can be built into each flight plan. In similar fashion, the Van Allen radiation belts around the earth and the possibility of solar flares require understanding and planning to avoid exposing the crew to an excessive dose of radio-activity. Besides, what is an "excessive dose" or indeed a "design dose"? Bill Anders looked after our interests in these areas. Since he holds a master of science degree in nuclear engineering, his assignment here made sense.

Charlie Bassett—training and simulators, operations handbooks Charlie was a very highly regarded member of our group, and his assignment here reflects the importance NASA has always placed on simulators, those multimillion-dollar machines which duplicate, insofar as possible, the actual spacecraft. In many cases it is more difficult to simulate something than it is to *do* it, and in some ways it was more difficult to get the simulators "flying" than the real McCoys which followed a couple of years later. In this kind of situation, it is vital to have someone with balanced judgment, who has flown a variety of new machines, to say, "Yes, we must have this effect" or "No, we can get by without that one." When you look out the CSM simulator window at an approaching lunar module, for example, must you see an actual three-dimensional replica of the LM, or will a televised picture of it be a suitable substitute? When you attempt to dock with it, must your simulator actually move in response to your controls and jar you at the instant of contact, or are these unnecessary and unduly complicated frills? As the simulator designs proceeded, so also did the training documents which the crew would need to study before flight and in some cases carry on board for possible reference during flight. What was the best way to organize this immense library, to make the miles and miles of electrical wiring somehow understandable, for maximum retention and quickest access? An electrical engineer and a very bright guy, Charlie wrestled with these problems.

Al Bean—recovery systems From the time the parachutes open until the crew is released from quarantine, an elaborate series of events must take place, involving not only NASA but the U.S. Navy and the Air Force to a lesser extent. A Navy carrier pilot, Al Bean had to make sure all the equipment design and attendant procedures made sense from a crew point of view. A chain being no stronger than its weakest link, it was entirely possible that the Apollo chain might break at the instant some rusted, antiquated crane attempted to grapple Columbia, the gem of the ocean, up on board.

Gene Cernan—spacecraft propulsion and Agena Agena was the auxiliary engine which was launched unmanned and whose power was later used by a Gemini which had docked with it to push the two vehicles together into a higher orbit than would have been possible using Gemini's energy alone. More important, it was the Gemini target vehicle which proved that rendezvous and docking in space were practical. The Gemini astronauts could send commands to the Agena, docked or undocked, and force it to do a variety of maneuvers and tasks, like a well-trained dog. Spacecraft propulsion refers to all the rocket motors with which Gemini, command and service modules, and lunar module were adorned, and which were used to propel them from one orbit to another, to land on or take off from the moon, etc., etc. Some of these engines are quite large, others tiny, but all are of interest to their operators. Gene's master's degree in aeronautical engineering concentrated on rocket-engine technology, which made him especially well suited to the job.

Roger Chaffee—communications, DSIF DSIF stands for deep space instrumentation facility. The DSIF consisted of tracking stations around the world, including three very powerful installations—one near Madrid, one in Australia, and one in the Mojave

Desert.* As early as 1963, the DSIF had bounced radar signals off the planet Mercury, over 60 million miles away, so we weren't too concerned about its power, but its accuracy was of vital concern to those of us who would be navigating with its help. On board the spacecraft, communications consisted of a wide array of transmitters and receivers, covering various portions of the energy spectrum from low frequency to the superduper high-frequency range called S-band.† At one time we even planned to have a teleprinter on board, so we wouldn't have to write down messages. With characteristic energy and enthusiasm, Roger plunged into the arcane world of band widths and Doppler shifts, making sure the complex equipment was going to do all it was advertised to do and that it was simply and sensibly designed from an operator's point of view.

Walt Cunningham—electrical and sequential, non-flight experiments If anyone got the short end of the stick, Walt did, because these areas are simply not the most interesting ones available. The sequential system, as its name implies, was designed to control certain events which took place in unalterable sequence, one after the other. On Gemini, for example, one gets out of orbit by firing the retrorockets. In sequence, this consists of: (1) maneuvering to retrofire attitude, (2) switching power to four main batteries, (3) activating the re-entry control system, (4) firing a guillotine to chop fuel lines leading aft to the adapter section, (5) firing another guillotine to chop electrical lines leading aft, (6) separating the spacecraft from the now unneeded adapter section, (7) firing the retrorockets, and (8) jettisoning the retrorocket package. The sequential system served as aide and monitor in the

---

* Located on Goldstone dry lake near Barstow, California, this station occupied familiar territory, as it had previously been the site of an Air Force bombing range, and I had spent many hours at Goldstone while stationed at nearby George AFB in 1954. A decade later, no planes were permitted to fly over Goldstone, the DSIF radar transmitter being supposedly so powerful that its energy could explode pyrotechnic devices in airplanes passing overhead. The reason for stationing the three huge antennas in the United States, Australia, and Spain is that one of the three will be pointed toward the moon at all times, ensuring continuous communications.
† The region around 3,000 million cycles per second, or 3,000 megahertz, as it is called.

process by illuminating amber lights in sequence at the proper time: (1) IND RETRO ATT, (2) BTRY PWR, (3) RCS, (4) SEP OAMS LINE, (5) SEP ELECT, (6) SEP ADAPT, (7) ARM AUTO RETRO, and (8) JETT RETRO.

The main thing here was to ensure reliability, which usually meant redundant circuit design, and, whenever possible, to think up swift alternate courses of action in case of malfunction. The electrical system is only slightly less boring; at least it includes the fuel cells, which miraculously generate electrical power by combining oxygen and hydrogen, getting water as a by-product. It's the opposite of the old high-school science experiment, wherein an electrical current is passed through water (H_2O), causing it to separate into a bell jar of H and a bell jar of O. The water produced by a fuel cell is supposed to be pure, but on Gemini it was contaminated by organic particles aptly nicknamed "furries." These furries turned the water the color of strong coffee and raised havoc with iron-stomached men. On Apollo a different type of fuel cell actually did produce potable water, a vital saving of weight on a lunar mission since a separate drinking water supply was not required.

Donn Eisele—attitude and translation controls Tough to explain. Webster says attitude is "the position of something in relation to a frame of reference"; by far the most common typographical error in aeronautics or astronautics is "altitude" for "attitude," and vice versa, so apparently the secretaries don't have it figured out. Attitude really means which way something is *pointed*. Up, down, toward the sun, or what? Remember doing an aileron roll in a T-38 a couple of chapters ago? If done properly, this roll would result in no change in pitch or yaw attitude (the nose would stay fixed on a point), but the craft would change 360 degrees in roll attitude. Translation, on the other hand, means *moving* through space, be it up or down, left or right. If you throw this book against the far wall, and I wouldn't blame you at this stage, you would have caused it to translate laterally or longitudinally (however you define it) that distance, and then translate vertically down to the floor. Inside a spacecraft, one flies with both hands. The left grasps a translational hand controller, the right an attitude hand controller. Between the two, all maneuvers are possible. The left controller sticks out of the instrument panel like a T-handle. It can be pushed in/out, up/down, or left/right, and the spacecraft will respond by *moving* in the corresponding direction. This is done by firing thrusters (small rocket engines) on the appropriate side of the spacecraft to cause this motion. As long as the left stick is held out of neutral, the selected thruster will fire. The right-hand controller is more sophisticated. It controls attitude, or which way we are *pointed*, and it is connected to the thrusters by means of an elaborate switching panel, which allows various options to be brought into play. The right-hand controller, rather than being a T-handle like its left counterpart, is more like the stick on an airplane: pull back to pitch up, push forward to rotate the nose down; left to roll left, right vice versa; all this just as in an airplane.

In addition, a spacecraft is too crowded to afford the luxury of rudder pedals, so the third (or yaw) axis is also built into the right-hand controller in such a fashion that rotating your hand left or right causes a yaw motion similar to kicking left or right rudder in an airplane. So far it's fairly straightforward, but now to discuss the "elaborate switching panel." Here is where the pilot leaves and the electrical engineer takes over, or where the test pilot/engineer truly comes into his own; here we start attaching rate gyros, and dead bands, and so forth, to the control stick in such a fashion that the pilot can choose to be more or less a part of an elaborate auto-pilot network. For example, he can arrange the switches in such a fashion that when he removes his hand from the right-hand stick he commands a zero rate, and the spacecraft will stay fixed in inertial space, gyro-stabilized, firing its thrusters as necessary to continue pointing in exactly the same direction, within the limits of the selected dead band. If the switches are not so selected, the unattended spacecraft will wander off and point wherever the laws of physics say it should. The first scheme is better for control, but at the expense of fuel consumption; in some situations, tight attitude control is a necessity, in others a luxury. It is the pilot's job, before all others, to determine how best to put these conflicting requirements together, how to fly to the moon with a minimum of thruster firing (saving fuel for rendezvous), while still retaining the option for tight, precise control systems during critical mission phases such as docking. Donn Eisele did these jobs for us.

Ted Freeman—boosters Boosters have several aliases: launch vehicles, rockets, missiles. It is true that they do use rocket engines and have boosted missiles, but probably launch vehicle is the most accurate label to pin on these huge slim cylinders we have all seen rising majestically up our television screens. They are mostly empty after their fuel burns out, but historically they are of supreme importance because without them there would be no space age. Tsiolkovsky, Goddard, Von Braun—these chaps never dreamed about simulators or recovery systems or mission planning, but about the raw rocket power which would be required to put men into orbit, onto the moon, out beyond Jupiter, or wherever. In my own view, the pendulum has swung; rockets are about as interesting as the powder which propels a bullet. To be fascinated by them is to prove McLuhan's point that the medium is the message, or perhaps it is a more Freudian thing. From the crew point of view, one rides the thing rather than flying it, and the trick is to know when to get off, or when to shut it down, if it turns out to be ailing. The boundaries of leaving or staying are generally defined by two sets of limits: if it wanders off past a certain *angle*, which might cause you to point in the wrong direction, or if it changes direction at an unacceptable *rate*, indicating a predilection to tumble ass over tea kettle. So maximum angles and rates are analyzed and memorized by the crew, which generally has time to respond to such excursions. On the other hand, if the thing blows up, it usually does so with an instantaneous finality which makes crew action futile. Ted Freeman died in an airplane crash within four months of his assignment to this difficult area.

Dick Gordon—cockpit integration   Everything the pilot does, he does in the cockpit, using the switches and controls available there. Most of his information comes to him through cockpit displays. In a machine as complicated as Apollo, the selection and arrangement of these tools is of paramount importance. In the first place, more information is available than can possibly be presented to the pilot at any one time, so each subsystem must be analyzed to determine what its essential measurements are. To measure the health and happiness of a fuel cell, should one check its voltage, or the amount of current it is producing, or the amount of hydrogen and oxygen it is consuming, or the purity of the water coming from it, or all these things, or none, or what? How often will the pilot be concerned about fuel-cell health? Certainly during the ascent into orbit he is more apt to spend his time scrutinizing the huge rocket to which he is attached, making sure it is not deviating from course. Yet the launch vehicle is quickly emptied and discarded, leaving gauges which are now useless but which occupy valuable panel space for the rest of the mission. Perhaps the fuel-cell indicators should occupy that prime space.

In addition to deciding what information is needed when, and in what order of priority, the cockpit designer also faces a host of other problems. All switches and controls must be accessible, some more than others. In the event of cabin-pressure failure, when the oxygen atmosphere has leaked out, the astronauts must work inside their rigid, confining, pressurized suits—and that makes things much more difficult. A simple task such as changing a lithium hydroxide canister (necessary to prevent a dangerous carbon dioxide build-up in your breathing oxygen) can be done one-handed in two minutes in shirtsleeves, but can become an arduous fifteen-minute wrestling match in a pressurized suit. Other switches and controls have to be operated, not only in a vacuum, but during the "high G" phases of the mission, during launch and re-entry, when acceleration forces cause an outstretched hand to weigh six or eight times as much as normal. The question of visibility also becomes very important; lighting within the cockpit is always a real problem, and as acceleration forces build up, pilots get tunnel vision, losing some of their peripheral visual acuity. Labels must be logical and legible.

The cockpit designer must weigh all these considerations, and many more, and come up with a foolproof design. One small example: that T-handle, the translational hand controller, the one in your left hand, also doubles as the control for aborting the flight. If the rocket starts to catch fire on the launch pad, for instance, the commander need only rotate his left wrist counterclockwise 30 degrees to initiate a series of events which will ignite escape rockets to pull the command module free from everything behind it, lift it out of the danger zone, open parachutes, etc. That's all it takes, one small movement of the wrist! It has to be that quick and simple, for there is no time for more elaborate switching arrangements. Yet consider the consequences of poor design. "Oops!" says Neil Armstrong, as he drops his pre-launch check list, bumping his arm against the translational hand controller as he attempts to retrieve it, aborting the first lunar landing attempt, making him the laughingstock of three billion people. A classic case of poor cockpit design is the ejection procedure which used to be in one Air Force trainer. It was a placard listing half a dozen important steps, printed boldly on the canopy rail where the pilot couldn't miss seeing it. The only flaw was that step 1 was "jettison the canopy." As the most experienced test pilot of the fourteen, Dick Gordon was well equipped to avoid problems such as these.

Rusty Schweickart—future programs and in-flight experiments I really don't know what Rusty did about future programs, but there was plenty to keep him hopping in the area of experiments, especially on the Gemini program. Going to the moon was almost an experiment unto itself, but on Gemini there was a review board in Washington which gave all comers their opportunity to explain why fertilizing sea urchin eggs in weightlessness was going to result in a great scientific breakthrough, or whatever. Once an experiment was approved, it was assigned to a specific flight, and then a long and sometimes tortured process took place, to convert something that seemed like a sound idea on paper into a practical reality in flight. Experiments were divided into categories and assigned coded letters: D for Department of Defense experiments, M for medical, S for scientific, and MSC for Manned Spacecraft Center ideas. The long-duration flights, such as Gemini 7, were heavy on the medical side (eight experiments) while the other experiments were sprinkled around, based on competing objectives, weight limitations, other demands on the crews' time, etc. Some met ignominious fates: Because the activating handle broke when Gus Grissom twisted it, the sea urchins never got to do their thing. Some, like the synoptic terrain photographs and the micrometeorite measurements, seemed to yield plenty of good, useful data. I don't know of any that were spectacular successes. On Apollo the same pattern of assigning experiments to specific flights was continued, up until the time of the spacecraft fire at Cape Kennedy that killed Grissom, White, and Chaffee. Then it was determined that building a new machine to tight specifications and sticking to them was difficult enough without the added complication of an array of experiments which added wires, cables, black boxes, and a variety of other equipment to the inside of a command module. Hence a moratorium was declared on experiments for the early Apollo flights, a move applauded by most of the astronauts.

Dave Scott—guidance and navigation (G&N) Important stuff, the spacecraft's brain really. Dave had done his master's thesis at M.I.T. on interplanetary navigation, which was especially appropriate since M.I.T. was designing the Apollo G&N equipment. Heart of the Apollo system is a lump of metal slightly larger than a basketball, called the inertial measuring unit, or IMU. The IMU contains three gyroscopes mounted at right angles to each other and connected to the spacecraft by gimbals. When the gyroscopes are spinning at full speed, they keep the IMU pointed in one direction ("fixed in inertial space"). Think of it as a "stable table" around which the spacecraft turns. Now if we knew which way the IMU was pointed, in comparison to the stars for example, then we could measure which way the spacecraft was pointed, relative to the same stars, by measuring the angles between the spacecraft axes and the IMU axes. These three angles (pitch, yaw, and roll gimbal angles) tell you which way you are pointed, relative to the stars, which are fixed in inertial space. Gyroscopes precess, or wander off, so the IMU must be restored to precise stellar alignment periodically by the astronaut, who selects two stars, peers at them through a telescope or sextant, superimposes cross hairs on them (one at a time, of course), and then pushes a button at the instant of perfect alignment. He then tells his computer which two stars (by pre-stored numbers or celestial coordinates) he has selected. Then the IMU is torqued around to its correct new alignment.

All this keeps your attitude information up to date, but does not help in telling you where you are, only which way you are pointed. The *where* part comes from a mathematical thing called a state vector, which is stored in your computer. A state vector has seven parts: three tell you position, three velocity, and one time. Position is expressed by three components $(x, y,$ and $z)$ of distance from a reference point, while velocity $(\dot{x}, \dot{y},$ and $\dot{z})$ is the rate of change of those three distances. Time (t), of course, pinpoints the time at which you were at any one spot (x, y, z) heading which way $(\dot{x}, \dot{y}, \dot{z})$. The computer knows where it was when it started (Launch Pad 39 at Cape Kennedy) and keeps book on all subsequent travel. During times of engine firing, accelerometers are turned on to record changes in velocity, but in coasting flight the computer keeps book by an involved mathematical computation of the gravitational pull exerted by sun, earth, and moon upon the spacecraft. As time goes by, the state vector loses accuracy and it must be updated, either by the ground computers, which can radio a new state vector up to the spacecraft, or by the astronauts, who can measure the angle between a selected star and the earth or moon horizon. If we want to change course, it's as simple as ABC. We just get Roger Chaffee's tracking network to send up a state vector and a couple of other things to Dave Scott's computer, which feeds pointing commands through Dick Gordon's instrument panel to Donn Eisele's controls, which cause Walt Cunningham's electricity to power Gene Cernan's engines, which fire, to get us out of Bill Anders's radiation zone into the position called for by Buzz Aldrin's flight plan. The rest of you guys must be loafing!

C. C. Williams—range operations and crew safety Cape Canaveral had been shooting rockets off for a long time before man got added on board, and when he was, it created a few problems. Rockets are fired toward the east, and as long as they keep going that way, no sweat! But when guidance systems go awry, and tons of explosive liquid propellants start wandering around the sky, then the range safety officer's finger gets itchy on his DESTRUCT button. If unmanned, again no sweat, but how about a Gemini or Apollo? What warning should he give before destroying, and what to do if the crew fails to heed, or does not receive, the warning? How much time to allow, how many times to try, how to balance the safety of two or three up there against entire communities down here? C.C. had these problems with which to wrestle, as well as the complicated crew procedures at the Cape, be they the normal ones of loading men aboard, or the emergency ones of unloading in a hell of a hurry. If one wanted to depart a Gemini gantry in a big hurry, one rode the slide wire, which simply meant that you clipped your parachute harness onto this steel wire and jumped off. You then slid down this wire, first almost vertically, then at an ever more shallow angle, until you reached the bottom, unhooked, dropped to the ground, and took off running. On Apollo it was a bit more complicated, as nearly everything was on Apollo. The slide wire had attached to it a small cable car which you entered and then released for the long slide down. At the bottom you got out of the car and jumped into a dark slippery tunnel, sliding along beneath earth and concrete, until finally you were spit out at the "rubber" room. This chamber was shock-mounted so as to survive the earth tremor caused by an exploding Saturn V, and its interior was literally built out of rubber, including rubber floors and rubber chairs to further protect the occupants from vibration. All this was C.C.'s domain.

Mike Collins—pressure suits and EVA Last and hopefully

not least was Mike Collins with pressure suits and EVA (extra-vehicular activity). EVA came in two varieties, Gemini space walks and Apollo moon walks. The equipment for each was different, the Gemini suit being made by the David Clark Company of Worcester, Massachusetts (Goddard's home town); oxygen came from the spacecraft via an umbilical, and then went through a chest pack built by a division of the Garrett Corporation in Los Angeles. The Apollo suit came from International Latex Corporation in Dover, Delaware, getting its oxygen supply from a back pack built by Hamilton Standard in Windsor Locks, Connecticut. There were also many other pieces of equipment for possible EVA use, the most notable being the astronaut maneuvering unit, almost a separate spacecraft in itself, which the Gemini astronaut was supposed to strap to his back and then use to propel himself in good Buck Rogers fashion through the sky. It was made by Ling-Temco-Vought in Dallas. My job was to monitor the development of all this equipment, to make sure that it was coming along all right, that it was going to be safe and practical to use, and that it would please the other guys in the astronaut office.

It was a tough job, because of geography if nothing else. I could get a lot done in Houston, working with NASA engineers in the crew systems division, but most of the serious design reviews and equipment inspections took place at the contractors' plants, which meant my old T-33 was kept busy jumping from Connecticut to California to Delaware to Texas to Massachusetts, trying to keep up with developments. And developments were not all rosy, especially in the Apollo suit department. In the first place, there is a kind of love-hate relationship between an astronaut and his pressure suit. Love because it is an intimate garment protecting him twenty-four hours a day, hate because it can be extremely uncomfortable and cumbersome. Generally, as time goes by, the emphasis shifts from hate to love, so that by flight day the astronaut has a garment which he regards as an old friend, one which has been meticulously tailored to his every dimension, one which has been worn long enough to be comfortable, but not long enough to suffer undue wear and tear. In fact, each crew member has three suits tailored to his individual measurements. The first, called the training suit, is manufactured as soon as he is assigned to a flight, generally a couple of months before public announcement of said fact. At MSC this was the prime source of rumors as to who was going to get which flight. "Hey, Charlie's got a trip to David Clark scheduled for next week. What's up?" To avoid leaks, the crew system division tailored Gemini suits to fit "Castor" and "Pollux" rather than putting any actual names on the documentation.

The training suit was well named. It was worn in simulators, centrifuges, zero-G airplanes, etc., whenever the realism of having the crew suited was required. It got battered and beaten by several hundred hours of hard use, but then it was never intended for flight. The second suit was the flight suit, the third the back-up suit. These two suits were intended to be identical, one of the two simply being designated as the prime choice (perhaps because it fit a tiny bit better) and the second its substitute. The flight and back-up suits were generally worn just enough (perhaps twenty hours each) to get them "broken in," in conformance with the infant-mortality theory popular in reliability engineering, which says that a component has a fairly high probability of initial failure, and then is at its all-time best (most reliable) immediately after this first danger zone. Since these suits were frightfully expensive, around thirty thousand dollars for a Gemini suit and more for an Apollo suit,* three suits each may seem too many, but the argument was that some malfunction like a broken zipper (it happens!) on launch morning could not be allowed to delay or cancel a multi-million-dollar flight.

If it is true that the astronaut ends up loving his pressure suit, it is equally valid to note that shock—if not downright hatred—is

---

* We used to joke that the suits cost a thousand dollars a pound, which is not far off the mark, with the heavier extravehicular suits being considerably more expensive than their lighter counterparts, which were designed solely for wear inside the spacecraft.

usually the first reaction to wearing one of these tailored gas bags. Hear Pete Conrad on the subject of an early Apollo suit, in a memo dated January 28, 1965: "Based on the zero-G work last week and some work . . . this week, I've concluded that the ILC suit is *useless* and should be abandoned. Bob Smyth has a custom-made suit ($65,000) and he looks like a pro-football player cast in cement. My suit . . . wasn't any better. I would like to take a Gemini extravehicular suit up there* for direct comparison."

In order to make sense out of all this, it is necessary to explain what a pressure suit is designed to do, and why that is so difficult. First and foremost, it must be airtight, so that it can be pumped up ("pressurized") to protect the astronaut from the vacuum of space. Without gas pressure around him, he would have nothing to breathe, and even if you solved this by giving him an oxygen mask, he would still die quickly because the fluids inside his body would vaporize—his blood would literally bubble. So we must begin with a gas bag which can be pressurized to around 3.7 pounds per square inch, which is about what you need if you use 100 percent oxygen as we did.† The gas bag, or bladder as it was called, was made of

---

* "Up there" is the Grumman plant at Bethpage, Long Island. Bob Smyth is a Grumman test pilot who, more than any one individual, was responsible for the lunar module cockpit layout, a superlative job according to LM pilots. Bob left the Apollo program long before the LM flew, in order to do the experimental test flying of the Gulfstream II, Grumman's jet transport.

† At first glance, something seems wrong here, because we humans are accustomed to 15 pounds per square inch of atmospheric pressure. But remember that we breathe air, 80 percent nitrogen and 20 percent oxygen, so that the partial pressure of oxygen in our lungs is only 20 percent of 15 psi, or 3 psi. This pressure is sufficient to cause plenty of oxygen to pass through the membrane of the alveoli into the bloodstream. Hence, breathing 100 percent oxygen at 3.7 psi is also sufficient. In addition to forcing enough oxygen into the bloodstream, 3.7 psi must also be sufficient pressure to hold all body fluids and gases in their usual state, to prevent painful or fatal gasifying. It turns out that 3.7 psi is O.K. in all respects except one. Nitrogen which has been previously absorbed in the body tissue comes out of solution at 3.7 psi, causing the "bends," usually with symptoms of painful joints, especially knees and elbows. On Gemini and Apollo we solved this problem simply by "pre-breathing," breathing 100 percent oxygen for several hours before the flight to purge all nitrogen from our system. This pre-breathing is the reason you always see astronauts with helmets on and locked, carrying a suitcase full of 100 percent oxygen, as they depart for the launch pad. They have been pre-breathing pure oxygen for an hour or two, and if ordinary nitrogen-contaminated air got inside their suits by some accident, then the launch would be delayed until the pre-breathing ritual had been repeated.

thin neoprene rubber, contoured to fit the body. The bladder material was soft and pliable, moving easily with the astronaut, provided the suit was not pressurized. When it was pumped up to 3.7 psi, it was a different matter. The next time you stop by your neighborhood filling station, watch them pump up an inner tube. It comes out of its box soft and floppy, but as the air pressure inside it increases, it becomes a fairly rigid doughnut. Same thing with a pressure suit bladder, even though we are only pumping to 3.7 psi instead of the inner tube's higher pressure. To prevent ballooning of the inner tube, it is restrained inside a relatively inflexible tire casing; in similar fashion, a pressure suit bladder is contained by a "restraint layer." Here is where black magic enters the rational world of design. The restraint system must allow the pressurized suit to bend as the astronaut bends, twist as he twists, and in general act as a tough extra layer of skin. It's not too difficult to design a flexible joint for a knee which only moves back and forth in the same plane, but how about a shoulder joint? It is capable of incredibly complex motions, and the suit must be able to follow them all, without undue effort by its occupant, without leaking, without changing shape, without zinging back to some awkward position the instant the astronaut relaxes. In a Gemini suit, the restraint layer was made of a cleverly woven net which allowed expansion in one direction, but in only one. By sewing net sections together with the direction of expansion carefully matched to the normal movements of the human torso, it was possible to achieve a fair degree of mobility inside the pressurized suit. At 3.7 psi, the Gemini suit wished to assume some neutral position, just as the inner tube wished to become a doughnut, but the woven net restraint kept its shape pretty well and allowed the astronaut to bend and flex the suit within reasonable limits. It wasn't without effort, however, and whenever you relaxed, the suit popped back to its neutral point.

The Apollo suit had a more sophisticated setup. Instead of having a simple restraining net, it controlled the shape of its inflated bladder by a complex array of bellows, stiff fabric, inflexible tubes, and sliding cables. In theory, these devices permitted much more mobility with much less effort than the Gemini suit, but in the early models it was not always so, as Pete Conrad pointed out. The Gemini suit was also a lot more comfortable unpressurized, because the net was soft and pliable, whereas some of the stiff joints and cables of the Apollo suit got in the way even when the suit was not pressurized.

Inside an airtight bladder, be it Gemini or Apollo, the astronaut would dissolve in a pool of sweat were there not some way to keep him cooled, so a complex array of ventilation tubes had to be added to torso, helmet, arms, and legs of the suit. These vent tubes all gathered into a manifold, allowing oxygen to be piped in and out of the suit through two large circular metal connectors on the chest. In the case of an Apollo moon-walking suit, there are four such connectors, one set attached to the spacecraft, the other to the back pack. In addition, there is an electrical connector which provides a path for radio signals to reach earphones. Biomedical information is routed from four sensors (which are taped to the chest), through belt-mounted electronic signal conditioners, into the same electrical connector. On Gemini one wore simple long johns, but on Apollo a more efficient cooling scheme employed water-cooled underwear into which tiny plastic pipes were sewn. They in turn were manifolded out through still another connector on the chest, routing the warm water to the back pack, where it was chilled and then pumped back into the underwear. Then there was a triangular plastic bag into which the penis was inserted, for urination during those parts of the flight when you couldn't get out of the suit. The urine could be dumped into the spacecraft's plumbing lines by another connector, mounted on the thigh.

Gloves, boots, and helmet had to be connected to the torso— no problem in the case of the boots, which were simply added over the bladder. Gloves and helmet, however, added another host of design problems. The gloves had to be thin and flexible to allow manipulation of switches and other delicate controls, such as the rotational hand controller, even when pressurized. They also had to be removable. They had to permit wrist bending and rotation, opening and closing of the palm, freedom for the thumb and fingers to grasp strongly or touch lightly, and so on. The helmet had to be strong, light, comfortable, contain earphones and microphones, provide excellent vision, keep out noise, and provide plenty of circulating oxygen to prevent a build-up of carbon dioxide from exhaled breath.

If things seem to be getting complicated, remember that all we have done so far is to allow a man to operate in a vacuum. Now we must add the other hazards of that vacuum: the blinding energy of the sun, the subfreezing temperature of any object shielded from the sun, and the possible penetration of the suit by tiny projectiles called micrometeorites. Fortunately, heat, cold, and micrometeorites could all have a common solution—a thick cover layer which would be a superb thermal insulator and an adequate shield against high-velocity impacts. The thermal part of the problem was straightforward in that we had definite design inputs: the temperature in darkness was expected to be −250 degrees F, in direct sunlight +250 degrees F, and at the bottom of a lunar crater +310 degrees F. Hence the thermal limits of the Gemini cover layer (−250 degrees F to +250 degrees F) and the Apollo cover layer (−250 degrees F to +310 degrees F) were well defined; it was simply a search for the most efficient materials, which turned out to be multiple layers of thin Mylar. The micrometeorite problem was different. Micrometeorites are the sharks of outer space: normally absent, usually harmless when present, but on rare occasions terribly destructive. How do ocean swimmers prepare for sharks, or space walkers for micrometeorites? We tried a mathematical approach, based on the number of micrometeorite impacts measured by previously recovered space vehicles and by theoretical computations. Our mathematical model showed very simply that if you stayed out in space a long time, your chances were excellent of getting hit by a small micrometeorite and poor of getting hit by a large one. In this kind of situation it is impossible to design 100 percent protection; common sense must take over. Since Mylar was not a good energy absorber, we added one layer of a felt-like material, which in conjunction with a tough nylon outer covering gave us good protection against the small, dust-like particles which make up the vast majority of the micrometeorite flux. At first we did not know how to package this thermal and micrometeorite protection. We tried overcoats and capes and other schemes, and finally decided to build the protection directly into the suit cover layer. This arrangement was far simpler once you were outside the spacecraft, but of course it meant that whenever you wore the pressure suit inside the spacecraft, you were saddled with a cover layer of extra bulk and clumsiness. On Apollo it did not matter too much, as the plan was to take the pressure suit off whenever it was not needed, but on Gemini it was not possible, because of the tiny crowded cabin, to remove the suit. Therefore, on the EVA flights (Geminis 4, 8, 9, 10, 11, and 12), the commander in the left seat wore a thin cover layer and the EVA man in the right seat a bulky one. When it came time to sleep, an argument usually ensued about the cabin temperature, which was too cold for the commander and too warm for the space walker.

There were other differences as well. The space walker's eyes needed special protection against the sun's rays, especially the ultraviolet component of sunlight, which can be very damaging. A special coating was added to the basic helmet visor, and then a second visor, tinted like sunglasses, was clamped on in such a way that it could be raised or lowered as desired. For Apollo, fifteen or twenty different schemes must have been tried before we finally settled on the gold-covered, double-visored outer helmet that moon walkers wear.

For me, late 1964 and early 1965 was a time for wrestling with all these design problems, helping the crew systems division (CSD) engineers evaluate new designs, talking to other astronauts who, like Pete Conrad, had complaints or suggestions, and keeping the entire astronaut group informed as to new developments. The common method of getting the word around was the memo, and as we fourteen dug deeper into the various specialty areas assigned us, the memos began to pile up by the dozen in our IN baskets. Some memos were written to ask opinions of the astronaut group, others to dispense information, announce decisions, or simply to vent the author's spleen. "Look what the bastards want to do now!" and "Look what I have saved us from!" were two favorite messages, implying that all was not well between the astronaut office and the rest of the space program. Such was not the case. As individuals and as a group, we astronauts were well received by other NASA and contractor groups, our ideas were carefully considered, and we enjoyed access to the top level of management. In fact, our opinions were solicited by so many people in so many different technical areas that we were run ragged. In my own area, there would be meetings, design reviews, simulations, and other activities at two or more different places simultaneously; I spent more time just getting there, and saying what crew members could reasonably be expected to do (and not to do) in flight, than I did writing memos, but occasionally I did feel the need to call the group's attention to developments in my area of "vicious specialization."

MEMORANDUM

DATE: 7 Oct. 64

TO: Astronauts
FROM: Michael Collins
SUBJECT: Extravehicular Ground Rule

Recent Gemini ingress tests in the zero-G KC-135 indicate that the extravehicular astro requires all his strength and agility to get back inside the spacecraft and to wedge himself far enough down in the seat to close the hatch. If he is incapacitated (dead, unconscious, badly hypoxic, or exhausted), there is no way for the second astro to get him back inside. So far all design effort has been directed toward preventing him from becoming incapacitated, and there are no plans to attempt to design any pulleys, etc., to reel him back in. In my opinion, this is a sound decision which realistically acknowledges the fact that you can't stuff an immobile object as large as a man in a pressurized suit into a space as small as the Gemini cockpit.

However, it is a harsh ground rule and one which means that, for a successful re-entry, the left-seat astro must disconnect his buddy's hoses and lines, close the hatch, and leave him in orbit. There are obviously many implications here, and I would like to hear any ideas you have on the subject.

A similar situation exists on Apollo, in regard to retrieving an astro incapacitated on the lunar surface. For example, a recent decision within CSD deleted any "buddy hose" connectors which would allow the astro inside the LEM* to rescue the man outside by hooking both up to the same back pack (PLSS). This is part of a developing philosophy which says "the man is one component of the extravehicular system which will not fail." Again, I think this is the way to go, but I would appreciate any dissenting comments, because now is the time to state our position.

Michael Collins

Generally the reaction to memos such as this was a deafening silence, because the other astronauts were too busy with their own special problems. In the case of this particular memo, I remember only one comment, Ed White's strenuous objection that it did not require "all his strength and agility" to get back inside the Gemini. In Ed's case, I'm sure it didn't, as he was a superb athlete, strong as a horse, but I was really shocked at the difficulty I had in getting that damned pressurized gas bag wedged far enough down in the right-hand seat to provide enough head clearance to close the hatch. It was easy enough on the ground, with an assist from gravity, but in the zero-G airplane I kept popping out, like a cork out of a bottle, and I was worried about it. Ed's worry was, I guess, that my memo might result in his EVA being canceled on his upcoming

* LEM—lunar excursion module, the old name, which later became shortened to simply LM.

General Pershing and my dad in Mexico, 1916

My family,
the day they
learned I was
to be an
astronaut

The T-38: our teacher, transportation, and toy

No food, but plenty of chiggers

Charlie Bassett and I, headed for Reno

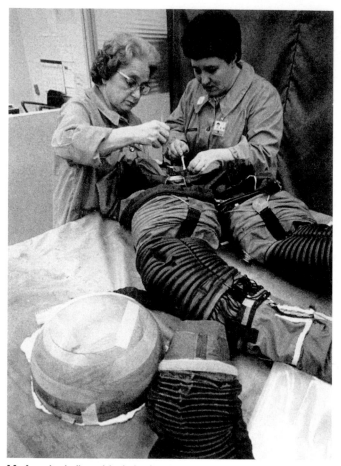

My favorite ladies with their gluepots

Whizzing across the slippery table

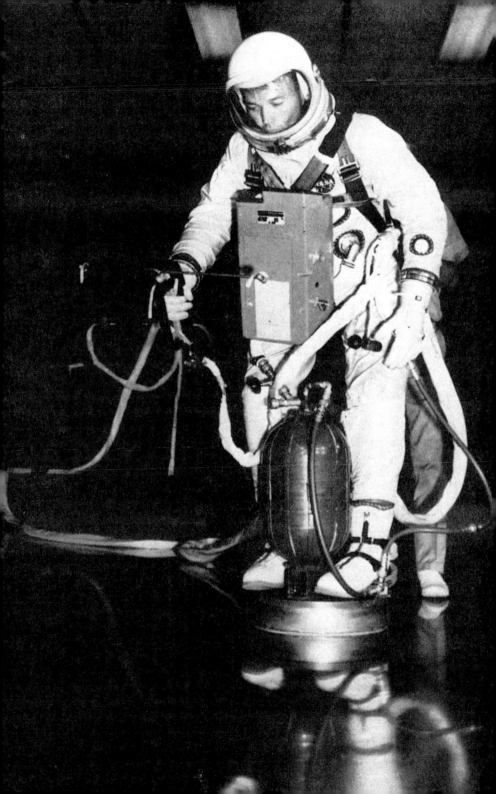

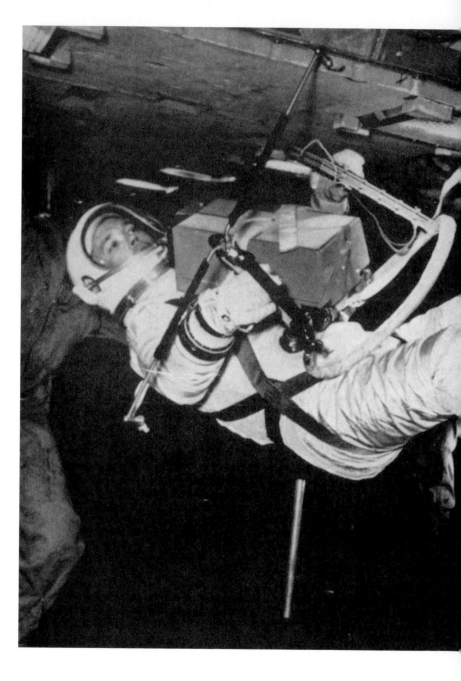

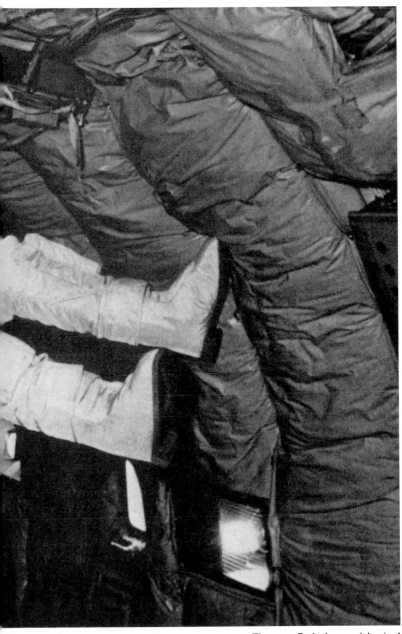

The zero-G airplane—sickening!

The Gemini: a tight squeeze, even with everything neatly packed

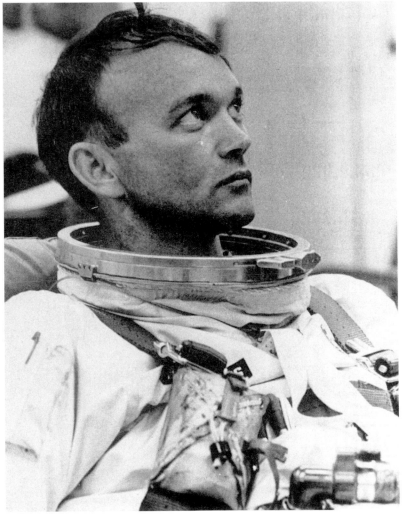

All set (I hope) for Gemini 10

Our Agena, parked and waiting

Here we go, trailing one extra wire

Wheeling into position, just before docking

The **Big** Engine lights, and kicks like a mule

What was once a pretty cockpit belonging to Grissom, White, and Chaffee

Arlington Cemetery: LBJ shakes hands with Roger Chaffee's son

Gemini 4 flight. As it turned out, neither of us needed to worry. Ed became the country's first space walker, and had no difficulty whatsoever in getting back inside, although he did have a bit of difficulty with a binding hatch. For all space walks after Ed's, including my own, the right-seat pan was cut down slightly to permit easier access and the hatch was loosened, so things worked out for all concerned. Flying in space turned out to be easier than flying in the zero-G airplane.

It was also a lot more pleasant. The zero-G airplane was a converted Air Force KC-135 (essentially a Boeing 707) which had had all the seats removed from the cabin and the interior padded. It reminded me strongly of a padded cell. The airplane usually flew out of Wright-Patterson AFB in Dayton, Ohio, although it could be summoned on special occasions to Houston or the Cape. It could imitate weightlessness for slightly over twenty seconds at a crack by flying parabolic arcs. The pilots would dive it down steeply, pull up fairly sharply (2 or 3 Gs), and then push the airplane over the top at exactly the right rate to keep our bodies suspended between ceiling and floor in the aft cabin. At least that was the theory. In practice, even the smoothest pilot wiggled a little bit, and we tended to bang around in the back. The first time I had flown it, back in 1963 at Edwards, it had been a ball. It had been a simple indoctrination flight, and our ARPS class piled on board with great enthusiasm, dressed in thin cotton flying suits rather than the cumbersome pressure suits. We laughed ourselves silly trying to drink water out of paper cups (in zero G you really can't do it, it floats up out of the cup in great spherical blobs before you can get it into your mouth), and did a hundred other silly things not possible here on earth. The climax of the flight was a carefully staged act featuring Ernie Volgenau, one of our instructors. A collegiate wrestling champion, Ernie looked exactly like Clark Kent, and his wife had sewn—especially for this occasion—a Superman suit for him to wear. It was made out of long-john underwear dyed sky blue, with brief red shorts over it, the great Superman S insignia on the chest, and a long flowing cape. Two of the biggest, fattest students (since they would weigh nothing, it didn't matter) were picked to be "criminals," and as soon as the pilot got us all weightless on the next parabola, Ernie uttered a scream of triumph, picked up one crook with each hand, examined them with disgust as he held them at arms' length, then knocked their terrified heads together, and hurled them the length of the fuselage! Talk about high camp, and all of it recorded for posterity by movie cameras.

So much for fun and games. The zero-G airplane I later learned to hate was altogether a different matter. First, we were there to work, not play. Second, we nearly always wore pressure suits, which were rarely well ventilated, meaning we were hot to begin with and got hotter yet. Third, we always had more things to find out than we had time to do, so we worked frantically during each parabola. Fourth, we flew so many parabolas! A few were fun, but after that it becomes more and more like drudgery. And finally, people would always get sick and throw up, usually right at the beginning of each flight. There is something very unsettling, even for experienced aviators, about repeating forty or fifty parabolas, the body alternating constantly between zero and 2 Gs. A typical plane-load included cameramen, engineers, and various supporting personnel—most of whom were not accustomed to this kind of thing and who were almost sure bets to throw up within the first half dozen parabolas. That didn't help. Some astronauts threw up. I never quite did, but I was close enough at times to be utterly miserable. I found that it helped to minimize my head movements, but frequently that was impossible as we thrashed about trying to retrieve make-believe micrometeorite packages from make-believe Agenas, or whatever the task of the day was. The strange thing about the zero-G airplane was that instead of becoming acclimated to it, my tolerance seemed to decrease. Toward the end of my Gemini career, one had only to say "KC-135" to make me feel quite queasy. Thank God, we didn't use it much for Apollo crew training.

A much more unsettling, and potentially very serious, problem came to my attention during my pressure suit work: under certain conditions, I got claustrophobia. One doesn't really wear a pressure suit, one gets inside it. Once gloves and helmet are locked in place and the suit is pressurized, you peer out at the world but are not part of it; you are trapped inside this rigid cocoon, as dependent on it as any chrysalis, dependent especially for an abundant flow of air or oxygen to breathe and to provide a cooling flow over the body. So far so good, no feeling of uneasiness yet, just as I have never felt uneasy strapped into a tiny cockpit. But now let a strenuous task be introduced, such as removing a bulky and balky hatch from a darkened tunnel, and as my breathing grew deeper and faster, an unfamiliar feeling of panic started edging into my consciousness. My God, I'm not getting enough to breathe, I've got to get out of here! Of course, I didn't dare to let these feelings be known, and somehow I always managed to finesse it. "Say, out there, could you turn up the flow and check the hoses to see if they're being pinched somewhere? I'll just rest here a minute." Somehow I was always able to muddle through, but on more than one occasion I found it necessary to raise my visor to get a whiff of the outside world, a solution obviously not available in the vacuum of space (or altitude chamber, for that matter).

I gave the matter deep and serious thought. It was certainly true that a lot of the tests I was conducting were very strenuous, and that the air supply, running through abnormally long hoses from test console into a mock-up or simulator, might not be as cool or swiftly flowing as in the real spacecraft. It was a well-known fact that restricted flow could result in a stagnation of exhaled breath, rich in carbon dioxide, within the helmet. This, in turn, could cause a feeling of malaise for valid physiological reasons. On the other hand, I seemed to be having trouble when others were not. Certainly I shouldn't need more oxygen than everyone else, and if other people seemed to be perfectly happy with the same hot, stale air supply that terrified me, then clearly something was wrong. Delicately I pursued the subject with some of the experienced pressure suit engineers. "Say, do people ever get claustrophobia when they get into one of these things for the first time?" "Oh yeah, I remember one guy went positively bananas first time we lowered the visor on him. Had to get him out right away. Couldn't stand it." "No kidding!" I tried to analyze just when that awful trapped feeling would begin inside me. All was well as long as I felt cool, could see well, and had freedom of movement. However, when the workload increased and I began to perspire heavily, then I became vulnerable, and if I was working in a dark corner or became entangled somehow, or if the visor became fogged with moisture, then I nearly ceased functioning, failing to respond to instructions for a minute or so at a time. Sometimes I could force myself to sit quietly and overcome the panic, the desire to get out of there at any cost, but at least once I had to find some pretext to end the test.

After these episodes I considered going to Deke, confessing all, and leaving the program. But of course, I didn't, because I wanted desperately to stay, and because I felt I could talk myself out of this ridiculous condition. After all, I had worn pressure suits on a dozen occasions at Edwards, including the F-104 zoom flights, and had never had the slightest tendency to feel trapped inside them. Furthermore, the claustrophobia, or whatever, was an infrequent visitor, and I could sail through most tests with no difficulty. Granted it would take only one such incident in flight to put me and my fellow crew members in grave peril, but on the other hand, on about half the planned flights the pressure suit would not be used at all, except in the highly unlikely event of cabin pressure failure. Somehow I could, I would, overcome this absurd impediment to flying in space.

As time went by, things got better. I learned to request maximum cooling flow at the beginning of a test sequence, even if it meant I half froze. I also learned to concentrate on my breathing, and could convince myself that all was well, that I really was getting enough to breathe. Finally, as I got to know the equipment better, and as the really tough tasks were re-engineered to make them simpler, I found fewer and fewer occasions when I would be gasping for breath. I never did discuss the matter with anyone, so I don't know whether I alone had the problem, but I must say I've never seen or heard of another claustrophobic astronaut. I must also say that my own love-hate relationship for my pressure suits is more strongly developed because of all this. I did have a highly successful space walk on Gemini 10, wearing pressure suit number G-4C-36. It was a beautiful suit, with special modifications to give me extra visibility and mobility; it fit perfectly, and I felt very much at home in it. Its paper-thin bladder, glued together with consummate care by some nice ladies in Worcester, was all that kept my 3.7 psi oxygen from escaping into the zero psi infinity surrounding me as I dangled from the end of my fifty-foot tether. For G-4C-36 I felt as close to love as one can feel for metal, fabric, and rubber; but when I think generically of pressure suits, I think, Ugh, nasty little coffins. I'd be pleased never to get inside one again.

In addition to looking after the development of pressure suit design, I was the astronaut office boy in charge of all the miscellaneous equipment to be used during extravehicular operations. During Gemini EVA, one wore a chest pack, on Apollo a larger back pack. They were engineering marvels, crammed full of oxygen supply bottles, radios, heat exchangers, water boilers, pumps, fans, and the other claptrap necessary to keep that suit pumped up and its occupant safe and comfortable.

There was also an experiment to be flown on two Gemini flights which was really far out. Called the astronaut maneuvering unit, or AMU, it was really a little spacecraft in itself. It was housed inside the rear of the Gemini, in the adapter section as it was called, and in order to don it, the astronaut had first to depressurize the cockpit, get out, pull himself along handrails, trailing an umbilical, and swing around into the adapter section. There he backed up against the AMU, attaching himself to it by pulling down an encircling arm on either side and by connecting his pressure suit to its electrical and oxygen supply, as replacements for the umbilical cord. Then it was possible, in theory at least, to cut the AMU free from the Gemini and, by firing small gas jets mounted on it, to fly out of the adapter section. The space walker, or space scooter, or whatever he would be called, could then "fly" himself around in front of the Gemini, where his buddy inside could watch as he went through a series of pirouettes, bows, zigs and zags, and other maneuvers calculated to show that a man could operate in space free of his mother craft. If the idea could be developed, the Air Force (which sponsored this equipment) foresaw a long list of practical tasks man might perform in orbit, such as inspecting and repairing friendly satellites or neutralizing hostile ones. Fascinating as the concept was, it never got a fair trial. On Gemini 9, Gene Cernan couldn't get the arms down in place, and worked himself up into such a sweat that Tom Stafford had to recall him to the cockpit. Because of Gene's difficulties, and some problems Dick Gordon had on Gemini 11, a second attempt on Gemini 12 was abandoned, much to the displeasure of Buzz Aldrin, who was scheduled to give it a workout on the final flight in the Gemini series.

The AMU was a huge, cumbersome affair, necessarily so because of the features it contained. It had gyroscopes to hold itself stabilized and a host of thrusters which could be fired to make it go up, down, forward, backward, and so forth. Within NASA, it had another serious flaw in addition to its complexity: NIH, or "not invented here." NASA's version of this Air Force invention was the hand-held maneuvering unit, or gun. In this scheme the astronaut held in his right hand a chubby handle with two triggers, from which extended two arms with a jet on the end of each, pointing to the rear, and a third jet in the middle, pointing forward. The theory was that if the astronaut wanted to go some place, he simply pointed the gun at his target and squeezed the GO trigger, firing the two jets. When he got there, he squeezed the STOP trigger, causing the third jet to fire. Since there were no stabilizing gyroscopes in this system, the astronaut had to prevent himself from tumbling or spinning by pointing the gun in a direction to oppose any unwanted body motion and squeezing the appropriate trigger. The gas supply for the three jets could be built into the gun, a very limited quantity of course, or it could come from a much larger tank in the spacecraft and be routed to the gun through an umbilical line. Ed White on Gemini 4 used the first system, and I used the second on Gemini 10. Apart from its simplicity, the gun had the added advantage that it could be stored inside the cockpit, and the space walker could connect himself to the umbilical cord and make other necessary preparations inside the cabin, prior to dumping cabin pressure, rather than fumbling with connectors inside a pressurized suit back in the adapter section, a tricky business at best. However, once donned or mounted, the AMU with its stabilization system was clearly superior in ease of control to the simpler gun. It's too bad the two were never compared directly, in a Rolls-Royce vs. Hot Rod flying competition.

In addition to worrying about the pressure suits and other extravehicular hardware, I spent a lot of time trying to define what tasks man might reasonably be expected to do *outside*, flying along as a human satellite. Could he really do serious repair work, or lasso other spacecraft, or retrieve packages? Even if he could, should he; or was there a better way of getting the job done? Of course, my background as a test pilot was useless in this work, but then it was such a far-out field that no one else had any practical experience to call upon either. It was simply a case of hopefully reasonable men making hopefully reasonable guesses, which could be verified only by in-flight experiments. In retrospect, I think we made only one fundamental mistake: without gravity to hold you in one place ("down"), you can use all your attention and energy simply maintaining the status quo. It was no problem on the moon, with its gravitational field, or strapped securely inside a spacecraft, but during a weightless space walk your body is constantly floating off to some unwanted position and you (unless you have been clever enough to think about footholds or short leashes or something) must constantly fight to get back where you want to be. Pretend that you are sitting in your favorite armchair and that someone suddenly turns gravity off. For simplicity let's assume the chair is bolted to the floor. At first absolutely nothing happens because you haven't moved a hair. But the slightest body movement will start things going: if you straighten just a bit, your thighs will push against the chair seat, it will push back ("For every action, there is an equal and opposite reaction") and you will slowly begin to rise. To counteract this, you reach between your legs and grab the seat with one hand. This action causes a torque on your body which starts rotating your head down. To stop this, you grab the chair arm with your other hand and apply a bit more torque in the opposite direction. This causes your fanny to bang down into the chair seat, a bit harder than you intended, which of course means that you pop back up much more quickly than your first floating motion. Soon you find yourself wrestling with that chair, with both hands, just to sit in it. A little practice teaches you to damp out all motion, at which point hands can gingerly be released, but any subsequent contact (which inevitably occurs when you are doing even the simplest task) will start the whole cycle all over again.

We noticed a little bit of this in the zero-G airplane, but in only twenty-some seconds it was not really possible to assess or even appreciate the problems which might result. Furthermore, we blamed a lot of our body-positioning difficulties on the mistaken notion that the pilot was flying a less than perfect parabola and was throwing us around artificially. On Gemini 4, Ed White wasn't out long and didn't have any meaningful tasks to perform, so we still didn't appreciate the problem. It took the combined experiences of Gene Cernan on 9, me on 10, and Dick Gordon on 11 to get a proper catalog of complaints and possible solutions. It then became Buzz Aldrin's duty on 12 to check out a variety of foot restraints, waist tethers, and other helpful devices, rather than zooming around the sky attached to the AMU, as previously planned.

If the EVA planning was the most stimulating and innovative part of my specialty area, then the pressure suit testing was the most dreary and tiring. As Pete Conrad intimated, all was not well with the Apollo suit and I found myself in the middle of a brouhaha concerning its shortcomings and how they could be over-come. The Apollo suit maker (ILC) had produced some very mobile early suits, which had attracted NASA's attention in the first place. However, it seemed that the harder they worked on refinements and improvements, the worse the suit got. Its most annoying feature was a shoulder joint which seemed to derive its mobility from the fact that it dug a trough an inch deep in the flesh of the deltoids. Its helmet was uncomfortable and offered poor downward visibility. On the other hand, the Gemini suit (by David Clark) was coming along nicely, and if its link net mesh system was inherently limited when it came to joint mobility, well at least it was comfortable. NASA began putting the squeeze on ILC, by putting pressure on Hamilton Standard, which had the contract to provide the entire extravehicular package, including back pack, thermal and micrometeorite protection, etc. ILC was a subcontractor to Ham Standard. Ham Standard responded by hiring some suit engineers of its own, and began producing its own experimental suits as a possible replacement for the ILC design. David Clark, not to be outdone, promptly produced its own Apollo suit, with an experimental sliding plastic shoulder ring to help the link net along. NASA encouraged these efforts, and it was not long before the situation was formally recognized by a three-way compe-tition to decide, once and for all, which company would build the all-important lunar suits. Each company was to enter one suit in the competition, and each was sized for Mike Collins. To be scrupulously fair to all, as well as to provide the maximum amount of usable information, a detailed and highly structured test plan was written. Everything that possibly could be done with a suit would be, not once but three times, under identical circumstances, and all would be duly recorded and documented. Many of the tests, such as leak-rate measurements, could be done without my participa-tion, but Bob Jones, the energetic young Ph.D. psychologist run-ning the tests, had a long list of items where "crew input" was required. So off we went, to North American in Downey, Cali-fornia, to check each suit's reach, visibility, and general mobility inside the command module mock-up, and then to Grumman at Bethpage, Long Island, to do the same with the LM.

In addition, I was reintroduced to that diabolical training and research device known as the centrifuge. At this time (June 1965) the MSC centrifuge wasn't working yet, so we packed our three suits off to Johnsville, Pennsylvania, and borrowed the Navy's. This made my third trip to "the wheel," as it was called, the first coming as part of our ARPS course at Edwards and the second during Basic Grubby Training. If the zero-G airplane performs its task only briefly and with doubtful accuracy, then the high-G wheel is exactly the opposite—it does its very specialized chore with deadly accuracy for as long as its passengers can stand it. Its job is to imitate the acceleration one experiences riding a rocket into space or the deceleration caused by re-entering the earth's atmosphere. It does it by swinging a small gondola, or imitation cockpit, around in a circle, on the end of a fifty-foot arm. As the arm spins faster and faster, centrifugal force pushes the occupant deeper and deeper into his seat or couch. The force can be measured in Gs, one G being the acceleration constant we are most accustomed to, the 32.2 feet per second per second acceleration of the earth's gravitational field. On Gemini the peak G was reached the instant before the Titan rocket's second-stage engine was shut down upon reaching orbit. Here the G was 7.5, or the body "weighed" 7.5 times its normal amount. During Gemini's atmospheric re-entry, the G level would usually reach 4. On Apollo the situation was different. The Saturn V's ride was "softer" than the Titan's, and only 4.5 Gs were experienced at first-stage burn-out. However, coming back from the moon, zinging back into the atmosphere at 25,000 miles an hour, really put it to you! Routinely 7 Gs were reached during the Apollo lunar return, and if your angle was slightly steeper than normal, it could easily soar to 10 or 15. Hence 15 Gs was our limit on the wheel, and it was no fun at all.

At anything over approximately 8 Gs, I start feeling very uncomfortable, with difficulty in breathing and a pain centered below my breastbone. By 10 Gs the pain has increased somewhat and breathing is nearly impossible; in fact, an entirely different breathing technique is needed at high Gs. If you breathe normally, you find you can exhale just fine, but when you try to inhale, it's impossible to reinflate your lungs, just as if steel bands were tightly encircling your chest. So you have to develop an entirely new method, keeping the lungs almost fully inflated at all times, and giving rapid little pants "off the top." In addition to breathing problems, at high Gs the vision starts to deteriorate, and darkness closes in from the edges toward the center. The intermediate stage is called tunnel vision, and this is where the cockpit designer must have been careful to group all necessary instruments for that stage of the flight directly in front of you and not out in the periphery where you are now blind. Of course, we have all heard of fighter pilots blacking out as they abruptly pull out of dives, but astronauts are much less prone to do so. It's not because the people are any different, but simply that the direction of the G is different. The easiest way to define direction is by reference to which way the eyes sag. When the fighter pilot "pulls positive Gs," as he calls it when he pulls out of a dive, that is "eyeballs down" in astronaut parlance. A pilot's "negative G" is "eyeballs up." Too many positive Gs and the blood cannot reach your brain and you "black out." Too many negative Gs and too much blood is forced into your brain and you "red out," a much more hazardous condition. The astronaut works at 90 degrees to all this. Unlike a pilot's seat, his couch is oriented in such a way that all his Gs are transverse, or "eyeballs in." While "eyeballs in" can hurt your chest and impair your breathing, you can endure much higher levels without losing vision or consciousness.

All these things have a kind of sinister fascination the first time through, especially when the doctor checks your eyes for hemorrhages after each trip on the wheel and you discover at the end of the day that your back is uniformly discolored by red speckles known as petichiae, a jillion tiny hemorrhages of the capillaries, caused by ruptures of the vessel walls under the stress of acceleration.* But once is enough, and after that it becomes most unpleasant work, with a lingering hangover that can last a day or two, sometimes accompanied by dizziness when the head is turned suddenly. So I wasn't too happy to be going back to Johnsville for thirds, especially since with three suits I would have to go through all the tests three times, measuring comfort, reach, and vision under various conditions.

Despite my discomfort, however, the suit competition was a good idea, because it spurred ILC on to a higher plateau of performance, especially in the area of shoulder mobility and comfort. The ILC suit was clearly superior to the other two, and it got even better in the four years between the 1965 competition and the first moon walk. By the end of Apollo, astronauts were spending long hours in lunar EVA with no apparent discomfort, a fact beyond our wildest expectations during 1965, when we got our first look at the lunar EVA hardware. On Apollo 17, for example, Gene Cernan spent seven and a half hours covering some twelve miles, statistics which would simply not have been believable in 1965. At that time, some astronauts were even suggesting that development of the back pack be abandoned, that instead we should simply plan to work with umbilical lines from the LM, which would have

* I will never forget the time at Edwards when my Fighter Ops compatriot Bob Looney somehow got "volunteered" for a project on the Johnsville wheel. It involved a fancy new restraint system which allowed the pilot to work at very high G levels, eyeballs *out*. When Looney got back to Edwards, he was the worst-looking mess I have ever seen. Both eyes had hemorrhaged, turning the whites absolutely blood red. It took about a month for his system to completely absorb the trapped blood, and for his appearance to return to normal. In the meantime, none of us could stand to look at him, and it was an absolute dictum that he had to wear sunglasses at all times. Looney's misfortune was part of a larger problem at Edwards, which was that various researchers insisted they had to have experienced test pilots as guinea pigs for their projects. Fortunately, our bosses were able to fight off most of these madmen, who, I always suspected, just wanted to add a bit of stature to their reports. One, I clearly recall, had to do with blood-pressure measurements on the wheel. However, blood pressure measured by a conventional arm cuff would not suffice; it had to be taken at the source. Of course the procedure was simplicity itself. A small incision was made into a vein in the armpit, and a plastic tube with a pressure sensor on the end of it was then fed into the body until the tube's end found its way *into* the heart itself. Then jump onto the wheel, and away you go; Naturally, the subject had to be an experienced test pilot.

limited our radius of action to fifty feet or so. The reasoning behind this was that the back packs were becoming larger, heavier, and more complex; that they were making a small cockpit even more crowded; that they would never be as reliable as umbilicals; that their weight meant less fuel could be carried (and the fuel budget was very tight); and that all we needed from the moon was a few rocks, anyway. My own opinion was that we should continue working on the fancy gear and postpone decisions on its use until we saw how it developed. The fact that this was also an expensive course to follow did not worry me a bit. One nice thing about Apollo was that no one ever told us we were running the price up too high.

It is not worthwhile to go around the world to count the cats in Zanzibar.

—Henry David Thoreau, *Walden*

Oh no? Then for what purpose, to count the pulse of man? To record his demise, or at least his ultimate foolishness? Nineteen sixty-five was the year to make this judgment. If it started out poorly for me, hot and sweaty inside my pressure suit, then at least it ended as a cool year for NASA, a year for Gemini and its wonders, because suddenly we were flying again, after a layoff of nearly two years.* At least the real astronauts were, and Gemini looked good. After two successful unmanned shots, Gemini 3 was ready by March 23, when Gus Grissom and John Young lifted off aboard the *Molly Brown*. Gus named the spacecraft in expectation that it would be as "unsinkable" as its Broadway namesake (unlike his Mercury spacecraft *Liberty Bell 7*, which rests at the bottom of the Atlantic Ocean). Gus and John had a busy three-orbit flight

* Gordo Cooper had finished up the Mercury program with his twenty-two-orbit flight aboard *Faith 7* on May 15, 1963.

designed to give the spacecraft's components as tough a workout as the five-hour flight plan would allow. Some people, Gus and John among them, thought that it should be kept up longer, perhaps until some problems emerged, but conservative management prevailed, and the *Molly Brown* plopped into the Atlantic some sixty miles short of target after three hectic orbits. Gus had time to change orbits three times by firing his thrusters (a necessary prelude to rendezvous), and John had time to run through a variety of subsystem tests. He also managed to eat a corned-beef sandwich which Wally Schirra had caused to be placed in his pressure suit pocket—a fact unknown to NASA, which reacted somewhat hysterically. The medics claimed that somehow that sandwich had negated the flight's medical protocol, while the engineers claimed that crumbs from it could easily have invaded the guts of the machinery with catastrophic effect. Some members of Congress became apoplectic, charging NASA with having lost all control of the astronaut group. I think most of us could have strangled Wally for bringing a tornado of upper-echelon attention down on us because of something as trivial as a corned-beef sandwich. Deke was even moved to write us all a memo: ". . . the attempt . . . to bootleg any item on board not approved by me will result in appropriate disciplinary action. In addition to jeopardizing your personal careers, it must be recognized that seemingly insignificant items can and have affected the prerogatives of follow-on crews. Witnes [*sic*] this memorandum." It was also about this time that the practice of naming spacecraft was directed to cease,* a disap-

* The Mercury spacecraft had all been given names, followed by the number 7 to indicate they belonged to the Original Seven: *Freedom* (Shepard), *Liberty Bell* (Grissom), *Friendship* (Glenn), *Aurora* (Carpenter), *Sigma* (Schirra), and *Faith* (Cooper). The Russians numbered their spacecraft within programs. Roughly, their Vostok (East) corresponded to our Mercury, their Voskhod (Rise) to our Gemini, and their Soyuz (Union) to our Apollo. But the Russians, in addition to numbering spacecraft within a series, also used individual call signs, such as *Seagull, Hawk, Diamond,* and *Argon*. Probably because *Molly Brown* was deemed a bit too irreverent, we flew by number alone until Apollo 9, when Houston was required to radio instructions to two manned vehicles simultaneously. Since they could not both be called Apollo 9, the restriction was lifted; the command module was promptly dubbed *Gumdrop* and the LM *Spider*, both names jocularly but powerfully descriptive of the craft they represented.

pointment to Jim McDivitt and Ed White, who wanted to name their upcoming Gemini 4 *American Eagle*.

Whatever its name, Gemini 4 was developing well. Gemini 3 had gotten us off to a grand start, with assurance that there were no *major* redesigns ahead, although it had not been equipped with such things as fuel cells and rendezvous radar, yet to be tested. With the exception of its landing inaccuracy, Gemini 3 had accomplished virtually all its test objectives. It also came at a busy time, just five days after Alexei Leonov stepped outside Voskhod II for ten minutes to become the world's first space walker, and the day before Ranger IX impacted the moon. Ranger IX sent back thousands of photographs, including close-ups taken seconds before crashing into the moon. Scientists analyzing them "noted that crater rims . . . seemed harder than the plains but that floors of the craters appeared to be solidified volcanic froth that would not support a landing vehicle."* So the optimism we enjoyed in the aftermath of a successful Gemini 3 was sobered somewhat by the sure knowledge that the Russians were ahead of us and the possibility that trouble lay ahead on the lunar surface.

Gemini 4 went a long way toward dispelling any gloom that might have accumulated. It was a flight which seemed to have everything. In the first place, it was to last four *days* instead of a paltry three orbits. Second, the spacecraft was going to turn around as soon as it got into orbit, and photograph and fly in formation with the upper stage of its Titan II launch vehicle. Third, Ed White was going to go EVA, and might even use his gun to fly over to the Titan! Fourth, the crew was photogenic, chatty, and gregarious, and could be relied upon to emote, rather than emitting the usual laconic grunts from space. The matter of Ed White's EVA was to be a surprise, and was not announced to the world until after Gemini 4 got airborne. In fact, the decision to go EVA had been made only ten days before the flight, and was the culmination of frantic behind-the-scenes activity at MSC as a small team of specialists developed and tested the umbilical cord, the gun, and a

* NASA SP–4006, "Astronautics and Aeronautics, 1965," p. 149.

specially developed miniature chest pack. Although I was the astronaut office specialist in EVA matters, I was not included in this team, a fact which rankled considerably. Granted I was busy as could be with Apollo problems, such as the three-way suit competition, but nonetheless I felt left out in the cold as the Gemini 4 EVA design team went about its work with great industry and secrecy.

Finally June 3, 1965, arrived, the thirteenth anniversary of Ed White's graduation from West Point (mine too). I felt a twinge of envy (jealousy?) as the great flight unfolded. They had a minor problem at the very beginning as Jim McDivitt allowed the Titan to drift too far away and then had difficulty catching it again, a graphic illustration if there ever was one of the tricky nature of our old friend orbital mechanics. Without precise computerized catch-up instructions, it was clear that Jim was going to waste a lot of gas catching the empty launch vehicle, so the attempt was called off, and Ed White's EVA became a separate event.

My wife had gone to the Cape with Susan Borman (Frank was McDivitt's back-up) to watch her first launch. As soon as this trauma was over, she headed on back to Houston on a plane jammed with reporters hurrying to get to Mission Control at MSC to continue covering the flight. At four hours after lift-off, the word was announced on the plane that EVA was about to begin. Pat, completely flabbergasted, screamed, "My God, he's getting out!" in horror and fascination at this bizarre turn of events. She had come a long way since the days she used to *listen* to Mercury launches on the radio because she didn't want to *see* the thing blow up before her eyes, but she still had a long way to go. Anyway, with or without her approval, Ed White opened his hatch and floated out, permitting Jim to take some of the most spectacular pictures ever to come from the space program. The gun seemed to work well during its very brief trial, and best news of all, Ed had no difficulty whatsoever. In fact, he was euphoric, most reluctant to get back in on schedule, and had to be coaxed a bit by Jim and

Mission Control. When he did get back in, the hatch stuck open for a bad moment, but Ed was able to wrestle it shut. Once back inside, the flight was routine for the remainder of the four days, and the crew was in good enough shape after the flight to allow planners to double up for the next one, making Gemini 5 an eight-day flight. In fact, the crew was in better shape than I was. I was feeling more and more rundown and was unable to shake a persistent cough, which was not helped any by the fifty hours of high-altitude, dry-oxygen flying I did during June 1965 as I scrambled from one pressure suit crisis to another.

The back-up crew for Gemini 4 was Frank Borman and Jim Lovell. Under Deke's system, which was just emerging with some clarity, a back-up crew could expect to skip two flights, and pick up the third as prime crew (Gemini 7 in this case). Apart from this, the Fourteen wanted to know who the back-up crew would be. We were running out of Old Heads. Of the Seven, only three were then active (Grissom, Cooper, and Schirra), and they had their names on Geminis 3, 5, and 6. Glenn had retired, Carpenter was underwater, and both Slayton (irregular heartbeats) and Shepard (inner-ear infection causing dizziness) were grounded. The Nine were all on or slated for prime Gemini crews, in the following order: Young (Gemini 3), McDivitt (Gemini 4), White (Gemini 4), Conrad (Gemini 5), Stafford (Gemini 6), Borman (Gemini 7), Lovell (Gemini 7), Armstrong (Gemini 8), See (Gemini 9). It is interesting to note that Neil Armstrong was the last in his group (Elliot See having been killed) to fly. Were they saving the best for last, or was his selection as first human to walk another planet just a fluke? What *was* obvious was that help was needed to fill back-up positions, and it made good sense to start feeding the Fourteen into the system at this point, when the Gemini 7 crew was about to be announced. Presumably they would be presented in some order of merit, so that when I learned, late in June, that Ed White and I would be the Gemini 7 back-up crew, it more than made up for the feelings of uncertainty and doubt that I had been experiencing because of my pressure suit troubles and lack of participation in the Gemini 4 EVA plans. This was the SUMMONS, the big time, the first inkling of a promotion, or recognition, or even acceptance, in my eighteen months at MSC. I had not known whether I was making a good or bad impression on Deke and Al. My guess was that they were probably not aware of my work one way or the other, and would therefore favor those who seemed to hustle and bustle in and out of their offices constantly, seeking advice and decisions, or perhaps those with whom they traveled and partied. Being more of a loner, and given the isolated nature of my pressure suit work, I felt far out on the periphery, despite the fact that I was well accepted and—I thought—respected in my peer group.

At any rate, I had little time for speculation as I was immediately caught up in the bustle and excitement of being on a crew, even a back-up crew. I canceled a geology field trip with glee, dropped my EVA equipment work like a hot potato, and flung myself full-time into Gemini 7. I didn't even have time to daydream about Deke's "skip two" system, which would seem to put me on the Gemini 10 prime crew. Gemini 10 was at least a light-year away, but Gemini 7 was to be in December, time was running short, and I had a hell of a lot to learn. Since I had concentrated mostly on Apollo problems, I was woefully ignorant about Gemini basics, and the three gents I was working with (Borman, Lovell, and White) could not afford to tread water while I caught up, so I really had to hustle. To make things worse, my persistent cough had gotten worse, was now accompanied by nightly fever, and had been diagnosed as viral pneumonia. I was grounded, which meant that I had to trail ignominiously after the trio when we traveled, as they would fly direct in their T-33s or T-38s, and I would limp along behind as closely as airline schedules permitted. Fortunately, in the midst of all this, Borman decided to program our annual vacation, which was my salvation. Pat, the kids, and I spent a beautiful week at Padre Island, an unspoiled spit of land off the Texas coast just north of the Mexican border. As I baked like a lizard in the near-tropical sun, my cough disappeared, and I returned to Houston energetic and optimistic.

Around Houston, being assigned to a crew was a status symbol that carried tangible and intangible benefits with it. For instance, crew members had priority in scheduling airplanes for travel. Crew members were exempt from routine PR chores, such as the week in the barrel. Each crew was assigned a small support team of engineers, whose assistance was invaluable in tracking down bits and pieces of information the crew needed, representing the crew at meetings, and keeping on top of any problems with the hardware or planning for the upcoming flight. This support allowed the crew to concentrate on learning the machines and planning how best to use them. This work took place primarily in classrooms and in simulators. The crews had their own pecking order, the premier crew being the one assigned to the next flight; they had first priority in the struggle for simulator time. NASA tried as hard as it could to plan far ahead, but human nature dictated that the flight everyone was really worried about was the next one, a fact blindingly visible to crew members. If you were a crewless astronaut trying to point out a long-term problem of great consequence, you were less likely to be heard than if you were scheduled for the next flight up and had some trifling complaint. In fact, the next crew to fly was showered with help, pampered with assurance and good will, and in general made to feel like fatted calves. Within crews, priority went to the prime members, but the back-up crew shared on a nearly equal basis; the training of the back-up crew was in theory identical to that of the prime crew, and in practice it was nearly so, at least on Gemini. The Gemini foursome (two prime, two back-up) was an easy number to handle, and generally we flew in two airplanes on identical schedules. On Apollo, with six crew members and two different flight vehicles, the situation was more complex and crew unity during training was less intense. The situation was further complicated because prime and back-up were not allowed to fly in the same airplane, lest a crash wipe out the entire capability in one specialty. For example, Borman was the Gemini 7 crew commander and Lovell was the prime crew pilot; White was Borman's back-up and I was Lovell's. I could fly with anyone but Lovell.

I was especially pleased to be on this crew, not only because I was the first of our fourteen to be chosen, but because I had known Frank Borman and Ed White well for years. Borman and I had both come to Edwards together in the summer of 1960, and had shared adjoining desks in test pilot school. I had known White for seventeen years, as we had entered West Point together, although I hadn't seen much of him in the intervening years. Both were good friends and easy to work with, although Borman annoyed a lot of people by the crisp and arbitrary military precision with which he ran the operation, and the merciless ribbing he applied to any who might disagree with him. Lovell, a most pleasant and relatively easygoing fellow, completed the team. Traveling with them was a pleasure, especially when I considered that otherwise I might be in the barrel.

It was a special pity to be leaving the barrel just as I discovered the answer to the all-time favorite question astronauts are asked. No one would ever ask it in a crowd, but after the lecture was over, inevitably someone would sidle up, amid the autographers, and half whisper, "Say, what I have always wondered is how you fellows . . . you know . . . how do you, well, go to the bathroom up there?" Allow me to enter for the record the official, approved Gemini 7 procedure for going potty in space:

Operating Procedure
Chemical Urine Volume Measuring System (CUVMS)
Condom Receiver

1. Uncoil collection/mixing bag from around selector valve.
2. Place penis against receiver inlet check valve and roll latex receiver onto penis.
3. Rotate selector valve knob (clockwise) to the "Urinate" position.

4. Urinate.
5. When urination is complete, turn selector valve knob to "Sample."
6. Roll off latex receiver and remove penis.
7. Obtain urine sample bag from stowage location.
8. Mark sample bag tag with required identification.
9. Place sample bag collar over selector valve sampler flange and turn collar 1/6 turn to stop position.
10. Knead collection/mixing bag to thoroughly mix urine and tracer chemical.
11. Rotate sample injector lever 90 degrees so that sample needle pierces sample bag rubber stopper.
12. Squeeze collection/mixing bag to transfer approximately 75 cc. of tracered urine into the sample bag.
13. Rotate the sample injector lever 90 degrees so as to retract the sample needle.
14. Remove filled urine sample bag from selector valve.
15. Stow filled urine sample bag.
16. Attach the CUVMS to the spacecraft overboard dump line by means of the quick disconnect.
17. Rotate selector valve knob to "Blow-Down" position.
18. Operate spacecraft overboard dump system.
19. Disconnect CUVMS from spacecraft overboard dump line at the quick disconnect.
20. Wrap collection/mixing bag around selector valve and stow CUVMS.

The Gemini 7 insignia, designed by the crew, shows a hand clutching a torch (no, not a CUVMS) in a symbolic representation of a marathon runner. It was an apt design because Gemini 7 was to be a marathon lasting two weeks, unless men or machines broke before then. The two main purposes of the Gemini program were to prove that weightlessness could be endured for the duration of a lunar voyage (eight days) and that rendezvous in space was practical. Gemini 7 was designed to be enough longer than a trip to the moon to provide a reassuring safety factor, and was to follow two build-up flights: Gemini 4 for four days and Gemini 5 for eight. No rendezvous was planned; instead it was to be above all a medical flight, a medical orgy of experimentation. If a simple thing like taking a pee requires twenty steps, consider what might be involved in complying with the whole array of Gemini medical experiments, all of which were included on Gemini 7:

M–1 Cardiovascular conditioning
M–3 In-flight exerciser
M–4 In-flight phonocardiogram
M–5 Bioassays body fluids
M–6 Bone demineralization
M–7 Calcium balance study
M–8 In-flight sleep analysis
M–9 Human otolith function

In addition to running the medical gamut (I don't know what happened to M–2), there were also eleven scientific, technical, or military experiments to be performed. The medical experiments generally involved comparing data measured before, during, and after the flight to measure the effects of weightlessness. Some were designed simply to measure, others to counteract, the slowly deteriorating physical condition of the astronauts. One of the latter was M–1, in which Pete Conrad on Gemini 5 and Jim Lovell on Gemini 7 were to be equipped with inflatable rubber thigh cuffs which were programmed to inflate automatically at intervals, day and night, squeezing the thighs in a crude imitation of gravity, which might stimulate the autonomic nervous system to readjust more quickly upon returning to earth. Their cohorts, Gordo Cooper on 5 and Borman on 7, were not so equipped and would act as controls. Exercise was also expected to be beneficial. M–3 was a simple bungee cord, with a handle on one end and a nylon loop on the other. It took sixty or seventy pounds of pull to extend the powerful bungee a mere twelve inches, and by looping the cord around the legs and pulling with the arms, it was possible to give both arms and legs a fairly strenuous workout. During and after a calibrated number of pulls, the pulse rate was measured, as was its rate of return to pre-exercise levels. M–4 measured heart noises (and possible heart-muscle fatigue), while M–5 was concerned with measuring the body's output of fluids. The body loses fluid in space, as zero G has the effect of causing more blood to pool in the thoracic region, where stretch receptors are tricked into thinking that there has been an increase in total body blood volume and respond by increasing urine output. The mechanism for this conversion is a complex one involving secretion of hormones ADH and aldosterone. M–6 measured loss of bone density during weightlessness by comparing pre- and post-flight X-rays of hand and heel.

M–7 (calcium balance) was a real doozy. It required very precise measurements, before and during flight, of calcium intake and output. The intake was controlled by the fact that for several weeks before the flight we ate and drank only what had been given us by the medical experimenters, who, of course, knew how much calcium was contained therein. They even measured my martinis. If we didn't eat or drink everything, our leftovers were carefully measured and weighed. Output was more complicated, as calcium can be lost through urine, feces, or perspiration. We wore special long-john underwear, which we turned in after use. We bathed in distilled water, and kept the dirty tubfuls for analysis. We kept all urine and stool samples, carrying bottles and ice cream containers with us wherever we went. The medics even packaged this claptrap into special attaché cases which we carried with us on trips. Thank God, this was before the days of airplane hijacking, because I think I would rather have been arrested than reveal the contents of my attaché case to some airline inspector. In flight, M–7 meant a careful inventory had to be kept of each urination (witness the twenty-step check list) and each stool sample in its little plastic bag had to be packed away in a locker.

M–8 measured brain waves by electrodes stuck onto a shaved scalp. As one of the experimenters put it, "A cooperative research program . . . has been directed to the following practical and scientific questions: Can the electrical activity of the brain, as it is revealed in the electroencephalogram (EEG) recorded from the scalp, provide important and useful information concerning such factors as the sleep-wakefulness cycle, degree of alertness, and readiness to perform? . . ."* This was the kind of experiment designed to infuriate any crew. An already complicated situation was to be made that much worse by the addition of uncomfortable electrodes, wires, electronic amplifiers, etc., etc., so that some son-of-a-bitch on the ground could determine whether we were ready to perform! A classic case of the tail wagging the dog, with decisions to be made by the wrong people (the medics) in the wrong place (the ground) with the wrong information (brain waves). Needless to say, there was little enthusiasm for furthering anything that might lead to M–8's kind of decision-making. M–9 was an innocuous little experiment to measure the ability of the inner ear (our normal balance mechanism) to function in weightlessness.

At 9 A.M. on August 21, 1965, Gordo Cooper and Pete Conrad began their eight days aboard Gemini 5. In addition to its medical objectives (abbreviated versions, for the most part, of the Gemini 7 experiments), Gemini 5 was to test fuel cells and rendezvous radar for the first time. To test the radar, a special pod, equipped with radar transponder and flashing lights, was carried in the adapter section and released shortly after Gordo and Pete reached orbit. They were to perform an elaborate rendezvous sequence, chasing the pod through the sky, but unfortunately low oxygen tank pressure required that electrical power be curtailed, and the fuel cells were not permitted to produce enough current to keep the rendezvous radar working. Despite this, however, the radar did get a brief workout, the fuel cells performed well, and the oxygen tank pressure recovered nicely, permitting the flight to go the full eight days. Because I missed the excitement of Ed White's EVA, or because I was so keenly tuned to the upcoming Gemini 7, for whatever reason, I felt Gemini 5 was a ho-hum, dull kind of flight,

* NASA SP–121, "Gemini Midprogram Conference," February 23–25, 1966, p. 423.

and even Pete Conrad's customary ebullience was muted. The crew seemed tired, dehydrated, and somewhat weakened upon return, and all of us on Gemini 7 pondered what the difference between eight and fourteen days might be.

In the first place, no one who has never seen a Gemini can fully appreciate what it's like being locked inside one for two weeks. It is so *damned* small, smaller than the front seat of a Volkswagen, with a large console between the pilots, sort of like having a color TV set in a VW separating two adults. In the Gemini, if you are 5 feet 9 inches, you can fully straighten your body; any taller and you will find your head banging the hatch or your feet bumping the floorboards, or both. Proper body positioning during ejection requires a seat design that is as bulky and uncomfortable as a Louis XIII chair. You cannot get out of the seat because there is no place to go. On the ground, at one G, no one could sit in a Gemini for eight days, I am convinced. Long before that he would be so sore and stiff and crowded and cramped and miserable I think even the most highly motivated person would literally go out of his mind. I couldn't sit in the Gemini simulator more than three hours at a stretch, and the simulator was tilted back 30 degrees or so from the horizontal to place the seat at its most comfortable angle. Then how had Cooper and Conrad managed, and why were Borman and Lovell expected to last fourteen days? The answer is weightlessness, which allows the astronaut fanny to float free of the seat, restoring circulation, preventing bedsores, and in general allowing the two men to rattle around inside their Volkswagen without being constantly squashed against any part of it by gravity.

Despite the relief of weightlessness, however, it was no bed of roses inside a Gemini. The cockpit was tiny, the two windows were tiny, the pressure suits were big and bulky, and there were a million items of loose equipment which constantly had to be stowed and restowed. The stowage compartments were long, deep, and narrow, which meant that if what you wanted was on the bottom, you ended up with a cockpit literally full of floating debris, which had to be gathered up and stuffed back into its box, by which time you had probably lost what you were looking for and had to begin the Easter egg hunt all over again. The cockpit was office, study, kitchen, dining room, bedroom, bathroom, and laboratory all in one. There were two of you, and you were always aware of it. If he stank, you smelled it; if he spilled urine, globs of it ended up in your face as easily as his. You could never get properly clean, not even once. You could not shave or drink hot coffee. Sensors were taped to your body not to be removed. For Borman and Lovell this was to be home for fourteen days.

During these two weeks their physical condition was expected to deteriorate slowly and selectively, to what level no one knew, and there was no way to find out short of *doing* it. The moment of truth would begin at re-entry, when the earth reasserted its gravitational dominance over their two bodies. Then and only then could the medics start to measure how much the body had forgotten or how severely it had become weakened. Some facts were clearly known. First, the weightless state in itself was relatively benign; without the inexorable pull of gravity, the muscles of the body had much less to do. The heart muscle, especially, did not have the arduous task of pumping blood uphill any more, because there was no "up." On earth, when we stand, the heart must pump blood, not only along the easy downhill route to our feet, but also uphill to our heads, fighting gravity all the way, somewhat as water is pumped around the water system in a city with hills and valleys. At zero G the heart has to overcome primarily just the internal resistance of the arteries and capillaries (so-called "peripheral resistance"), a less arduous task than fighting gravity. Normally, on earth, our blood is prevented from falling back down into dependent parts of the body by the operation of strategically positioned, one-way flapper valves in the veins.* In our body's system for recycling "used blood," these veins ultimately carry blood back to the heart. The venous

---

* A curious comparison can be made between man and giraffe. The giraffe has flapper valves in the two jugular veins in its neck, which close when it bends over to drink or feed, to prevent a gravity-assisted gush of blood from rushing down and possibly rupturing blood vessels in its brain.

valves are controlled and "fine-tuned" by the autonomic nervous system, the one over which we have no conscious control. In space, these flapper valves are somewhat less useful because there is no tendency for the blood to slip back "down" into the legs, so after a couple of days the autonomic nervous system says, "To hell with it," and reduces its fine control over the valves. That is well and good as long as weightlessness continues, but during re-entry, gravity comes back with a vengeance and the flapper valves are quickly needed once again. Unfortunately, it takes the body a little while to readapt to this task and in the interim the astronaut finds himself lightheaded, especially if he stands, for blood tends to pool in his lower extremities and it's difficult for the heart to pump it up to his head without more responsive help from the flapper valves.

In addition to forgetful flapper valves, the body in several other ways deteriorates without gravity to fight against. Gravity is the force we work against as we tote "heavy" packages "up" stairs; without such loads we do not need dense or strong bones or well-conditioned muscles, just as people lying in bed do not need them. In fact, from a physiological point of view, being weightless is closer to lying prone than any other terrestrial condition, and bed-rest patients are studied to measure bone demineralization and muscle atrophy, in an attempt to extrapolate this information to space flight. Under these circumstances, how should one prepare his body for space flight? Theories within the astronaut group ran the gamut. Bon vivant Wally Schirra allowed as how the best way to prepare for a restful experience was to rest. Mathematician Neil Armstrong suggested that a person was given only a finite number of heartbeats in this life, and he was not going to hasten his demise by asking his heart to speed up during exercise. In the opposite corner were the jocks, chief among them Ed White, who might begin a typical day by joyfully running three miles and end it with half a dozen games of squash and handball. In between, inconspicuous under the dome of the bell-shaped curve, cowered the majority. They felt exercise was helpful, pleasantly or odiously so, but they weren't militant about it, and they pursued whatever sports or exercises they felt fulfilled their individual needs. Surprisingly, NASA had no formal physical-conditioning program, and we were free to do a little, a lot, or nothing. There was a very nice gymnasium at MSC reserved for our use, with two handball and two squash courts. It got a lot of use, especially during the noon hour, when astronauts doubled up on their virtues by working out *and* skipping lunch.

My own view is that, since space causes the body to weaken, one should start strong—at least in a cardiovascular sense—so that in a given period of time one runs a lesser risk of reaching a dangerous level. If the Dow Jones must plummet a hundred points, I would prefer that it start from eleven hundred, not eight hundred. A highly conditioned cardiovascular system can be achieved and maintained only by exercise, sustained exercise which causes a reasonably heavy load on the heart muscle. This load is best measured by the pulse rate, which normally (for me at least) is around sixty beats per minute at rest and around 180 when I am working at full capacity. Games like handball are a lot of fun, are strenuous, and are good cardiovascular conditioners, but they cause the pulse rate to shoot up and down erratically, in response to the pace of the play. A more efficient way is to put a calibrated, steady load on the heart, and the best way to do this, in my humble judgment, is by running.* If I devote just one hour a week to running, I can stay in good shape. I do this by running four times a week on my local track, two miles per outing, and timing myself. In my Gemini days I could make two miles in thirteen minutes, but my time has gradually crept up, and today I feel as if I am going to cough up a lung if I do it in fourteen. This pace can be examined from an Olympic or a geriatric point of view, and is either tortoise-like or spectacularly fast. Personally, I think it's pretty goddamned swift, for one who has never enjoyed track as a sport and who used to think up elaborate excuses to offer to the high-school football coach when he ended practice with, "O.K.,

---

* Or jogging, or just walking. Before anyone takes to cross-country trail or local track, however, a green light from the family doctor is a *must*. The National Jogging Association, 1910 K Street N.W., Washington, D.C., is most helpful in putting out a variety of literature, which contains sane advice for the middle-aged sedentary who wishes to recapture the spring of youth.

everybody two laps around the field." My plan is to continue as I grow older and slower until it takes fifteen minutes for two miles, and after that I am going to run, walk, stagger, lurch, or crawl as far as I can in fifteen minutes. One hour a week is enough.

Since I freely admit that running is simultaneously boring and painful, why do I do it? For one reason, I used to be in good shape and I don't like to contemplate my body's transformation into a wheezing, jelly-like lump of protoplasm. I'm willing to do what I can to slow this process, and an hour a week seems a small investment. Further, I find that endurance on the track somehow converts to endurance behind the desk, and I feel less tired after a day of paper work if I work out regularly. The medics also say your chances for avoiding, or at least surviving, a heart attack are better if you are exercising regularly. The reason for this is that a heart attack, or myocardial infarction, simply means that part of the heart muscle has been deprived of its blood supply and the tissue has died. The runner has a more muscular heart with a better developed network of blood vessels feeding the heart muscle itself, so the consequences of any one vessel becoming clogged are lessened. My good friend, fellow jogger, and occasional cynic Jack Whitelaw says that a strong heart merely serves to prolong the agony of terminal cancer patients, a theory I try not to think about as I puff around the track. But I guess the basic reason I run is that I think a body, like a brain, should be used, stretched, forced to its limits. Just as the concept of celibacy is abhorrent to me, so do I feel sorry for those who will never know the desperate pound of 180 beats per minute, or the golden afterglow of recovery from it. The body machine, like a flying machine, can be operated with greater enjoyment, poise, and confidence if one has explored its limits.

I hate to mention it, but closely tied in with all this health and exuberance of the jocks is smoking. My life so far has been divided into three equal parts. For the first third, I was too young to smoke. For the second, I smoked—heavily—and enjoyed it immensely, although less and less each year as the enormity of the lung cancer situation became ever more clear. Self-inflicted cancer —what an obscenity! For the most recent third of my life, too old to be deceived by the golden lie of the perfect people in the cigarette ads, I have not smoked. Quitting was a most difficult but satisfying milestone for me. One Sunday in the spring of 1962, I got up badly hung over with a throbbing head and a throat like an old flue, drier and dirtier than the Mojave Desert sand outside my door. I barked at Pat, "All right, goddamn it, when this pack of cigarettes runs out, that's it!" It didn't take long, and by evening I was empty-handed, fidgety, angry, and determined. The next morning, nervous beyond description, I reported in to Fighter Ops as usual, and was reminded that this was the day I had promised Ted Sturmthal, one of the B-70 test pilots, that I would ride in the right seat with him on a routine test of a B-52 with new engines. Ted needed someone qualified to maintain a body temperature of approximately 98 degrees to throw a few switches he couldn't reach from the left seat. The only hitch was, we were to keep the lumbering beast up for four hours, hardly enough to get a bomber pilot limbered up, but at least twice as long as any self-respecting fighter pilot likes to stay airborne, and of course an eternity for one who expects the imminent onset of delirium tremens as the least ominous symptom of nicotine withdrawal. But a promise is a promise, and besides I might as well log flying time as I twitched and fidgeted, so off we went. For four of the most miserable hours I have ever spent, I convinced Ted that fighter pilots were weird indeed. Like a teething baby, I slobbered and gummed my fingertips, my pencils, the corner of my handkerchief. I blew imaginary smoke rings, I inhaled mightily and exhaled in staccato little puffs. Ted peered at me strangely, and I realized he was motioning to one of my switches. I threw it and three or four more like it. I screwed up everything I touched, and I'm sure Ted was as happy as I was to get back on the ground.

Things slowly got better as the weeks went by, and finally I could make it through an entire day without thinking once of cigarettes. Three months after I quit, I went to Brooks Air Force

Base for the second time, and the physical exam included a repeat of the previous year's uphill stroll on the treadmill. I noted, to my amazement, that my endurance had improved by about 20 percent, despite the fact that I had not changed weight or exercise habits or done anything different except give up my two packs a day.

It was not long after this that I really got serious about NASA and its astronaut program, to the point that I began running, the first time in my life I had done so voluntarily, without some coach yelling at me. I got so gung ho I even started running up the mountain in the desert behind the Edwards base housing area on 100 degree afternoons, which is the safest time to do it because rattlesnakes are sensible creatures and never venture out until it cools off in the evening. My theory was that the mountain, which got steeper and steeper as you went up, was good practice for the treadmill, which was adjusted to get a few degrees steeper as each minute elapsed. So when I got to Brooks for my third and final exam, I was really set for that treadmill. I was going to destroy it, reduce it to a useless pile of rubble and all its keepers to uncomprehending, incoherent babblers. Well, folks, that's one I lost, because although my time did show fair improvement, the increase was actually less than it had been the year before. I conclude from all this that not exercising is bad, smoking is even worse, and he who would fly in space should know it and prepare himself accordingly. Those who don't fly in space might think it over too. You don't have to be a zealot about it.

Borman and Lovell were both good non-smoking runners and were in great shape by flight time. So were Ed White and I, or at least Ed was, and I did my best to trail along after him. Most of our time, however, was devoted not to things physical, but to making sure we understood that big, complicated pile of machinery, that we knew how each hour would be spent in orbit, and that we would be prepared for any possible eventuality. In previous flights the crew had split their sleep periods, so that someone would be awake at all times. But, however considerate the awake partner might be, he still made enough commotion talking on the radio and performing experiments to interrupt his partner's sleep, and part of Conrad and Cooper's post-flight fatigue was attributed to the fact that neither had slept well for eight days. On Gemini 7 the system was changed to simultaneous* sleep periods, the theory being that the equipment had been sufficiently well tested to be left unattended, so that the crew could power down the spacecraft, put blinds on the windows, and get eight uninterrupted hours of rest per day.

It also made good sense from a physiological point of view to have one eight-hour period in twenty-four devoted to rest, because that of course is what we are accustomed to here on earth. Our internal body clocks follow this biological activity, a rest-sleep cycle, known as the circadian rhythm. This twenty-four-hour rhythm is keyed to the earth's rotational period, and causes us to become sleepy after dark and awaken the next day. It takes roughly a week to change it, so that if we fly from Washington to Tokyo we arrive to find ourselves wide awake at night and drowsy in the daytime. But what happens in orbit, when we go around the earth once each ninety minutes, covering one solar "day" and "night" in this period of time? Well, the body ignores this fact and clings to its established twenty-four-hour circadian rhythm, so that when it is night back home (Cape Kennedy, since that is where the crew has been living for more than a week prior to the flight), the crew becomes drowsy, even though it may be rapidly changing from dawn to noon to dusk in orbit. Hence we kept our watches on Cape local time and tried to plan orbital activities accordingly. On later flights we changed our watches to Houston local time for simplicity's sake in talking to people in Mission Control; since Houston and the Cape are only an hour apart, it didn't matter much. In addition to our wristwatches and two conventional clocks on the instrument panel, the Gemini had a digital elapsed timer, which started counting at lift-off and continued throughout the flight. It measured GET (ground elapsed time), as we called it, and

---

* We had to be careful of our terminology, as some outsider might not understand if we said, "Oh yes, that crew is sleeping together." Simultaneously, please.

it meshed with our flight plan, which recorded major and minor events in terms of hours and minutes after lift-off.

The flight plan for Gemini 7 was fairly straightforward, because there was no rendezvous involved, and with only one vehicle sailing through the sky, timing was not particularly critical. On the rendezvous flights, split-second timing was required for the various maneuvers designed to bring Gemini and Agena, its target vehicle, to the same spot at the same time and at the same velocity. On Gemini 7 the main things we had to practice were launch, re-entry, how to conduct the various experiments, and emergency procedures. Emergency-procedure drills are a never-ending part of any pilot's training. They have to be reviewed periodically, no matter how experienced the pilot may be in his machine, and kept fresh enough in his mind so that he can react quickly and accurately no matter what the circumstances.

In my own case, the value of this kind of practice had been dramatically demonstrated one pleasant sunny day near Chaumont, France, way back in the summer of 1956. It was a Saturday afternoon and I was flying my F-86 on a NATO exercise, low over the woods and farmland near Chaumont, looking for "friendly" troops below. If I found any friendlies, as indicated by prearranged signals, I was to relay their whereabouts to a group of helicopters waiting nearby, who would airlift them out from behind "enemy" lines. Suddenly I felt a sharp thump, and the cockpit began to fill up with light-gray smoke. I called my wingman, Second Lieutenant Charles E. Sexton, to come look me over at close range, and at the same time began a gentle climb back toward the nearby airbase at Chaumont. By the time Sexy got abeam of me, the smoke was getting so thick that I was having trouble seeing the instruments, but I could still see that the fire warning lights had not come on. The F-86 had two fire lights, a forward and an aft, and the forward one is especially bad news, as the machine has been known to blow sky-high a second or two after it has come on. I checked both bulbs by a push-to-test circuit and they both passed. Fortunately, Sexy had by now spotted a raging fire just aft of the cockpit and started yelling over the air, "You're on fire. Get out, get out!" As quickly as I could, I doubled over and actuated the canopy jettison mechanism (if you didn't bend over, the canopy hit you in the head as it departed), then sat erect and ran through the steps necessary to fire the explosive charge beneath my seat, which shot the seat out along rails, with me in it. I had no sensation of motion: one instant I was inside the cockpit and the next I was tumbling end over end in a fierce wind blast. My helmet flew off. I reached down and unbuckled my seat belt, kicked free of the tumbling seat, and immediately yanked the D-ring mounted on the left side of my chest. For what seemed like a long time, nothing happened; then there was an almighty jerk and I was swaying back and forth under an orange-and-white canopy. Looking up, I discovered to my horror that the canopy was shot through with small holes! Were they getting bigger? My attention was so focused on measuring hole diameters that I let go of the D-ring, a good-luck souvenir, and didn't even notice the ground rushing up at me. Finally, at the last instant, I made a vain attempt to assume the proper position, hit like a sack of cement, and tumbled over backward into the soft plowed dirt of a farmer's field.

I staggered up, gathered my parachute in my arms, and looked around in vain for my helmet and D-ring. Sexy had called the waiting helicopters, and before anyone on the ground appeared, I was whisked on board one and deposited in the courtyard of the Chaumont AFB hospital a few miles away, as it is a prescribed ritual to have a physical exam after an ejection. I waved at the helicopter pilot, he waved at me, he zoomed off, and I tried the hospital door. Locked. A small clinic on a small American base on a weekend, but *someone* should be around. Finally, on the other side of the building, I found an unlocked door, a darkened corridor or two, and finally one lighted office. There I politely inquired about seeing a doctor, but the orderly on duty seemed quite agitated and kept shouting, "No doctor! Big crash and doctor's out looking!" When? "Just now. Big crash!" But that's me. "No. Big crash!" Finally I convinced the young man, but it really didn't matter, as only the errant physician could sign the necessary papers and he was out caroming about the countryside in a truck converted into a four-wheel-drive ambulance, looking for bodies at the crash site. By the time he returned, hours later, at least I had filled out all the necessary forms. He examined me and duly noted on his report: "Patient has had adequate sleep and nutrition for two (2) days prior to the accident. He is on no medication and denies ingestion of drugs and excess alcohol. Last physical exam October '55."

My home base sent a T-33 to pick me up, and it wasn't more than a couple of hours before I was ensconced in the Chambley Officers' Club bar, ingesting excess alcohol and reliving the experience before a small but select audience, including my best girl, Pat Finnegan. I was relieved to find out that the F-86 had crashed into a forest, injuring no one, and my worst complaint was a skinned knee, although I have been tempted to blame spinal disk problems in later years on the tremendous compressive G forces (eyeballs down) experienced during this ejection. I learned that the small holes in the parachute canopy were fairly routine and were caused by the heat of friction during deployment, which actually melted the nylon. I later found out that the fire was caused by radio gear igniting a pool of leaking fuel. More important than these things, however, were a couple of other lessons I learned. First, and most reassuring, was that I had reacted as I had practiced. I had done a lot of things right, such as waiting long enough to gain sufficient ejection altitude, but not long enough to risk an explosion. My memorized procedures had been clear in my mind when I needed them, and I had followed them swiftly and accurately. In short, I gained a lot of confidence in my own ability to practice emergency procedures and to use them if necessary.

In the F-86 days, we had no simulator, but simply practiced in the classroom by imagining the necessary switches and motions, or else we went out onto the flight line or into the hangar and got cockpit time, memorizing each switch's position and function. The most important ones we learned to locate blindfolded. In Gemini days we would have scoffed at such crudity: the simulators were exact replicas of the spacecraft cockpit, and the various dials, gauges, and switches were hooked to a computer which caused them to give realistic readings and the proper response to the pilot's switch changes or stick movement. Gemini 7 was a good flight for my initiation to the simulator, for with no rendezvous to practice, I could concentrate instead on the basic machine and its normal and emergency procedures.

As time went by, and I came to learn the Gemini machinery, I fell in love with Gemini 7. I really wanted to fly that flight, and was half convinced right up to launch day that one of the prime crew would be taken ill and that Ed and I would make the flight. Mind you, I didn't wish Frank or Jim any bad luck, I was simply preparing myself for a possibility I felt was not too remote. In retrospect I can't imagine why I wanted to be locked up for fourteen days in an orbiting men's room, with no rendezvous or EVA, instead of waiting contentedly to be assigned to one of the later, shorter, more exciting Gemini flights.

Just as August 16, 1956, was the day F-86F 52-5244 had chosen to catch fire, so was October 25, 1965, the day for Agena 5002, Gemini 6's assigned target vehicle, to blow sky-high. The Gemini 6 crew watched the Agena's flawless launch atop an Atlas booster, then busied themselves with preparations for their own launch. But six minutes later came the awful news. No sooner had the Agena's main engine, or PPS (primary propulsion system) ignited, than telemetry suddenly ceased and the Cape radars began tracking five or six targets instead of one. Apparently the PPS had gotten a hard start and caused the whole thing to blow. The Gemini 6 crew, Wally Schirra and Tom Stafford, had been eager to make the first space rendezvous and docking, but with no target vehicle, they were all dressed up with no place to go. Some quick planning was done, and Gemini 7 was pressed into service as a target vehicle for Gemini 6. Of course, no actual docking, or touching, of the two vehicles could take place, but since rendezvous was much the more complex part of Gemini 6's mission, it was decided to press on and get some rendezvous data by using Gemini 7 as a target rather than waiting for another Agena. It didn't change Gemini 7's plans much; it simply made us change places with Gemini 6, which would follow us from the same launch pad, as soon as the ground crews could repair pad damage and erect Gemini 6 in our place. It looked as if we might get both vehicles airborne, with a ten-day interval between, before the end of 1965.

Gemini 6 was to be as short as 7 was long. Borman was determined to stay up fourteen days; Schirra was determined to get the rendezvous done and land as soon as possible. Tom Stafford wanted to go EVA, but Wally would have none of it. Scientists pleaded for experiments to be carried, but Wally just laughed. He wanted simplicity, brevity, and success—the last would follow surely on the heels of the first two. He had the world's greatest human computer, Tom Stafford, to analyze the rendezvous problem, and once that was over Wally was going to put his BEAT ARMY sign in the window, play a few bars of "Jingle Bells" on his harmonica for the benefit of Frank and Jim, and then come home and have a cigarette. Cornbeef John (Young) would have gotten skinned alive for such pure corn, but Wally somehow carried it off with élan.

As the great day approached for Gemini 7 (December 4), I reluctantly concluded that Jim Lovell was really going to make the flight after all, so like a good back-up crew member, I did everything I could to be helpful, right down to setting up his switches and getting his side of the cockpit shipshape prior to launch. I even had a tiny sampler embroidered, which he could stick on his instrument panel if he wanted. HOME SWEET HOME, it said, and for two weeks it would in fact be home, sweet or not. Noon on the fourth of December found Ed White and me dressed in white smocks and white caps, standing around the "white room," the enclosed system of portable platforms built around the upper end of the Titan II rocket that served as loading dock, communications center, and the only access into the Gemini. When Jim and Frank appeared, we couldn't say a word to them, as they were sealed inside their pressure suits, denitrogenating in 100 percent oxygen, but there was much hopping around, little visual jokes were made, everybody clapped everybody else on the back, and then we loaded them on board, pushing them far enough down in the seat so that the hatches could close freely without bumping their heads. As soon as Jim's shoulder harness, seat belt, oxygen hoses, and communication wires were secured, there was time for one last handshake; then the hatch was latched behind him, and I retreated to the gantry, with the others, and the white room was dismantled.

With nothing else to do, feeling completely useless all of a sudden, Ed White and I retreated to a safe vantage point, wooden bleachers located outside the launch control center a couple of miles away. As the 2:30 P.M. lift-off time approached, I felt increasingly nervous. If an Atlas-Agena could blow sky-high, why not a Gemini-Titan? The fact that one never had almost made it worse, as there was always a first time for everything, and did not the flawless launches of Geminis 1 through 5 mean poorer, not better, odds for those that came after? Such logic, or lack of it, is prevalent at launch time. All the months of calm dispassionate analysis give way to a few minutes of emotion, an outburst of hope and horror as the inert beast comes alive for the first time, shakes itself and its new-found tail of fire, and starts slowly—so slowly—to move. For the first few seconds it is purely a spectacle, for with the eyes alone involved, one can see but not succumb. But when the great crackling Mach 1 roar arrives, and the very ground under you shakes, then you are there, you are part of it, and you laugh or cry or yell or whisper. For six months I had consumed Gemini 7, and now it was consuming me. It was going without me, or at least with only a part of me; at least it appeared to be going, and it was getting help from the crowd. "Go! . . . go! . . . go! . . ." they cried rhythmically. "Go . . . go," I whispered, out of phase and with a lump in my throat. "*Please* go." And ever more swiftly it did, until finally it was but a dot in the sky and then gone altogether, whereupon everyone congratulated his neighbor and said, Well, that's over. But of course for the crew it was just beginning.

Ed and I flew back to Houston and visited the Mission Control Center. The back-up crewmen are celebrities of a sort around MCC during the time "their" flight is up, and are called upon for advice, especially if things go wrong, as they know the machine intimately. However, they usually have no regularly assigned duties, as the astronaut office is represented in MCC by the CAPCOMs,* who generally work around the clock in three shifts and who do all the talking to the spacecraft. Since Gemini 7 was going well, there really wasn't a hell of a lot for me to do, and I got into a daily routine of briefing Susan Borman and Marilyn Lovell on the progress of the flight. Susan, in particular, showed determined interest in the flight plan, and I would update it daily for her and show her on a map what part of the globe she could expect Frank to be over at any time. As the wearing days went by, this little ritual grew more and more strained. I think, naturally enough, all Susan really wanted to ask was whether Frank was going to make it home again. None of us knew the answer to that, which Susan also knew, so she didn't ask the question.

Eight days into the flight of Gemini 7, Gemini 6 was ready to join it, and Launch Complex 19 at the Cape again had a Gemini-Titan poised for launch. Wally Schirra and Tom Stafford had no Agena to worry about this time, and all appeared normal during the countdown. In fact, they got engine ignition and an indication of lift-off—then a sudden silence. They had about a second to review two scenarios: (1) the engines for some reason had shut down after lift-off, and they were now on the brink of disaster and would either settle back down or topple over, requiring immediate ejection to avoid the ensuing holocaust; (2) the engines had shut down the instant *before* lift-off, in which case they were still firmly bolted to the launch pad and could stay put unless some new danger developed. The design of the hardware spoke for option 1

---

* A hangover from the Mercury days, when spacecraft were called capsules and those who communicated with them were therefore CAPsule COMmunicators.

(supposedly the lift-off signal in the cockpit was foolproof), but the seat of Wally's pants spoke for 2. It felt solid under him. A panicky type might have ejected anyway, and had Wally and Tom done so, certainly no one familiar with the hardware would have blamed them. But supercool Wally kept his head, picked the correct option, and Gemini 6 was saved to fly another day. It turned out that an electrical plug had vibrated loose after engine ignition but a split second before lift-off, causing a spurious shut-down signal to be sent to the Titan's engines. The following day it was announced that "a plastic dust cover carelessly left in a fuel line would have blocked the Gemini 6 launch even if an electrical plug had not dropped out of the tail and shut down the Titan II engines . . . The device apparently had been installed at the . . . plant and was not removed due to 'human error' . . ."* Wally and Tom must have been living right.

Their third launch attempt, on December 15, 1965, was finally successful, and after a four-orbit chase, as planned, Gemini 6 pulled up alongside Gemini 7 and flew in formation with it for five hours, during which time the two crews chatted and took pictures of each other. Then it was time for Wally and Tom to separate, grab some sleep, play "Jingle Bells," and re-enter, leaving Frank and Jim to plod along for another couple of days. Finally December 18 came and it was Frank's turn to assume the retrofire attitude and prepare the electrical circuitry which would enable the four solid retrorocket motors to fire. That is, if they *would*, if they hadn't become too cold-soaked or otherwise damaged by over three hundred hours of exposure to the vacuum of space. Firing the retrorockets was the only way to get Gemini 7 down before it ran out of oxygen, so all of us in MCC breathed a great sigh of relief as Frank reported a successful retrofire. Now we had a shorter list of worries. Would they come down in the right place, would the parachute open, would they be in good physical shape? It wasn't long before our TV screens showed the two stepping out (yes, they

* NASA SP–4006, "Astronautics and Aeronautics, 1965," p. 548.

could walk) onto the carrier deck. They looked pale, tired, and unshaven, but happy, smiling, and not in any obvious physical distress.

Post-flight physical results confirmed the fact that they were in good shape, in some ways less the wear for their fourteen days than Cooper and Conrad had been after eight. Borman had lost ten pounds and Lovell six. Lovell's pulsating thigh cuffs had had no detectable effect. Borman's brain waves had shown that he slept fitfully for a couple of hours the first night, and well for seven hours the second night, which Borman had long since reported himself. The third night was not recorded, since the EEG scalp electrodes had been prematurely jerked loose, whether accidentally or on purpose only Borman knew. The crew's tolerance for exercise, as measured by their hearts' response to pulling on the bungee cord, had not changed appreciably over the fourteen days. The bone density X-rays were a pleasant surprise, since a loss of 3 percent, roughly, was noted—only one third that of the Gemini 5 crew. The results of the intricate calcium-balance studies I have never seen, nor do I have any burning desire to search archives for them. Nor can I report how long it took the autonomic nervous system to relearn its leg flapper valve duties, but one has only to remember the televised view of Borman and Lovell's confident walk across the carrier deck to realize no serious problems existed. The two main factors which explain why the Gemini 7 crew fared so well compared to 5 are that 7 had less bulky pressure suits and were also allowed to remove them for most of the flight, and that 7 employed the more restful routine of simultaneous sleep periods. I think a third reason was Frank Borman's disciplined determination. Goddamn it, his flapper valves would too close, and Lovell's had better also. Otherwise court-martial them!

With both Gemini 6 and 7 on the ground, five of the Gemini program's ten manned flights were completed, and more than half of the Gemini questions had been answered. The two main ones were man's durability for a weightless lunar voyage and the practicality of a lunar strategy which required a space rendezvous for its successful completion. True, an actual docking had not taken place, and despite Ed White's twenty-minute foray, there was no solid data to prove man could operate effectively outside his space-craft, but these were considered to be fairly minor problems. The single big unknown remaining was the whole rendezvous question. Rendezvous is an incredibly complicated game, with more variables than one likes to list. Just because Wally and Tom had done it once, in earth orbit under carefully controlled conditions, did not mean it could be done in lunar orbit with all the permutations and combinations of lighting, timing, and spacing—not to mention all the possible equipment breakdowns ("degraded modes," they were called) which might still make rendezvous possible but which would require the use of different techniques. Hence the remaining five Gemini flights were all to feature rendezvous, docking, and EVA, with as many variations on the rendezvous theme as could be squeezed into five earth orbital flights.

The next flight up, Gemini 8, belonged to Neil Armstrong and Dave Scott, which meant that Dave would be the first Fourteen to fly, despite the fact that I had been the first to be assigned to crew duty. Dave on 8 and Charlie Bassett on 9 had sneaked in ahead of me, bypassing the back-up crew apprenticeship, a fact which right-fully put me in my place. When the 8 crew had been announced in September and the 9 crew in November, I had been happy for them, and felt that this was further proof that NASA was a good outfit and knew what it was doing. But now, as 1966 began, I was getting a bit nervous. For six months I had figured that Gemini 10 would be mine, but now Ed White told me it wasn't going to be his, that he had the word from Deke that he was moving on to Apollo. He was ambivalent about it, wanting to fly another Gemini but also wanting to get in on the ground floor of Apollo. Per-sonally, I thought Apollo made more sense for him at that time, and I thought he might very well end up first on the moon, as he had a lot of things going for him. He had projected exactly the right image as the nation's first space walker, and why not do the same as the first moon walker? At any rate, it was clear that our back-up crew would not fly Gemini 10, so when I heard from John Young (he and Gus Grissom had just come off the Gemini 6 back-up crew) that he and I were to fly 10, I was overjoyed. I would miss Ed, but I liked John, and besides I would have flown by myself or with a kangaroo—I just wanted to fly. All that stuff about crew psychological compatibility is crap. Almost anyone can put up with almost anyone else for a clearly defined period of time in pursuit of a mutual objective important to each. So John and I started getting it all together for 10, which would be launched in July 1966.

7

I keep six honest serving men
(They taught me all I knew):
Their names are What and Why and When
and How and Where and Who.

—Rudyard Kipling, "The Elephant's Child"

The first task John and I faced was to review the plans for Gemini
10, because neither of us knew much more than the bare fact that
it was one of the "standard" flights remaining in the series. Each of
the five remaining was to last three days, and would feature rendezvous, EVA, and a variety of scientific, technical, and medical
experiments. Each would be different in that the specific rendezvous techniques would be as varied as earth orbital conditions
permitted, and each particular EVA objective would be changed,
as would the experiments. When we added everything up, John
and I were immediately sobered by the magnitude of the job asked
of us. In three days, we were going to conduct two rendezvous,
with two different Agenas in two different orbits; two EVAs, each

quite different from the other; and fifteen scientific and technical experiments. In addition, we were going to be navigational guinea pigs and were going to compute on board our spacecraft all the maneuvers necessary to find and catch our first Agena, instead of using ground-computed instructions to get us within range of our own radar. This technique had not been perfected, but an expanded on-board computer memory was being designed, called Module VI, and it was left to us to figure out the best procedures for using it by trying out various methods in the simulator.

The first thing we figured out was that we were really going to be rushed to get all these things done in three days. Since the spacecraft had enough provisions on board for four days (oxygen being the critical item), why not stay up an extra day? For some reason this idea did not appeal to Chuck Mathews, the Gemini program director, or to his staff, and our request for an extension was denied.* We gave brief thought to requesting that some of the experiments be deleted, as one or two of them seemed a bit silly to us, but they had all been officially sanctioned after a long review process and we didn't want to waste too much of our time banging our heads against the bureaucratic wall. We were all in favor of rendezvous and EVA and didn't wish to curtail either, which left us with only Module VI as a possible candidate for deletion. Since it was in its infancy, I think that if John and I had said that it simply wasn't far enough along, management would have assigned it to a later flight. In retrospect, I think that would have been the best thing to do, or perhaps it should have been deleted entirely. Module VI used a very cumbersome procedure, whereby I was to use a portable sextant to measure the angle between a star and the earth's horizon (repeating the process at carefully prescribed intervals with stars selected for their positions relative to our orbit) and feed the angles and times into Module VI of the computer. Combining Module VI data with a variety of charts and graphs carried on board, we would be able to determine our orbit and

---

* It was not until the final Gemini flight that NASA management relented, for some unknown reason, and allowed 12 to stay up four days.

predict where we would be at a given future time relative to our Agena target, whose orbit we knew. The silly part of all this was that Module VI's *modus operandi* was really not applicable to Apollo, which employed a far more sophisticated and accurate navigational scheme. Therefore, we would be spending valuable simulator time developing a technique which appeared to be a technological dead end. Module VI would actually decrease the chances of our making the rest of the flight successful, by robbing us of valuable preparation time on the ground and by jam-packing the first few hours of the flight itself. On the other hand, was this not a noble cause, to build an autonomous capability, to allow a manned spacecraft to roam free of ground control, to compute its own maneuvers? Was not the very name of the game, in manned space flight, to put the pilots in control? We kept Module VI.

With all fifteen experiments, both rendezvous, and Module VI securely on board, all with their own powerful constituency around MSC, the two EVA periods became prime candidates for cancellation in an effort to simplify the flight. I remember Dick Carley, Module VI's strongest supporter, taking me aside one day shortly after I had been assigned to the Gemini 10 crew and extolling the virtues of on-board navigation. Dick started with the Arabs, tracing the progress of civilization in terms of the distance and accuracy of man's travels. "Out of sight of land finally," he whispered in sepulchral awe at the audacity of the early great navigators. Dick would make me a modern-day Magellan. Dick would delete both EVAs and make Gemini 10 a navigational orgy, repeating orbit determination and prediction exercises over and over until they had been perfected. Of course, he was talking to the wrong guy. After two years at MSC, specializing in pressure suits and EVA, this was like trying to take the cubs away from Mama Bear.

The EVAs were the last things I wanted to delete; all Dick accomplished (besides causing John to start calling me Magellan) was to organize our thinking a bit, and we immediately compiled a list of cogent (we hoped) reasons for including two periods of

EVA. During the first EVA, a simple stand-up-in-the-hatch exercise, we could do some solid science, getting the ultraviolet signatures of selected stars, information which astronomers on the ground are denied because of the heavy filtering effect of the earth's atmosphere.* Then on the second EVA, if the proper equipment were developed quickly, I could make a really interesting space walk, using a gun to propel myself over to an old Agena (the Gemini 8 target vehicle) and removing from it a micrometeorite-measurement device which had been exposed to the hazards of space for four months. There were also a variety of supporting technical reasons why this EVA plan made sense, but it did require some new equipment to be developed swiftly, and I knew Chuck Mathews, who had to keep the lid on changes, would not be pleased by this. Fortunately, Chuck is a patient man who listens. I have seen him sit hour after hour in meetings, heavy-lidded but wide awake, impassively chain-smoking, nodding occasionally—a professional decision-maker who lets the conversation ramble just long enough to get all the facts out on the table.

On this occasion I was as nervous as Chuck was calm, and when, after politely hearing me out, he was still noncommittal, I figured Carley had won and was going to make a Magellan out of this poor soul who got lost every time he visited the Pentagon. It was not until a couple of days later that I learned Chuck had agreed to proceed with the necessary hardware development. The basics were all there already: pressure suit, chest pack, maneuvering gun, umbilical cord. I simply needed a greater radius of action and more fuel for the gun than had been provided so far, which meant a new fifty-foot umbilical cord had to be produced, and the gun's propellant gas would be stored in bottles added aboard the spacecraft (instead of in the gun itself) and piped out to the gun through the umbilical cord. Thus the cord, in addition to being

---

* It was not enough simply to lift the camera above the earth's atmosphere. It had also to be moved outside the spacecraft, because the protective glass of the spacecraft windows screened out most of the ultraviolet rays which the astronomers wished to measure. Hence EVA was necessary, to expose the camera lens to an unfiltered view of space.

long, was going to be thick: inside it would be one hose containing oxygen to breathe and another containing nitrogen for the maneuvering gun. Also inside it was a tether, a stout nylon line to keep me securely attached to my mother ship no matter how hard I jerked against it, and finally a bundle of electrical wires which allowed me to converse with John and allowed the medics to monitor my heart rate. I could also talk to the ground if John pushed the cockpit microphone button for me. The umbilical, to be protected from solar heat and the hot exhaust gases of the Gemini's maneuver thrusters, was encased in a heavy nylon sheath about two inches in diameter. The whole thing, to borrow my daughter's vocabulary, was GROSS!

Having willingly agreed to do everything on everybody's list for Gemini 10, John and I now had the task of organizing it into a workable flight plan, admitting secretly to ourselves that we were trying to cram four pounds into a three-pound bag. For instance, our load of fifteen experiments exceeded that of any other Gemini flight except the two rendezvous-less, EVA-less, long-duration flights of eight and fourteen days, when they had little else to do. In terms of time available, we were to devote 37 percent of our flight time to experiments, the highest figure in the program (Gemini 12 was runner-up with 30 percent). The first rendezvous was straightforward in that we would take four orbits ("M = 4" in MSC parlance*) to catch our Agena, which would be launched an hour and forty-one minutes ahead of us. But of course the addition of the Module VI on-board computations would absolutely saturate these first four orbits to the point that there would be literally no time to look out the window except to take star-horizon mea-

---

* Early in the program, that is, during Mercury and early Gemini days, "orbit" was the only word used to describe going around the earth once. Later "revolution" or "rev" took its place. Technically, an orbit describes an event in earth coordinates, a rev in inertial coordinates; by that I mean a rev is one 360-degree trip around the spinning globe, with no concern for what piece of real estate is below, while orbit defines the interval between "overhead" passes relative to a spot on the ground. If the earth did not rotate on its axis, an orbit and a rev would be identical, but since the earth turns toward the east and spacecraft are launched in an eastward direction, the period (or time) of a rev is slightly less than that of an orbit. M = 4 really means four revs to catch the Agena.

surements. The second rendezvous was a different matter, because it would be made with a different target, the Gemini 8 Agena, which had been waiting for us in orbit for four months, after it had been abandoned by Neil Armstrong and Dave Scott. So what? Well, the big difference was that the 8 Agena had long since "died," as its batteries had discharged, which meant that it had no radar transponder. This in turn made our radar useless, because when it called, the 8 Agena could not answer; hence, we would have to employ purely optical means for rendezvous. Furthermore, when the Agena died, its system for stabilization was incapacitated and we might very well find it spinning like a top, or tumbling end over end, in which case how did I get that micrometeorite package off it? Pessimists (realists?) envisioned a colossal snarl, with Gemini, Agena, and me all knotted together by the infernal fifty-foot umbilical cord.

So we—the crew—had a lot to do to organize all these tasks into some semblance of order. In good bureaucratic fashion we gathered our allies and held meetings. We had a lot of help, in the form of a support team of engineers headed by Ed Hoskins. We also had a back-up crew of Jim Lovell and Buzz Aldrin, who were less than delighted with their assignment, but who nonetheless pitched in with good solid suggestions. The reason they were unhappy was that under Deke's crew-assignment system, they could expect to skip the next two flights (Geminis 11 and 12) and pick up the third. But there wasn't any Gemini 13, so they were lame ducks, wasting time on Gemini when they should be pressing on with Apollo.

By this time, January was turning to February and we were starting to make good progress in organizing our flight. It shaped up like this: a late-afternoon takeoff was necessary, because of the orbital plane of the 8 Agena, following which we would spend five hours catching and docking with our own Agena, Module VI permitting. Then we would fire its big engine (another first) and use its power to boost us into a higher orbit, which would start us closing on the 8 Agena. Then we would sleep. On the second day, a fourteen-hour work schedule included a number of Agena engine firings, experiments galore, and a stand-up-in-the-hatch EVA. The third day spanned fifteen working hours, and included the intricate series of maneuvers needed to join Agena 8, plus experiments along the way, the EVA itself, a subsequent hatch opening to jettison unneeded equipment, and descent into a lower orbit. The fourth day covered eight brief hours, finishing up the experiments in time to prepare for retrofire and splashdown into the Atlantic seventy hours and twenty-seven minutes after lift-off. Sandwiched in as best we could were meals, fuel-cell purges, platform alignments, radio conversations, and all the other little things vital to keeping men and machines operating. In theory, of the seventy hours, we worked forty-six and had twenty-four off, but in practice it didn't quite turn out that way.

By this time, my life was assuming a personal urgency I had never felt before. This wasn't pressure suit design for someone else, or back-up duties for someone else, this was for Gemini 10, and it was going to fly in a few short months and I was going to be aboard it. In a way it was eerie, planning this flight; for instance, when we discussed it in meetings, John and I were referred to as "the crew." We adopted this third person convention without question, and I would find myself saying, "Yes, the crew will . . ." do this or that zany thing, and then suddenly realize with a start, "that" wasn't the crew, "that" was *me!* It was useful, however, in giving one a certain detachment in discussing and analyzing what the crew might do in some theoretical tight squeeze, such as getting one's umbilical ensnarled with a tumbling Agena.

I kept a seven-by-ten-inch black notebook divided into six sections, as follows: (1) Schedule, (2) Systems Briefings, (3) Experiments, (4) Flight Plan, (5) Miscellaneous, and (6) Open Items. Section 6 meant problems of which I became aware as we went along, and which were duly listed by number. As long as they remained unsolved, or open, I reviewed them periodically and bugged the appropriate people for solutions. As they were solved, they were closed, and I drew a line through that number. By the morning of launch, I had 138 items, and all 138 had been crossed out. If this process was a bit scary and time-consuming, it was also immensely satisfying. It was going to be one hell of a flight, if only I could figure out . . . Whip out the notebook and write it down before I forgot it. The flight was constantly on my mind. When the notebook was not handy, I scribbled on napkins, bits of paper from my wallet, or whatever was available. I can remember sitting in the Holiday Inn in St. Louis one night after dinner, nursing a beer by myself in the bar, and getting a bright idea which I duly recorded on a tiny scrap of paper. The bored bartender ambled over and said, "Oh, making plans, eh?" "Oh yes," I trilled. "You see, there is going to be this rocket circling the earth, named Agena, and I am going to go in orbit after it and visit it! I was just thinking I might carry an American flag along with me and tie it to the Agena's antenna. Don't you think that would be a nice idea?" "Sure, buddy, sure. You bet," he mumbled, beat a hasty retreat to his cash register, and stayed there. He didn't even offer me a second beer. But I didn't need one, for what the bartender realized better than I was that I was already high, on a five-month high. Never one to keep a diary, I did jot down one day how I felt about things in general as a result of this peculiar ambiance. The gist of it was that, contrary to what I would have guessed, my all-consuming devotion to Gemini 10 was *not* all-consuming at all, but was causing me to live on a plane of heightened awareness. Life was good. Music sounded better than it ordinarily did. Food tasted better, and so did wine. Supposedly mediocre wines revealed nuances of flavor I had never noticed before; they approached greatness. I was working long, long hours, but they were not grinding me down.

Unfortunately, into this euphoria plunged a T-38, carrying Elliot See and Charlie Bassett to their deaths. The accident happened in St. Louis, where I had been spending so much time. Elliot and Charlie were the prime Gemini 9 crew and crashed early one Monday morning at the end of a flight up from Houston. Charlie had been neighbor, classmate, tutor, and friend. He was a remarkable man—one who seemed to have everything: extraordinary intelligence, steely determination, infectious good humor, good looks—and then he was decapitated in a parking lot. Elliot I hardly knew, except that he was a friendly and gentle man, and now both were gone. In addition to the great personal loss felt by their friends, Elliot and Charlie left a void in the Gemini schedule, which was quickly filled in by calling upon the back-up crew system. Tom Stafford and Gene Cernan, who had been back-up, would now fly 9, and our back-up crew of Lovell and Aldrin was assigned to them. We were given Al Bean and C. C. Williams. Since all the surviving group of Nine were to fly one or more Gemini flights anyway, it didn't change their fortunes as much as it did those of the Fourteen. The original flight sequence for our group read Scott, Bassett, Collins, Gordon, and Cernan. It now became Scott, Cernan, Collins, Gordon, and Aldrin.

At any rate, no sooner had our new Gemini 10 team of Young, Collins, Bean, and Williams been assembled than Deke gave us a peculiar assignment: we were to sit with him and Warren North on a selection board to pick a new group of astronauts. This was completely incomprehensible to me for two reasons: first, we were going full blower in our training for Gemini 10, with an extremely tight four months ahead of us; and second, we already had about thirty astronauts still alive and kicking and on flying status. Who needed more, and why bother us with picking them? John remonstrated, but Deke was firm about it, so we dropped everything for the first week of March and picked nineteen astronauts. It was a fascinating time, even if it did put us a week behind in our training, a week we never regained.

The first thing I did was to find the personnel office, which had conducted the preliminary screening. By a simple check of age, education, and flying experience, they had reduced several hundred applicants to around thirty-five. Each of these finalists was accompanied by an impressive dossier. If one spent the time, and I spent the weekend at it, it was possible to get a pretty fair picture of each man: this one seduced compatriots' wives, that one was apparently a latent homosexual. This one flew well, but that one's boss was eloquent in what he did *not* say. This one had a string of college professors whom he had impressed; that one had slipped through school unnoticed. This one had acquaintances among the present astronauts; that one was too young to know any of us. One insisted upon $25,000 a year, but apparently no one else cared about money. I harked back to my own traumatic days as an applicant, or supplicant, and vowed to do as conscientious a job as possible to screen these men, to cull any phonies, to pick the very best. There were no blacks* and no women in the group.

I think our selection board breathed a sigh of relief that there were no women, because women made problems, no doubt about it. It was bad enough to have to unzip your pressure suit, stick a plastic bag on your bottom, and defecate—with ugly old John Young sitting six inches away. How about if it was a woman? Besides, penisless, she couldn't even use a CUVMS,† so that system would have to be completely redesigned. No, it was better to stick with men. The absence of blacks was a different matter. NASA should have had them, our group would have welcomed them, and I don't know why none showed up. Perhaps there simply weren't any who had the flying/educational backgrounds required, or perhaps they were more interested in other careers. I only know that no one was eliminated because of color, as there was no place on the application form to record color or creed, and the first chance any NASA official had to find out was when the finalists appeared for their physical exams. Be that as it may, and I think it was unfortunate, it was our board's job to select, not recruit, and we set about organizing an equitable system for grading applicants. Deke proposed a system which had been used in previous selections, and with minor modifications we agreed. It was a thirty-point system divided equally into three parts: academics,

* The closest this country has come to having a black astronaut was the selection of Major Robert H. Lawrence, Jr., on June 30, 1967, as a member of the Air Force Manned Orbiting Laboratory astronaut group. A Ph.D. chemist in addition to being a qualified test pilot, Lawrence was killed on December 8, 1967, in the crash of an F-104 at Edwards AFB. In mid-1969, the Manned Orbiting Laboratory program was canceled.
† See page 144, step 2.

pilot performance, character and motivation. "Academics" was really a misnomer, as an examination of its components will reveal: IQ score—one point; academic degrees, honors, and other credentials—four points; results of NASA-administered aptitude tests—three points; and results of a technical interview—two points. Pilot performance broke down into: examination of flying records (total time, type of airplane, etc.)—three points; flying rating by test pilot school or other supervisors—one point; and results of technical interview—six points. Character and motivation was not subdivided, but the entire ten-point package was examined in the interview, and the victim's personality was an important part of it. Hence, of the thirty points (the maximum a candidate could earn), eighteen could be awarded during the all-important interview. My recollection is that we spent an hour per man, using roughly forty-five minutes to quiz him and fifteen in a postmortem. We sat all day long in a stuffy room in the Rice Hotel, interviewing from early morning to early evening, for one solid week.

Our board, which was examining candidates from the Air Force, Navy, Marines, and civilian life, had representatives from each. Warren North was the civilian, John Young was Navy, C. C. Williams was a Marine, and I was Air Force. Deke (civilian, ex-Air Force) was our chairman. Before I started working for NASA, I thought the matter of service affiliation was quite important, and guessed that astronaut selection and assignment to specific crews was predicated in large part on balancing Air Force vs. Navy vs. civilians. When we fourteen were selected, at the very beginning we tended to be a bit cliquish; however, it wasn't long before bonds were developed which were much stronger than color of uniforms (which we didn't wear anyway), and I soon realized that I didn't care, indeed didn't even think about, who belonged to which service. We were all in it together, unlike the generals and admirals in the Pentagon who have to keep squabbling about bombers vs. submarines to keep each other honest. Nowhere was this NASA, or at least astronaut, lack of concern over origins more apparent than in this selection board. We had no quotas, and I frequently found myself arguing for a Navy man or a civilian, and against an Air Force candidate. And the others did the same. Actually, there was a lot less arguing than I thought there would be: the winners and the losers generally sprang out at us, from paper and in person, and it was not difficult to award our points and to rank everyone in order.

I suppose the most difficult part of our job was to measure motivation.* Did this gent really want to come to work for us? Did he want it badly enough to keep his nose to the grindstone through the long years of apprenticeship? This was a tough judgment to make, and in evaluating eighteen out of thirty points strictly on the basis of an interview, we ran the risk of getting snowed by some smooth talker. But what impressed us the most was not the glib expression of devotion to our cause but the solid evidence that the man had been interested enough to do his homework, that he had studied Gemini and Apollo, knew something of their problems, and proposed reasonable solutions based upon his experience. We pushed their knowledge to its limits, and there the smart ones said, "I don't know" and the others tried to bullshit us. In retrospect, I think our board did a pretty good job, and there are no real turkeys among the Nineteen. I recall that Fred Haise came out number one and of course Tail End Charlie wasn't selected; we had unanimous agreement on those who were picked.

The only astounding thing, to John, C.C., and me, was that Deke said, "We'll take as many as we consider qualified," and when that turned out to be a grand total of nineteen, Deke was happy to hire all of them. What the hell was he going to do with nineteen more astronauts? Aside from the "pull up the ladder, I'm aboard" syndrome, we honestly didn't think any more than half a dozen would be required, but Deke was worried about meeting the most optimistic future launch schedules. At that time (March

---

* I will never forget the immortal words of Bob Smith, an Air Force buddy of mine who was seeing me off to my own interview a couple of years earlier. "Now, don't forget, Mike," he said, "when they ask you why you want to be an astronaut, tell them it's because of all the money and ass you can get."

1966) NASA was considering a program called Apollo Applications (the progenitor of Skylab), which would overlap flights in earth orbit with the lunar Apollo launches. In fact, toward the end of the decade, if you believed the optimists, the Cape launch schedule would be frantic indeed, with either an Apollo or an Apollo Applications flight going nearly every month. With the prime and back-up crew arrangement, six astronauts would be tied up for a year or so in preparation for each flight, and that meant Deke needed lots of people. So he hired them.* The sad exodus, years later, of highly qualified, highly motivated, expensively trained young men who had never flown in space were the result of this miscalculation. I guess Deke played it the way he had to, the conservative way, which says you plan for what *can* happen, not for what you guess *will*, but it seemed to me at the time that nineteen was a bit much. John Young promptly dubbed them the Original Nineteen, in parody of the immortals, the Original Seven, but it was not until the next group was picked a year later that the nomenclature matched the facts. They knew enough to call themselves the Excess Eleven.

After this week of fun and frolic at the Rice Hotel, the Gemini 10 crew barely had time to crawl back into the simulator and wrestle a bit with Module VI before it was time to crawl back out and watch Gemini 8 fly. I "watched" the launch from Mission Control in Houston, watched as the dot representing the Titan launch vehicle climbed smoothly up its assigned line on the wall chart, watched as Neil Armstrong and Dave Scott separated their Gemini from the Titan and began the four-orbit chase of their Agena, which had been launched earlier in the day and which seemed alive and well, for a change. The M–4 rendezvous proceeded smoothly to its climax, and by suppertime the Gemini had docked with the Agena. This docking, or actual locking together of two vehicles in space, was the first in history, and was important

---

* At its peak, in the summer of 1967, the astronaut office numbered fifty-six.

because it was a necessary part of returning two Americans from the surface of the moon. It was also 100 percent manually flown, not unlike mid-air refueling of airplanes, and it made us pilots feel good to hear Neil report that it had been easy, with no surprises. So I headed on home for a martini and dinner, assured that 8 was off to a rollicking good start.

No sooner had I walked in the front door than I got quite a shock. Pat was upset by a cryptic phone call she had just gotten from the Scott residence, asking would we please look after Dave's kids for a couple of hours so they would be out of the way until "they brought them down." While I was hurriedly telephoning Mission Control, Dave's father-in-law arrived with the two children, and between the two conversations I gathered that 8 was indeed being ordered back to earth, after having tumbled wildly out of control. That damned Agena, I thought, and drove back to Mission Control in a hurry. There a totally different story unfolded. All had gone peacefully for a half hour after the docking, at which time the crew had noticed some yaw and roll rates developing. This indicated a malfunction in either the Gemini's or the Agena's attitude-control system. Which was it? Neil found he could manually fire his thrusters to counteract the unwanted motion, but it started again as soon as he let go of the attitude hand controller. From this he concluded the problem was probably with the Agena. Being over the Pacific Ocean out of radio contact with all ground stations, Neil and Dave were unable to get any help from the information being telemetered by Gemini and Agena to the ground. Next Neil decided to undock, a logical, analytical step, because he could then move a short distance away from the Agena and study each vehicle independently of the other. He did so, fully expecting the yaw and roll motion of the Agena to accelerate, while his own quieted down. Exactly the reverse happened! The Gemini began to spin like a top, and Neil and Dave soon found themselves tumbling at 300 degrees per second, a rate which was uncomfortably fast and which was getting still swifter as each second went by. There was no time to ponder the problem at length, and no ground station to consult. Chances were it was a thruster that had stuck open and was firing continuously, but there were sixteen thrusters, and the motion—combining as it did components of yaw, pitch, and roll—made identification of the guilty one nearly impossible. They did the next best thing. They shut down the entire orbital attitude maneuvering system. Now they were mercilessly under the command of Newton's second law: unless acted upon by external forces, an object at rest will remain at rest, while an object in motion will remain in motion. In other words, the stuck thruster was silenced and could not cause the rate to build up past 300 degrees per second, but without bringing other forces into play, the 300 degrees per second would continue indefinitely and that was intolerable. Intolerable because they might drift into, and strike, the nearby Agena; intolerable because they could not continue the mission that way; intolerable because they might develop vertigo and nausea if it kept up. So Neil was forced to play his ace in the hole and activate the attitude-control system that was reserved for re-entry. Using this second group of thrusters, he was quickly able to stop the tumbling, so that by the time Gemini 8 arrived within radio range of the next ground station, the excitement was all over. The whole thing had lasted perhaps ten minutes, but they were the hairiest ten minutes in the space program so far. The malfunction required that the Gemini 8 flight be ended right away, since it was a firm Mission rule that any time the re-entry attitude control system was activated, re-entry must follow promptly.

As soon as I determined from overhearing various conversations in Mission Control that everything was under control, I made the short trip home again, to brief Pat and General Ott, Lurton Scott's father and an old pilot himself. The Scott children were too young (Tracy not quite five, Doug two and a half) to have any real appreciation of what was going on, and by now it was getting past their bedtimes. We wanted to ease them home (a block away), but about that time the word got out that Jan Armstrong was on her way over to Lurton Scott's house, and the front yard was ablaze with television lights and excited reporters. Fortunately, all the activity was focused on the front door of the Scott residence, and we were able to tiptoe in the back way without attracting any attention, although obviously the kids knew something was up. When Jan arrived, she made a beeline for the front door, trailing a small wake of NASA officials who explained that she wanted to be with Lurton during this time of crisis and that she would discuss it with the press in due time. So the TV crews turned off their klieg lights and waited.

As it turned out, the re-entry was routine in all respects except that the premature landing was made in the Pacific rather than the Atlantic. But good old NASA, with its "What happens if?" philosophy, had made sure that the Navy stationed a destroyer there just in case. Therefore, it wasn't long before the safety of the crew was assured and the incident was over, except for the crushing disappointment for Neil and Dave and all who had worked so hard on those parts of the flight (such as the EVA) which were left undone. But at least Jan and Lurton's immediate worry was over. After following the flight intently for eleven hours, they were tired and Jan was ready to go home. The instant the Scotts' front door closed behind her, the yard came alive. Flashbulbs popped, the blinding TV lights came back on, microphones were stuck in her face, and twenty voices babbled out questions. The only friendly, familiar object Jan could see through the haze of lights was her car, and she made straight for it. Off she went, leaving a howl of disappointment behind her. She hadn't even gotten in the "thrilled, proud, and pleased" which is the astronaut wife's standard.

Neil and Dave weren't pleased either, especially Dave, who became the nation's first almost space walker, and I could certainly sympathize with him. He had spent months in intense, complex training for a very involved space walk using a newly developed chest pack and back pack, and now he would never have a chance to find out how the equipment would have worked, or how it felt to float free in space. Secretly, some of us were relieved at Dave's misfortune, because his EVA had seemed terribly complex and dangerous. The heart of the problem was that the Gemini cockpit was too small to store large pieces of equipment, such as the back packs scheduled on Geminis 8, 9, and 12. Therefore, the back pack was stored back in the adapter section, where it was exposed to the vacuum of space. The procedure was for the space walker to depressurize the cabin, open the hatch, and make his way on hand-rails back to the adapter, breathing and communicating through the umbilical cord leading from the cockpit. So far so good, but now comes the hairy part, because next he had to transfer his oxygen supply and voice link from the umbilical to the back pack. This required the disconnecting and reinstallation of oxygen and radio lines leading into the suit through the chest pack, and in Dave's case, an additional line to provide propellant gas for his hand-held maneuvering gun. Because of the limited mobility and visibility inside the pressurized suit, this changing of connectors was a difficult business, and one especially unforgiving of error, with no assistant present to point out tangled lines or any of a dozen other hazards lurking back there in the lonely vacuum of the adapter section. My own EVA scheme on Gemini 10 was far from ideal, in that I had to stuff everything into an already crowded cockpit, but at least I could make nearly all my preparations inside the pressurized cocoon, where John could help me if need be. Not so with Dave's complicated gear, which required the most involved procedures of any Gemini EVA, and before the flight I was frankly worried about it. I mentioned my concern to Neil a year later, after the end of the Gemini program, saying that the cancellation of Dave's EVA might have been fortuitous after all, but Neil didn't see it that way and snapped that if any of the EVA schemes had been wild, it was mine on 10, with an unstabilized Agena floating around. I guess one learns to become comfortable with the familiar, no matter how strange it may appear to others.

Gemini 8's problem was diagnosed as a continuously firing thruster caused by a short circuit in the wiring to the thruster. Suitable changes were made in Gemini 9 and subsequent space-craft, and the rapid pace of the program continued undiminished.

John and I spent nearly all of April and May in St. Louis, nursing Gemini 10 through its final assembly and subsequent tests, including "flying" it inside an altitude chamber, and running through every phase of the flight with test equipment plugged in to measure the correct response of each pressure switch, diode, thruster, parachute, computer, etc., etc. If something didn't behave exactly according to its specifications, it was replaced. The spacecraft spent most of its time inside a white room—a room kept meticulously clean, painted white, with filtered air. Everyone was dressed in white smocks and caps, and white booties. The white room ran on a twenty-four-hour schedule and usually had three or four spacecraft in it, mounted vertically with their noses pointed up in the air, as they would be on the launch pad. Hour after hour John and I lay on our backs with our legs up in the air, lazily flicking Gemini 10's switches as each test sequence required. Each test had a script, a huge volume listing the role of each supporting technician in the drama of make-believe flying. Some sequences were straightforward; others required a complex interrelationship between the test conductor, his supporting crew of test-equipment operators, and the crew in the cockpit. We perused the volume ahead of time, and red-lined any notation referring to our participation. Then, precisely at the correct instant in the test sequence (hopefully), we would activate the right switch for exactly the right interval. If we or anyone else on the team goofed, the test was repeated, sometimes three or four times. It was a very inefficient use of our time, but it wasn't bad training because we got to know *our own* machine intimately, and each one was slightly different, not only slightly different in design but different in the response of systems which were supposed to be identical in all spacecraft and in the simulator. These little differences were nice to find out about on the ground to prevent surprises in flight.

The white-room work was also good in that it drummed cockpit design and switch location into our heads under the mild pressure of the test sequences. If one mistakenly threw the computer switch to "Re-entry" instead of putting the attitude-control switch (directly below it) to "Re-entry," then no great harm was done; but faces got red, tempers got short (especially at 3 A.M.), and a long, involved test sequence was botched and had to be repeated at the expense of a bit of sack time. One also learned the precise location of each kidney, as the upended Gemini seats were really tough on the lower back, and after a couple of hours it was impossible to find a comfortable position. Also there is something about this abnormal feet-above-head position that seems to spur the kidneys into furious activity, which, coupled with a liberal coffee intake to fight drowsiness, meant hourly trips to the men's room, if the script allowed. If not, one wallowed in agony and self-pity, with the accursed script propped uncomfortably on top of a distended bladder, flipping switches interminably and rejoicing at the glamorous life of an astronaut.

When it was not the spacecraft, it was the simulator. The St. Louis simulator specialized in rendezvous, which meant that it could precisely imitate the response of the platform, radar, computer, and other gadgets necessary for rendezvous but not the operation of all the other systems, which we would learn in the Houston and Cape simulators. Since we were Module VI pioneers, we needed this St. Louis simulator badly because it came equipped with a highly skilled team of McDonnell engineers, the people who had invented Module VI, or at least implemented it. Day after day we sat in this infernal machine, drinking watery coffee out of thin tan plastic cups, watching a light on a darkened screen (the Agena) surrounded by other lights (the stars). The visual projection was crude indeed, but inside the cockpit precision was possible and the radar, platform, and computer all hummed along (unless they were deemed to have "failed") in perfect pretense as we chased our Agena around and around and around. Sometimes we caught it without difficulty and without undue expenditure of fuel; at other times, especially when we were denied the use of radar or platform or computer, we arced wildly through the skies, missing Agena entirely, or catching it only after our tanks had theoretically run dry. After each rendezvous we repaired to an adjacent room where we could see our imaginary path through the heavens plotted by a moving stylus on a large piece of graph paper. Here mistakes could be analyzed and suggestions made. The most ridiculous-looking plot on the paper was what we called a "whifferdill," a helix-shaped approach path which found the Gemini making ever smaller circles around the Agena before finally catching it. In addition to looking strange, the whifferdill used a lot of gas, and John and I were to discover later that the St. Louis simulator was quite accurate in its predictions of this fact. Whifferdills are caused by allowing the target to move relative to the stars (failing to null the inertial line of sight rate, as the experts would describe it) and by not making large enough corrections early enough to completely stop this unwanted motion.

By mid-May, I was sick of whifferdills, white rooms, and St. Louis in general, and was delighted to take a brief break by flying down to the Cape for the Gemini 9 launch on May 17. As usual, the Gemini launch was preceded by an Atlas, carrying the Agena as its top stage. In midmorning, it thundered aloft in good shape, apparently, but two minutes after lift-off one of the Atlas engines went out of control, pitching hard over and causing the vehicle to plunge into the Atlantic Ocean. Stafford seemed to be an Agena jinx, this being the second one shot out from under him, the first having blown up the previous fall as he and Wally Schirra were awaiting takeoff in Gemini 6. Now he was teamed with Gene Cernan, and the two of them had a two-week wait while a substitute target vehicle was pressed into service. After a second false start on June 1, caused by a balky data transmitter which refused to feed information into the Gemini computer, Gemini 9 finally made it into orbit on June 3, 1966—one year to the day after Jim McDivitt's and Ed White's Gemini 4. After a successful rendezvous, Stafford and Cernan found that the nose cone of their target vehicle had failed to come off, but was still attached, half open, making it look to the crew like an "angry alligator." Unable to dock with this beast, Tom and Gene were, however, able to proceed with EVA and their assigned experiments. The EVA was to feature, for the first time, the AMU. However, Gene worked himself up into such a lather during the early phases of the EVA that his visor completely fogged over, and both he and Tom wisely decided that this was no way to begin his flight as a separate satellite aboard the AMU. As has been discussed earlier, the heart of the matter was that without proper handholds, foot restraints, and other attaching devices, the space walker inside his pressurized suit burned up tremendous amounts of energy just trying to keep his body in proper position relative to the job at hand, with nothing left over for productive work. It became obvious that this heavy workload was a by-product or consequence of weightlessness, rather than a function of weightlessness itself, but nonetheless Gene's difficulty caused consternation among the equipment designers, and they cast a jaundiced eye on the EVA plans of later flights.

Suddenly Gemini 10 found itself the focus of attention, the favored son, as we were next up. Our training really hit high gear as operations shifted from Houston and St. Louis to the Cape. Spacecraft 10 had been shipped to the Cape in mid-May, as had our Agena 5005. There the two vehicles met for the first time, and "communicated" with each other in an elaborate test, in which John and I sat in the Gemini and radioed commands to the Agena, to check its responses. I sent these instructions by manipulating a device near my right elbow, called an encoder, a little box topped by two concentric wheels and a lever. I spoke to the Agena in three-digit words. For instance, 251 meant, Please turn your lights on. Firing the Agena's big engine was a bit more complicated: I had to tell it 041–571–450–521–501. The three-digit instructions always ended in either a one or a zero, and were formed by setting up the first digit on the outer wheel and the second digit on the inner wheel, and then transmitting all three by turning the lever from center to either the left (for zero) or the right (for one). The point is that the selection of the last digit was also the manipulation which transmitted the message instantaneously, so there was a built-in possibility for error, for the hand to move right instead of left, or vice versa, and then it was too late to change your mind. In fact, this particular test at the Cape had never been performed flawlessly before; at least one mistake had always been included among the many* messages, but I was determined it wasn't going to happen to me. In fact, I was willing to bet on it. It just so happened that at this time John Young owed me a couple of martinis, so we arranged a complicated betting scheme, doubling and redoubling as the test progressed. As each successful command went out, my martini ledger grew and grew, and my right arm got stronger and stronger as my concentration increased. It wasn't long before I held that encoder lever in a vise-like grip.† Finally, an hour or so into the test, I sent one last zero, snapping the lever smartly clockwise. It went clockwise all right, with a little sprouting noise far beyond its normal stop, and then it lay limp and could be twiddled in either direction with no resistance whatsoever. "John, I broke the goddamned thing." John was philosophical about it, even when faced with a ninety-two-martini loss. "Well," he said after a slight pause, "I reckon this is the best place there is to break it." So we beat a hasty retreat, leaving the engineers to redesign the encoder by adding metal barriers on either side of the lever to prevent some nutty tennis ball masseur from ripping it out of its socket again. On the ground, it only cost John some martinis but in flight it would have cost us a large portion of the flight, because as things turned out we needed that obedient Agena badly.

Being "in training" for a flight means, to the average layman, something vaguely like being in training for a prize fight. True, there are physical aspects to space flight, and I played the cardio-vascular game by running two miles a day, and the recreational game to release tensions by an occasional handball match, and the

---

* During the flight of Gemini 10, I sent the Agena 350 commands (all of which it obeyed), but these were spread out over a couple of days.

† My grip wasn't weakened any by the fact that by now I was carrying a tennis ball with me everywhere I went, absent-mindedly squeezing it between thumb and fore-finger whenever the situation permitted. This little Queeg-like quirk had nothing to do with the Agena, but was simply an attempt to strengthen my hands, as my EVA simulations had convinced me that my fingers—working inside pressurized gloves—were the first part of me to tire. This was especially true of my right hand (I am left-handed), which had to grasp the handle of the maneuvering gun in a rather awkward fashion between thumb and forefinger.

muscular game by exercising the only muscles which would really get a strenuous workout (those controlling my arms and hands) by squeezing a tennis ball and lifting weights. But all this was small potatoes—the real game was the mental one, and it was played in the simulators. Here the battle was lost or won; here the crew "flew" the critical phases of the flight over and over again until it seemed that every possible mistake had been made, and apologized for, and corrected, and that no further surprises in space were possible. This was the very heart and soul of the NASA system; this was where we spent our time above all other choices. Running on the beach is grand; learning geology is commendable; jungle living is amusing; the centrifuge hurts; the spacecraft ground tests are useful; but one is not to fly until the simulator has told him he is ready. No easy process this, for flying a simulator is in many ways more difficult than flying the spacecraft itself. The simulator must not only duplicate all the motions of the cockpit instruments, but in addition must feign a knowledge of the outside world and—via a huge computer—keep book on imaginary motions of earth, Gemini, Agena, and sun. In short, the simulator must be able to reproduce the total environment accurately at any future time, to place us in it at exactly the right spot, and to keep track of our subsequent travels. Mistakes must be documented, preferably in graphic form. "What happens if?" must be asked in a hundred different ways, and answered a thousand times. Only then, when the questions become repetitious and the answers uniformly correct, is the crew ready to fly.

We were nearing this point, John and I, in late June of 1966—although we still had a couple of rough edges to smooth out. One was the infernal Module VI, which still was pushing me to my limits of time and understanding. With C. C. Williams's help, I had "massaged" the procedures, simplifying the charts and star sightings to an irreducible minimum. It was still a formidable load, and one even less tolerant than sending commands to the Agena, for once a mistake was made, there was no way to crosscheck it. Instead all computations downstream of the mistake were poisoned, and the ultimate result was invariably incorrect. The second area of doubt was the EVA, although here I was (foolishly?) a lot more confident than I was about Module VI. If the on-board navigational ledgerdemain of Module VI proved faulty, then no physical damage was done, only mild embarrassment as Mission Control would override our plans and substitute, in lieu of our crude calculations, the considered opinion of a whole bevy of grounded experts, backed up by a whole basement full of computers. An EVA miscalculation, on the other hand, could result in death—my death. Of course the very heart of the matter is that NASA, which loves redundant systems, has never been able to figure out a practical way of providing redundant pressure shells. By that I mean that between the little soft pink body of the astronaut and the hard vacuum of space there is only *one* thin shell of aluminum (the spacecraft) or of rubber (the pressure suit bladder). To this basic fact add the complication of having to glue that rubber bladder together from an individually tailored array of body-fitting fragments. When you think of a space walker, you may visualize a chap confidently exploiting the most advanced technology which this rich and powerful nation can provide, but not me, friends. I see a covey of little old ladies hunched over their glue pots in Worcester, Massachusetts, and I only hope that between discussions of Friday-night bingo and the new monsignor, their attention doesn't wander too far. Maybe it's basic insanity to abandon hearth and home to dance in space on the end of a fifty-foot cord.

Some of the NASA hierarchy also thought it a bit mad and, in the wake of Gene Cernan's difficulties, became increasingly interested in the details of my EVA plans. Underwater simulation of weightlessness was just then coming into vogue, and proposals were made to duplicate all my EVA tasks in a water tank and analyze any difficulties which might develop. Fortunately or unfortunately, John and I simply didn't have time to drop what we were doing, with only a month left to go, and chase this red herring underwater. We pleaded our case with Mathews and his people. Our strongest argument was the simplicity of my equipment, all of which could be donned inside the cockpit before depressurizing. The only thing I had to do, in the way of preparation, after I left the cockpit, was to attach a nitrogen line to the side of the spacecraft. The connection was made just aft of the cockpit, and there was a handrail located conveniently nearby. I had only to release a cover plate, exposing the nitrogen valve, into which I plunged my umbilical connector. If I did it right, the mechanism locked itself into place; if I missed, the collar on the umbilical end snapped forward and had to be recocked—a two-handed operation. But this was the only tricky part, and even that was not a life or death matter, because the nitrogen was needed only to power the gun, and that was certainly not required to assure my safe return to the cockpit. Mathews said O.K., we didn't have to go underwater, but someone should check it out. So underwater simulations were conducted, and the conclusion was that I might have trouble with this task, and then again, I might not. John and I received this news with straight faces (sometimes I had trouble with it in the zero-G airplane, and other times I didn't) and pressed on with our simulator work.

In the matter of the gun, simplicity worked for us and against us. The gun was light and compact and could be stored inside the cockpit. Good. On the other hand, its effectiveness as a control device was seriously questioned, and there didn't seem to be any accurate way to check its performance on the ground. My training device was a supersmooth metal floor about the size of a boxing ring. On it could be placed a circular pad about the size of a commercial floor polisher. Gas jets inside the polisher were turned on, lifting the pad a tiny fraction of an inch off the floor. This system created as frictionless a bond between floor and polisher as was possible. Next I got dressed up in my suit, chest pack, and gun, and stood on top of the polisher. In this way, I could practice squirting gas jets from my gun. I could start on one side of the boxing ring and "fly" over to a target on the other side. As I went, my body would tend to twist sideways (yaw) one way or the other, and I learned how to damp out this unwanted motion and still arrive at my target. So far so good, but this simulation was pretty crude and could only address itself to one body axis at a time. By that I mean, if I stood up on the polisher I could learn about controlling the yaw axis, but only yaw; if I lay on my side on a couch mounted on pads, I could vary "pitch" as I moved across the ring. Roll was imitated by lying on my back, but in this position it was not possible to move from one side of the ring to another. It was never possible to fly up or down, only back and forth across the ring. In short, the entire control problem could only be examined piecemeal, and crudely at that. But to remove that micrometeorite experiment from the Agena, I might need tight control of roll, pitch, and yaw simultaneously, something I had never been able to practice. Even worse than that, the gun was not capable of generating pure roll, pitch, or yaw rotations, but as a by-product produced a force which caused the body to move off in some unwanted direction.

The opponents of the gun pointed this out, and were met with an emotional counterattack from the gun's defenders, who said that simplicity meant safety, that the competing AMU was a monstrosity, and that besides, the gun had worked fine for Ed White. I was caught in the middle of all this. My EVA research the previous year had convinced me that something like the huge back pack, the AMU, was needed to perform precise maneuvers, but on the other hand all I had was the gun, and I might just as well make the best of it. Perhaps great precision was not required in my EVA scheme. For instance, if John could fly close enough to the Agena, perhaps I could just float over to it, if it weren't tumbling too swiftly, if we could approach it from an angle where John could see it while I could reach it—if . . . if . . . if. It was getting too close to the flight to have such a long list of "ifs," but there really wasn't much we could do about it as EVA planning doesn't yield to the same precise mathematical analysis as does rendezvous. As June turned to July, John and I were spending five days a week at the Cape, flying the simulator, reviewing systems and emergency procedures, memorizing the flight plan, getting final briefings by experimenters, and helping put Gemini 10 through its last tests out on Launch Pad 19. Then on weekends I would fly home, to spend Saturday practicing EVA in Harold Williams's frictionless boxing ring, or slippery table, as I called it. I got fairly good with the gun, at least good enough that the MSC managers were content to let our EVA plans proceed without interference, despite the specter of exhaustion which Gene Cernan's EVA had raised.

Sunday was my day, and I fought to keep it clear. My 1966 calendar shows that in all our hectic preparations for Gemini 10, I worked parts of only three Sundays. I might have to get up at 4:30 Monday morning to make an appointment at St. Louis or the Cape, but Sunday was a day for relaxation with Pat and the kids, a day for balancing the checkbook and pruning roses and cooking exotic dishes (frequently disasters) and turning the garden hose on the dog. It was true that as July 18 drew closer, my thoughts were more and more preoccupied with the flight, but naturally my placid, even temper prevailed, and I recall thinking how grand it was to be able to share my upcoming experience with my family with such composure, equanimity, and good humor. Harking back to this same period, Pat says I resented interruptions and was preoccupied, distracted, and totally irritable! God bless her, she waited a couple of years to tell me this, so when I left the house on July 10 to go fly Gemini, I floated out on a wave of solid support and love and even partial understanding of this thing I was about to try. I certainly had doubts about successfully completing everything on our list, and I'm positive John had too, but by and large I was confident, at least about the big things. The danger was certainly there, and no one knew it better than those of us who had spent our recent careers studying various ways in which the treacherous machinery might destroy us. Beyond fallible machines lurked even more fallible humans. There were *so* many ways in which we could screw up, *so* many possibilities for error in the frantic three-day flight, that hardly an hour would pass without a fresh opportunity for disaster. This we knew, and in me the knowledge generated fear of a sort, but it was an intellectual fear, not an emotional one. I don't know if I can explain the difference, except to say that under the heading of intellectual fear I list such things as getting cancer (and I'm frankly afraid of that), while emotional fear would include watching the aerialists high-wiring at a circus without a net (and they scare hell out of me). Intellectual fear is an armchair admission of the frailty of human life, while emotional fear is a viscera-gripping, panic-inducing reaction of the brain and body.

While the upcoming flight of Gemini 10 certainly added to my list of intellectual fears (added 138 black book items?), I can honestly say that never before or during the flight was I afraid in the emotional sense. I think one reason for this is that old cliché about fear of the unknown. It is absolutely true: noises, smells, events whose origin and consequence are unknown are the ones that cause physical fear—and all our training was designed to prevent surprises of that sort. Finally, I must say that below the threshold of fear for life and limb there lurks a similar emotion, perhaps not fear but at least apprehension, a worry, not that you are going to be killed, but that you will be terribly embarrassed. In the case of professional pilots at least, "death before dishonor" is replaced by "death before embarrassment," and the accident files are bulging with cases in which a pilot's professional pride, or obstinacy, caused him to choose an ultimately suicidal course of action rather than admit a mistake. I claim no immunity from this disease.

Our last week of training at the Cape started out frantically and ended serenely. As a flight nears, the crew balances the time remaining against training objectives not yet attained, and no matter how diligent the crew, a point is reached where they must admit they are simply not going to get everything done that they had charted for themselves. At least that's the way it was on Gemini 10, but thankfully we had arranged our priorities so that the items left unstudied were in the "nice to know" rather than the "need to know" category. In my case, the pace of training had been building for months, and peaked about three days before the flight. At that point I simply said, "The hell with it, if I don't know it now I never will." Besides, the last couple of days before a flight are always too exciting to spend hiding in the simulator. Old friends send telegrams or call long distance to wish you bon voyage. Some part of the spacecraft always breaks, and the launch schedule is thereby always jeopardized. Everyone wrings his hands and scurries about and fixes the problem, except the crew, which by now has managed to reach a higher plateau of consciousness, one of lordly detachment from such unpleasantries as leaking fuel cells, which is the form the Gemini 10 devil assumed.

At flight time Cocoa Beach and the surrounding villages are pulsating with life. People start yelling "Go!" a couple of days ahead of time, throngs of tourists invade the beaches, the honky-tonks overflow, the locals party around the clock, and all is well with the economy—for a fleeting moment at least. The crew knows this but cannot partake of it, as it is locked up out at the Cape, tucked out of harm's way in crew quarters, a plush prison of pleasant, if somewhat antiseptic, furnishings and abundant, if somewhat bland, food. In the case of the Gemini 10 crew, our routine in crew quarters was more pleasant than most, because the position of the 8 Agena in the sky was such that we would not pass under its orbital plane until late afternoon. With our launch thereby scheduled for 5:20 EST we were able—indeed we were required—to adjust our own internal clocks by sleeping later and later in the morning, until finally, for the last couple of days, we were staying up until 3 or 4 A.M. and sleeping until noon. Granted we were staying up studying, but somehow the late hours carried with them a connotation of leisure and relaxation that was especially helpful as July 18 got closer and closer.

One of the questions an astronaut is most frequently asked is "How long does it take to train for a flight?" I hope you are sufficiently bewildered by the events described so far in this book that you don't know the answer, because I don't either. For six months, I had been training specifically for Gemini 10, with its peculiar

EVA and rendezvous problems, but this period had of course been preceded by six months of Gemini basics during my tenure as back-up for Gemini 7. Was the answer, then, one year? But how about all the Basic Grubby Training, with its emphasis on orbital mechanics, or the jungle training, or centrifuge, or zero-G airplane, or pressure suit work? Could I omit as training my experience in the test pilot and research pilot schools? Or what I learned test flying at Edwards, or ejecting from a smoke-filled cockpit years before? Or the math I learned in school, the foundation upon which orbital mechanics was built? I just don't know how long it takes, except that it had taken me thirty-five years to reach crew quarters at Merritt Island, Florida, and on July 17, 1966, I felt ready. I felt it had taken exactly the right amount of time. I wouldn't have wanted to launch a day sooner or a day later. I knew that tomorrow might bring some shocking surprises, perhaps even a fatality or two, but I felt we had a reasonable probability of success, a fair shot at it, and what more could one ask? I had a nice long chat with Pat on the phone, and then I fiddled with the flight plan for a while, comparing notes with John until 3 A.M. Then I hit the sack and slept like a baby until nearly noon.

Oh, I have slipped the surly bonds of earth
And danced the skies on laughter-silvered wings . . .

—John Gillespie Magee, Jr., "High Flight"*

The familiar pressure suit is there waiting for me, in a modified
house trailer where we will dress. It is not far from Pad 19, where
our Gemini-Titan stands. The medical technician makes annoying
small talk, which I must answer, as he shaves a clearing or two in
my chest hair and tapes his thin disk-shaped sensors to my skin.
Then into cotton underwear with a built-in belt of electronic signal
conditioners, which will amplify my heartbeat for relay to Mission
Control. I am plugged into a machine which certifies that an

---

* John Gillespie Magee, Jr., was an American who joined the Royal Canadian Air
Force and was killed at the age of nineteen, flying a Spitfire during the Battle of
Britain. He was born in China and lived in Washington, where his father was rector
of St. John's Church across from the White House. "High Flight" is without a
doubt the best-known poem among aviators and can be found, elegantly framed or
carelessly tacked, on home and office walls around most military airfields. The poem
in its entirety can be found on page 243.

acceptable signal is being received. While this is going on, a color TV set describes the upcoming rendezvous, using a "Gemini" toy electric train on an inner track catching up with a slower "Agena" train on an outer track. John and I chuckle at the announcer, who makes one absurd oversimplification after another. Then it's time to don a triangular yellow plastic urine bag by inserting the penis into a rubber receiver built into one corner of it. There are three sizes of receivers (small, medium, large), which are always referred to in more heroic terms: extra large, immense, and unbelievable. One selects based on bitter experience in ground tests, where a poor fit in the feet-above-head Gemini seating arrangement means urine trickling up the spine and pooling in the small of the back. "Wetbacks," the suit technicians call these neophyte astronauts, with displeasure.

Now into the suit: first the feet struggling with the turns and twists of the nylon inner liner; then jackknifing torso over double to get arms far enough into place to slip head into the neck ring; then standing and zipping up the back, with help. Next, tight, tight gloves are wormed on and snapped into wrist rings. Then the helmet gently, almost reverently lowered onto the neck ring until, satisfied with the alignment, it is brutally shoved downward, locks snapping and clicking into place. The transition into space begins when the visor is lowered and locked. From now on no air will be breathed, only pure oxygen; no human voice will be heard, unless electronically piped in. Through the barrier of the suit, the world can still be seen, but that is all—not smelled, or heard, or felt, or tasted. Today G-4C-36 feels good. No lurking terror, only smug satisfaction at its familiar clutch, with no lumps, no bumps, no blemishes—an old friend reborn for this occasion.

The outside world has already changed perceptibly. People are self-conscious; there is no spontaneity in their movements. They could flub their lines in rehearsal and laugh, but today the little chuckles are all programmed. While John and I are denitrogenating, regally slumped on brown overstuffed reclining armchairs, messengers file in and out with good tidings. The condition of the spacecraft is announced (superb, of course). The weather is cooperating, despite a little thunderstorm which seems to be drifting by to the north. Late afternoon in July in Florida: if we have only one thunderstorm in the area we are lucky—may it keep moving.

Time to leave for the launch pad now, in a little van; then into the grillework cage of a small elevator which laboriously creaks up the side of the gantry, into the white room. My mother is watching from somewhere nearby with my sister and my brother-in-law. Did they see me before I disappeared indoors again? In the white room all is ordered efficiency, hustle and bustle with a purpose, as preparations are made for its dismantlement. The customary joke is made. Guenter Wendt, the pad leader, presents us with huge styrofoam tools, wrench and pliers four feet long, in mockery of our having broken bits and pieces of the spacecraft over the months. He's not far wrong. Here I am, just a fancy heavy-equipment operator who couldn't fix any piece of this machine if it broke. Even my wife is a better mechanic than I.

Into the cockpit now, with willing hands making the necessary assist, shoving a bit, connecting oxygen hoses and parachute harness and communications lines, and finally lowering the hatches gently upon us. We are isolated at last, in our own little world, with only the crackle of the intercom and the hiss of oxygen for company. The spacecraft oxygen has a slightly different smell, clean and crisp and vaguely antiseptic. I look over at John and smile. Magnified inside his polished faceplate, his nose looks longer and more pointed than usual. It gives him a foxy, crafty look that I like. Perhaps he knows something about the next three days that no one else knows, and he is pleased about it. We talk now, to each other and to the people in the nearby blockhouse and in faraway Mission Control. It is technical chatter, busy work designed to make sure that everybody can hear everybody else, and that all is well with the machine. Final preparations include gimbaling, or swiveling, the two Titan engines which will start us on our way; despite a warning, this test causes a ripple of surprise, for it is unlike any we have done before. After long months, the slumbering beast is finally awakening, with a shudder easily felt as ninety feet below us the two engines dip and sway. My God, it moves! The next surprise is not pleasant. Scanning the busy array of gauges in the tiny cockpit, my eyes stop suddenly on the propellant quantity gauge over in front of John. It's flat on zero! How can that be? With all the checks, and counterchecks, and tests and verifications, could someone have forgotten to fill the tank with our rendezvous fuel? Impossible, it must simply be that something is wrong somewhere in the sensing system. What to do now? I reach over and tap the glass and peer at John. He nods curtly and goes on about his business. I decide to tough it out; if John is willing to launch without this measurement, so am I. To hell with it. It's not five minutes after this momentous decision that I am awash with embarrassment, as suddenly the gauge springs into life and properly shows a full reading. It is simply that, unlike in the simulator, this measurement is not activated until just before launch. I should have known this.

Our Atlas-Agena is off now, we are told, and we breathe a sigh of relief as the Agena reaches a good orbit. We will need it. Although it has departed from a pad a mere mile or so from us, we cannot see it or any other part of the outside world, save a tiny patch of blue sky overhead. It's 81 degrees and the wind is blowing at sixteen knots, but you couldn't prove it by us. We are indoors, lying on our backs looking straight up, with the Atlantic just under our right elbows, our feet pointing north. Our trajectory will carry us straight up for a while; then we will lazily arc over to the east (our right) and will achieve orbit one hundred miles above the Atlantic, lying on our right sides. Time to climb, five minutes forty-one seconds; distance traveled downrange, 530 miles; velocity at engine cutoff, 17,500 miles per hour. Or at least, so we have been led to believe. Right now we are velocity zero, distance traveled zero, prospects unknown. Days have turned to hours, and hours to minutes. No simulation this, no ride back down on the elevator, no debriefing over coffee. Our primary instrument becomes the clock, and finally the excited voice on the radio, trying to sound bored, reaches the end of its message: $10 - 9 - 8$ . . . grab the ejection D-ring between your legs with both hands; one jerk and our seats will explode free of this monster . . . $7 - 6 - 5$ . . . it's really going to happen . . . $4 - 3 - 2 - 1$ . . . engines should be starting—IGNITION—pay attention to those gauges—LIFT-OFF!

A barely perceptible bump, and we are airborne. Fairly high noise level, but we *feel* the machine, rather than *hear* it. Down below the engines shift back and forth in rapid little spastic motions, keeping the cigar-shaped load poised in delicate balance despite gusty winds and sloshing fuel tanks. Up on top we feel this actively in the form of minute sideways jerks. There is absolutely no sensation of speed, and only a moderate increase above one G as we are gently pushed back into our contoured seats. I am dimly aware that a thin overcast layer above us seems to be getting closer when *pow* we burst through the wispy clouds in brief but clear contradiction to the seat-of-the-pants feeling of standing still. Goddamn, we are moving out! As the G level begins to build, so does a choppy, buzzing vibration, not side to side now, but fore and aft. This is the so-called POGO motion, and we are expecting it; it is no surprise and no discomfort, causing only a high-frequency quivering of body and instrument panel, which makes the dial faces appear slightly out of focus. In fifty seconds we pass our ejection seat limit and I loosen the death grip on my D-ring. Noise and vibration increase sharply as we approach Mach 1; then there is an abrupt smoothing effect as we reach the supersonic domain in the thin upper atmosphere.* The G level is getting noticeable now as the first-stage fuel tanks are nearly empty, but the two first-stage engines are still churning away at full thrust. "Staging" (the shutdown of the empty first stage, separation from it, and ignition of the second-stage engine) nears, as the clock approaches two and a half minutes and the G meter creeps up over 5.

Staging is a shock. Too many things happen too swiftly for the

---

* For the technical reader, the typical Gemini-Titan reaches a maximum dynamic pressure (MAX Q) of 750 pounds per square foot, at an altitude of forty thousand feet and a speed of 1.5 Mach number, one minute and twenty seconds after lift-off.

brain to render a verdict. The eye barely has time to register catastrophe and rescue: the G load abruptly ceases, and I feel myself flying forward against restraining straps. The window is instantaneously full of reds and yellows and bright particles and whizzing pieces of debris, and then, as quickly as chaos has come, it evaporates, leaving black sky and quiet ride as the second stage hums serenely along. On the ground Pat watches her TV screen and thinks that the vehicle has exploded. She is right. An instant after the two stages separated, the first-stage oxidizer tank ruptured explosively, spraying debris in all directions with dramatic, if harmless, visual effect. Back in the cockpit we have no time to discuss the matter. John, up for his second time on a Titan, knows this one is different, but not me—I luxuriate in my ignorance and begin to enjoy this ride.

We are now far above any aerodynamic disturbances, and are in fact "in space," except that we don't yet have the speed to keep us up here. If the engine quits now, we will ignominiously fall back down into the ocean. But the speed is coming, and the G forces herald its approach, as we build up past 5 and 6. At 7 Gs the discomfort begins, with a giant hand pushing hard against my chest, but it doesn't last more than a few seconds, because we have arrived. Precisely on schedule the second-stage engine shuts down (SECO); we separate from the spent Titan and check our velocity. We are a paltry twenty miles per hour slow and we immediately correct for it by using our own thrusters to make up the difference. On the ground our CAPCOM, Gordo Cooper, throws the wrong switch and broadcasts to us a conversation intended to stay within Mission Control. "All personnel, we're going to debrief in the Ready Room." With what I hope is the proper tone of sarcasm I say, "Sorry we're going to miss that debriefing down there." Gordo ignores me.

John and I have our check lists out now, and we swiftly run through the routine we have practiced so many times in the simulator, converting our Gemini from a passive payload to an active, orbiting spacecraft. We must hurry, because in seven minutes we will be in total darkness and the first night is jam-packed with star sightings to use in our orbit determination (Module VI) calculations. The test pilot in me winces at the necessity to rush at this stage of the game: here we are with a brand-new machine airborne for the first time and we don't even have time to make sure it is operating properly. Everything is rush—rush—rush to get ready for Module VI. The tourist in me is equally outraged, for outside my tiny window is the most glorious spectacle of sea and sky I have ever witnessed, and I have no time for it. I press my nose briefly against the glass, but John—half joking, half serious—allows me only a few seconds before ordering me back to work.

Inside the cockpit I get my first manifestation of weightlessness: tiny bits of debris, washers, screws, pieces of potting compound, dirt particles—a miniature armada—are floating aimlessly about. In an hour or so they will be gone, sucked into the inlet screen of the ventilation system, but for the time being they are an amusing oddity as well as a sober reminder that this machine has been assembled by fallible hands which drop things which find their way into inaccessible and possibly dangerous crevices. My body doesn't feel much different now that it is finally weightless. A slight feeling of fullness in my head, and a tendency for relaxed arms to float up in front of my face, praying-mantis style. That's about all, confined as I am inside the tiny cockpit.

We have taken off from Cape Kennedy out over the Atlantic, with our first landfall the coast of Angola, in Southwest Africa. We are far short of it, not yet across the equator, when we plunge suddenly into darkness. In orbit, traveling three hundred miles a minute, dawn and dusk come with a startling rapidity that compresses all the old familiar color changes into a very dynamic minute or two. Sunrise is the more dramatic of the two, as the eye which has become accustomed to darkness is forced into blinking attention by the sudden stabbing light in the east: first golden, then fiery orange, then incandescent white. But darkness can sneak up on you until, like an old lady fumbling with her knitting, you realize you can't see and have to turn on the light.

Darkness is also the signal for me to unstow my sextant and begin my tussle with Module VI. My first star is Schedar, in Cassiopeia—that W-shaped constellation in the northern sky. Schedar is the right-hand lower point of the W, and is no problem to find. My job now is to measure the angle between Schedar and the horizon half a dozen times or so. Each time I will tell my computer what the angle is, and the computer will learn from this the exact height of the visual horizon, as it appears to me, and use this number in determining more accurately what the diameter of the earth should be considered to be in order to make my orbit determination calculations as accurate as possible. My first shock comes when I try to find the horizon: somewhere in that homogeneous void, black sky changes to black water—that I can see, grossly, by the fact that the stars stop at a certain point, but they seem to ooze into oblivion without any sharp line of demarcation. A perpetual optimist, and only twenty-three minutes into the greatest adventure of my life, I am nonetheless compelled to face up to it. "John, I hate to be pessimistic, but I think this is going to be sort of bad." John gives me as good an answer as anyone can: "Well, do the best you can." After some experimenting I find I can watch the image of the star through my sextant as it approaches the horizon. There is a faint muddy band, the airglow, into which the star, descending, almost disappears, but if I watch closely enough, it re-emerges in a very narrow zone of visibility before being snuffed out by the atmosphere. It is this instant of snuffing that I must capture, as accurately as possible. When I do I yell MARK, and John pushes a button. Then we must turn up the lights long enough to read the sextant angle, and manually punch it into the computer, digit by digit. I am getting the hang of it, but the process is slow and we are running behind by the time we have finished with Schedar and are ready for Hamal, our second star. Hamal, in the constellation Aries (the brightest star in the ram's horn), has a nice ring to its name. I don't know what it means in Arabic, but I always confuse it with the horse thief of Kipling's "The Ballad of East and West": "Kamal is out with twenty men to raise the Border side, / And he has lifted the Colonel's mare that is the Colonel's pride: / He has lifted her out of the stable-door between the dawn and the day, / And turned the calkins upon her feet, and ridden her far away."

Now Hamal turns out to be as elusive as Kamal. I can find it, with my naked eye, but when I put my sextant on it and start working it toward the horizon, it mysteriously fades away. Valuable minutes are slipping by, and I finally get one measurement at the last possible moment. Then a quick and sloppy try at Vega and Altair before night ends and we break into brilliant sunshine near the western coast of Australia. John is jubilant. "Isn't that beautiful!" Deep in Module VI calculations, I can only grumble, "Don't even tell me about it, babe. I'll look tomorrow."

The ground isn't concerned about Module VI and they will never share John's view, but they have been following our progress and now remind us there is another dimension to the flight. "Your O_2 tank pressure is dropping a little bit. Try to keep an eye on it." They also want to know how our fuel cells are doing. I haven't had a minute to spend with these temperamental twins that magically convert oxygen and hydrogen into electricity and water, but a quick check tells me they are getting along fine without me. As usual, one seems to be loafing a bit, but the stronger of the two is capable of picking up the load. All the other systems seem to be equally healthy, which is fortunate as we have no time to troubleshoot systems if we are to continue doing our own navigating with Module VI. By the time we reach Hawaii, I have finished some preliminary calculations, and I compare my answers with teacher's. The ground-supplied solutions are close to mine in some cases but far apart in others. Generally I am predicting a series of maneuvers to take place at approximately the right times, with approximately the right magnitudes, to get us up to our Agena, but I am wrong on our lateral, or out-of-plane, position. Worse yet, I can't seem to find out why we are different, no matter how many times I go back over the calculations.

As we sweep in over the Baja California coast, I take a brief respite from my mathematical chores to unpack and install our cameras in their brackets, and take some planned photos of terrain and weather over the southern United States. With a solid undercast, they won't be worth much. Now, some hour and thirty-seven minutes after lift-off, we have completed our first orbit—and a busy one it has been. Instead of being back over Cape Kennedy, the world has slipped sideways a bit beneath us and we are over Key West as we begin our second orbit at dusk. Now back to the sextant with different stars (Fomalhaut and Arcturus—what mysterious names) but similar results. I have great difficulty manipulating the sextant, using the nearly invisible horizon. By now we have agreed to use the ground's computations rather than our own, in making the maneuvers necessary to overtake our Agena, but despite this defeat I am carrying on with my Magellan act, and I am still filling charts and graphs with scribbled numbers, and am frantically punching instructions into our computer. John decides I need cheering up. "I think you are doing a tremendous job." "Oh, shoot, I'm doing a lousy job," I reply, and mean it. John blames it on the universe. "What the heck! If you can't see the stars, you can't see them. I've been telling you this for six months."

At two hours and ten minutes we fire our thrusters briefly to get in proper phase with our Agena target; at 2:30 we fire again to change our orbital plane slightly to coincide with Agena's; at 3:48 we ease into a circular orbit fifteen miles below it. The ground has coached us on all these maneuvers. At 4:34 we make a major move on our own, using our radar and computer to calculate an intercept trajectory which will cross our path with the Agena's at a point in the sky one third of an orbit later. We correct our course twice along the way, at preplanned intervals, to stay on this path. Everything looks good as Agena grows from a mere flashing light to a barely discernible cylinder. John is flying and I am feeding him helpful tidbits of information.

The two things he is most interested in from me are the distance from us to Agena (range, or simply R) and our overtaking speed (range rate, or $\dot{R}$, pronounced R dot). R and $\dot{R}$ must each reach zero at the precise instant of docking; before that time, there is a complex relationship between the two. On the one extreme, if $\dot{R}$ exceeds R by an unacceptable margin, we will whiz by our Agena, unable to stop in time. At the other end of the worry list is the situation where $\dot{R}$ is allowed to get too small too fast, and this will cause us to screech to a halt short of Agena, at which point orbital mechanics dictates that we make a little pirouette in the sky, and the chase of the elusive Agena begins anew.

I'm not fond of either option, and I sing out the numbers as John calls for them. "What's $\dot{R}$ now?" "I have, right now, eighty-three [feet per second]." "What should it be?" "Should be about seventy." "O.K. We'd better take out some, right?" "Yes." "I'm going to brake, babe." "Yes, brake. Range is two miles. You've got a sixty-five $\dot{R}$ at two miles." "One point eight miles, $\dot{R}$ sixty-two. If the cockpit lights bother you, let me know. $\dot{R}$ forty-nine, John." "O.K." "Forty-four $\dot{R}$, and a range of 1.4 miles." "Forty-one $\dot{R}$ . . . and range a little over a mile." "O.K., $\dot{R}$ twenty-nine, range .8 of a mile." "Twenty-two $\dot{R}$, John. Nice magical number, right?" "Yes . . . what's $\dot{R}$?" "Eleven." "O.K." "Three tenths of a mile." "O.K." "What's the $\dot{R}$ now?" "Eleven still, holding eleven, nice number . . . $\dot{R}$ nine, seven." "O.K." "Nine, nine, .2 of a mile." "Seven hundred feet . . . 660 feet . . . holding 660 . . . six hundred feet . . . six hundred feet holding." Something is wrong. We are off to one side and turning, but not getting any closer. $\dot{R}$ vacillates in the vicinity of four or five feet per second. Good Lord, we are doing a whifferdill around the Agena, an out-of-plane whifferdill, just like the ones we used to do in St. Louis in the simulator. John doesn't like it. "Whoa, whoa, whoa, you bum! Is that darn thing rolling or are we?" "We are."

John and I both know what has to be done now. We have sinned. We have somehow wandered off our prescribed path of righteousness, albeit only a tiny little meander, but now we must pay the penalty, and that means fuel. With the physics of our motion relative to the Agena now working against us instead of for us, John has no choice but to hose out whatever fuel is necessary to cut short our helix-shaped circumnavigation of the Agena, driving us right up to it and stopping next to it. The "brute force method," John will call it at our post-flight conference, an apt description, but he will not be able to explain why this problem occurred. Part of the difficulty, we will find out later, can be traced to an improperly aligned inertial platform, giving us an initial out-of-plane error. At any rate, we are finally parked next to our Agena, right on time, but with only 36 percent of our fuel remaining instead of the 60 percent we should have at this stage of the game. This is bad news. We are not sure exactly what will have to be deleted or abbreviated, but clearly we will not be able to do everything we planned with only 36 percent remaining.

It is not more than a couple of minutes before the ground registers polite interest. "Could you give us a propellant quantity read-out please?" "We're reading 36 percent." "36 percent?" . . . "Roger." We are in no mood to elaborate on it. We don't like talking about whifferdills . . . "Thank you," they say, but finally they can't resist tut-tutting a bit at the heinous number. "O.K. You seem to have used a tremendous amount of fuel—the propellant between RKV and CSQ* . . ." Goddamn it, we know that, the question is where to go from here. How do we salvage all we can of our experiments and EVA activities and the rendezvous with our second Agena? This is the sort of problem that is right down Mission Control's alley, and we have great confidence that somehow they will sort it all out. Meanwhile, we are busy with other things: for the second time in history one spacecraft docks with another, as John smoothly and expertly guides our nose into the docking collar on the front of the Agena with nary a bump. He finds it as easy as Neil did on Gemini 8. While John is playing pilot with great finesse, I am pretending to be a digital computer with my Agena encoder: 030–021–250–140–211–070, I tell the Agena, a short eighteen-digit message. I will send it 144 more digits before

* RKV = Rose Knot Victor; CSQ = Coastal Sentry Quebec. These are two ships used as tracking stations in the worldwide Gemini net. They fill gaps which would otherwise exist in the South Atlantic and western Pacific.

bedtime. It is our engine now; we will need its fuel to boost us to its brother—the Gemini 8 Agena somewhere up there above us. The 8 Agena orbits inertly at an altitude of 250 miles. It has no lights and will not respond to our radar. The ground knows where it is, approximately, and we will use the power of our own Agena's engine, as calculated by the ground, to guide us to a spot where we can see it ourselves; then we will be on our own.

The first step in this overtaking process is an unusual one, and will take John and me higher than man has ever ventured before— 475 miles. The purpose of the maneuver, however, is not to set a new world's altitude record, but rather to establish the proper timing of our orbit relative to 8 Agena's orbit. We need to slow down a bit, for a while at least, and that means we must go high, because higher is slower, remember? The Agena engine, with its sixteen thousand pounds of thrust, requires only fourteen seconds of burn time to boost us up to 475 miles. The engine is mounted on the far end of the Agena (we are docked with the near end), so that we can't quite see it. We will recognize its presence, however, as its thrust will tend to push us up against our instrument panel, with an acceleration of one G, eyeballs out. The burn will take place over Hawaii, so that our peak altitude (apogee) will occur 180 degrees later, on the opposite side of the earth from Hawaii, over the South Atlantic. As it happens, this region, known as the South Atlantic anomaly, is where the inner Van Allen radiation belt dips down, and at 475 miles we will be grazing the bottom of it. This we have known for months, but how hazardous our foray will be is not known with precision,* so that the on-board measurement of our radiation dose rate and total dose will be items of great

---

* The medics have predicted a total dose of nineteen rads, which supposedly is safe enough. One big problem in predicting is that the intensity of the inner Van Allen belt is not constant, but has been decreasing at an unknown rate since July 1962, when a United States high-altitude nuclear test sent the count of trapped electrons soaring. Besides, what is "safe" or not is an exercise in statistics rather than any well-defined threshold of threat. The effects of radiation are so poorly understood, vary so much with individuals, and can take so long to develop that no one can say with any certainty how much harm is being done with a given dose rate. I only know that if I ever develop eye cataracts, I will try to blame it on Gemini 10.

interest. At seven hours into the flight I read our total dose as zero, in response to a ground request.

At seven hours and thirty-eight minutes we will light the Agena engine. In the meantime, we must stow cameras and other loose items so they will do no damage (to themselves or to us) when the engine lights. My job during the burn will be to monitor a row of instruments and lights mounted on the front end of the Agena and, if any dangerous abnormalities within the Agena are indicated, to shut down the engine by flipping a switch. I must also keep one eye on the clock, and if the engine doesn't shut down on time, I will immediately command it to shut down, because we don't want to go any higher than planned, mainly because it will foul up the timing of our strategy for overtaking the 8 Agena, and also because it would put us deeper into the Van Allen radiation belt.

As the time approaches, everything looks good, both to us and to the ground, which sends us one last admonition, in the form of a question: "Have you fastened your restraint harness?" "Roger, we're tightened down." This situation is the opposite of any other I have experienced, in that the engine is in front of us and will thrust toward us, pushing our seats out from under us. The engine start sequence is automatic, once I have observed the preliminary amenities by sending the Agena a long string of three-digit commands, ending with 501. Receiving these, the Agena goes into an eighty-four-second routine which will culminate in engine ignition, I hope. We are right on schedule as I send the necessary instructions, and then all we can do is wait. I alternate between checking the clock and looking out the window at the status display panel, as the lights and dials on the front of the Agena are called. At the appointed moment, all I see is a string of snowballs shooting out the back of the Agena in a widening cone. The unexpected white stream is quite pretty against the black sky. Aw shucks, I think, it's not going to light, when suddenly the whole sky turns orange-white and I am plastered against my shoulder straps. There is no subtlety to this engine, no gentleness in its approach. I am supposed to monitor the status display panel, but I cannot prevent my eyes from wandering past it to the glorious Fourth of July spectacle radiating out from the engine. Out of long habit, however, I check my instruments, and all seems to be going well inside the cockpit, which in this case simply means that fourteen seconds have not yet elapsed. We are swaying mildly back and forth (that is John's job to measure), and as the clock (finally!) passes through fourteen seconds, I send a command for this raucous engine to cease. At that very instant, it has come to the same conclusion, and we are now jerked back into weightlessness and treated to a thirty-second barrage of visual effects even more spectacular than the preliminaries. It is nearly sunset, with the sun directly behind us; it clearly illuminates each particle, spark, and fireball coming out of the engine, and there are plenty of them. Some are small as fireflies, others large as basketballs; some depart lazily, others zing off at a great speed. There is a golden halo encircling the entire Agena that fades very slowly. I whip out my camera and take a couple of shots,* and then compare notes with John. "That was really something!" He agrees. "When that baby lights, there's no doubt about it." Perhaps because of the bizarre arrangement—from a pilot's point of view—of being able to see your own engine light and push you off backward, or perhaps because all my senses were magnified by the day's activities, or perhaps because of the sudden onset of G forces after seven hours of weightlessness, I don't know the reason, but that fourteen seconds seemed an eternity, the acceleration seemed much greater than it was supposed to, and the slight change in direction appeared a major shift. It all added up to a very uncomfortable feeling, and I now convey this to John by saying, "I almost shut it down, I almost did." "No, you didn't." "I almost did. If you had said shit, I would have shut it down. Really." John,

---

* For some reason, the pictures do not show the globs and particles, but the generalized glow emanating from the Agena's engine is clearly visible and is quite pretty as it radiates out several feet before fading into blackness. It appears more platinum than the golden color I recall. As with a dying dolphin, the colors fade rapidly, and their subtleties, while clearly discernible to the supersensitive instrument we call the eye, are unfortunately not captured by the rather crude emulsion of the film.

who doesn't miss a thing, claims incompetence. "I was too busy hanging on the wall." We agree that the closest thing to this swift onset of thrust that we have experienced before is the afterburner of the J-57 jet engine. If improperly adjusted, one can get a "hard light" when the throttle is moved to the afterburner position. In some of the early F-100 Super Sabres, at least, the light can be hard enough to knock your feet off the rudder pedals.

But now we are high above the Super Sabres and are beginning a shallow, one-half-orbit climb from 180 miles to 475, all because of the energy added to our orbit during that fourteen seconds. The ground fusses at us for radiation meter readings and can't believe the tiny numbers we read to them. At eight hours and nine minutes we have accumulated only .04 rad; by 8:20 it's .18 rad; finally, at 8:37, they accuse us of having the device turned off. "We're wondering if your dosimeter is still snubbed." "No, it's not still snubbed. It's reading .23 rad." "O.K. It looks like the . . . rate is less by a factor of about ten, and there is no sweat down here on that." Good. They are happy, we are happy, this unbelievable day is drawing to a close.

It's two o'clock in the morning at Cape Kennedy, and it's time for us to grab some food and some sleep. However, we are a glum pair, for as the excitement of the wild ride on the Agena subsides, the ground puts things in perspective. "We'd like a propellant quantity read-out . . ." "Roger. It reads about 32 percent now." Two thirds of our fuel gone already, and so many things left to do! Perhaps we can cling to this Agena for longer than we had originally planned, and use its gas instead of our own, but some of our experiments cannot be done properly while attached to Agena, and besides, we have to get rid of it before we approach the Gemini 8 Agena, because its engine cannot handle the delicate and constantly changing demands for thrust during the terminal phases of the rendezvous. I have a sick feeling that one or both of my EVAs may be canceled, and that will really hurt.

Speaking of hurting, my left knee hurts, a throbbing ache that began a couple of hours ago, gradually worsened, and is now holding steady at a moderate but very uncomfortable level of pain. I think it is nitrogen coming out of solution in the tissues and creating little bubbles which press on my nerves. For some reason, joints—particularly elbows and knees—are the most sensitive to this form of the bends. The reason I make this diagnosis is that the pain is exactly like ones I have felt before in altitude chambers. True, I denitrogenated for a couple of hours before lift-off, but the time required varies considerably with the individual, and I must be more sensitive than most—or perhaps some air leaked into my suit on the launch pad as I transferred my hoses from the portable oxygen supply to the spacecraft oxygen supply. That's all irrelevant anyway; the point is what to do now. Discuss it or try to ignore it? I have a vivid picture of the avalanche of medical conferences one quick complaint will produce. It will cause everything short of a house call. At a minimum, it will result in a stream of excited radio transmissions lasting half the night. This I don't need. What can they tell me to do, besides take a couple of aspirin? I take two without calling. If this is the bends, then time is on my side, and within a couple of hours the nitrogen bubbles should disperse.

John is morose and even less chatty than usual. When he does say something, it is about the rendezvous and his perplexity over the excessive fuel consumption. He can't figure out why it happened, and I am no help, for two reasons. First, counting his stint as Tom Stafford's back-up on Gemini 6, John has forgotten more about rendezvous than I have yet been able to learn; and second, my contribution to the rendezvous—the keeping of an elaborate set of charts—is limited to the fore-aft and up-down domain, and does not include the left-right plane in which we made our whiffer-dill. Between John's rendezvous anguish and my fear of EVA cancellation, not to mention my goddamn knee, it's pretty glum up here. As I prepare the cockpit for our eight-hour sleep period, I think, Jesus, this really *is* a hostile environment whose laws are terribly unforgiving of mistakes, and Lord knows, we have plenty of opportunities for mistakes tomorrow and the next day.

In addition to my knee, which still hurts through the aspirin, there are two other impediments to sleep: my hands and my head. My hands dangle in front of me at eye level, attached as they are to relaxed arms, which seem to need gravity to hold them down. My worry is that just past my hands is the instrument panel, festooned with toggle switches whose position is important and must not be changed during the night. What if these ridiculous floating hands bump a switch or two? I must put these hands somewhere out of harm's way, but where? I try behind my back, but with a pressure suit on, that is uncomfortable. I consider stuffing them in my mouth. In the meantime, I am having head problems as well. It just doesn't seem right to go to sleep without a pillow or something substantial for the head to rest on; yet here I am floating free with only an occasional little bump as my head touches the hatch and bounces off.

John is ready to put the lights out, which, in addition to dousing the cockpit lights, means fastening a thin metal plate securely over each of the two windows, since over half of each orbit will be spent in blazing sunlight. We get a nice surprise as we unpack the plates and find some kind soul has pasted a photo on each, photos of two voluptuous, wildly beautiful girls. They are shocking intruders into this masculine little cubbyhole full of machinery, and they seem weirdly out of place; but if we can't have darkness without girls, well, that's life in orbit. Up on the windows go the girls, and in the pitch-black cabin I fidget around until finally I find a position that allows me at least to doze fitfully. I have turned slightly to my right, away from John, and have wedged my head high up in a corner in simulation of pillow pressure. I am still not happy about my dangling hands, but I decide my worries about inadvertently flipping switches are silly, and I try to ignore my hands and my knee. After a couple of hours of this, I pop two more aspirin and try again. This time I do drift off to sleep for perhaps two hours, and when I awake I note with relief that my knee pain has nearly disappeared. I can't go back to sleep, but there's no point in waking John, so I sit there and run over the coming day's activities in my mind.

Basically it is a day for experiments, some of which I will perform while standing up in the open hatch. There is also a lot of undocking and docking with the Agena, to give John and me practice, and there are a couple of maneuvers to adjust our orbit and align it more precisely with that of the 8 Agena. The docking practice is an unnecessary frill, and a gas waster, so we can delete that, but if we stay stuck to the Agena all day, some of the experiments will have to be modified. The ground has had all night to work on that problem, and I only hope their solution does not delete the stand-in-the-hatch EVA. John is clattering around over on his side now and is obviously awake, so we remove the window shades and peer out at the world. About all we can see is a tantalizing little slice of it, for the Agena looms large in our windows and blocks out nearly everything. To make it worse, the Agena rigidly aligns itself with the horizon, so that our heads are always pointed up, giving us a grand view of the black void of space but precious little opportunity to observe any real estate whizzing by below. As John aptly points out, riding the Agena is not unlike being stuck behind the engine of a train and having to peer down the track on either side of it.

By the time we have stowed and unstowed a few items of equipment, checked our systems, eaten breakfast, and washed up slightly, a couple more hours of our "sleep" period have passed, and it is now time for Houston to start calling. They do, and John answers. We are off to a brisk start. "Roger. Good morning, John. We'd like a crew status report and a radiometer reading, please." "Roger. Crew status is GO. Gun counter reads 335.* We slept pretty good last night.† Radiometer reads .78 rad and the dose rate is off-scale low." Point seventy-eight! That is much, much less than anyone would have guessed, and it won't go much higher than

---

* This refers to our water supply, which comes to us in half-ounce squirts through a water gun. Each time we press the trigger, to drink or to fill a dehydrated food bag, the counter clicks once. The ground wants to make sure that we are drinking enough water, but with only one counter for two men, it's impossible for them to tell who is doing the drinking.
† The liar!

that, because in two hours we are going to burn the Agena engine again and get back down out of this high orbit, back down below the Van Allen radiation belts, back down so that we may overtake our 8 Agena from below. Purely as an accident of timing, we have found it necessary to come up here to 475 miles, and as a casual by-product, John and I now hold the world's altitude record. I don't know whether to laugh or to cry, when I think of all the pioneer aviators who have aspired to this record and who have put their reputations, money, and lives into seeking it, and now John and I are handed it on a platter.

When the burn comes, again the Agena kicks like a mule, and John is still impressed. "It may only be one G, but it's the biggest one G we ever saw. That thing really lights into you!" Our new orbit is 185 by 240 miles, and before the day is over we will circularize it at 240, approximately eight miles below the 8 Agena, twelve hundred miles behind it, and closing slowly. We have deleted the docking practice, so our next task will be the stand-in-the-hatch EVA. The main reason for it is Experiment S-13, which involves taking the ultraviolet signatures of selected hot, young stars. Since the spacecraft window glass does not transmit ultraviolet light (to protect our eyes), we must operate our 70-mm. camera with the hatch open. There is a bracket into which I will snap the camera mount. Then a special timer will allow twenty-second exposures to be made with precision. Unfortunately, I tried to plan ahead and assembled the camera, mount, and timer before the Agena burn, assembled them and stuffed them down into the footwell, out of harm's way. However, when the Agena engine ignited, it plastered camera et al. up against the bulkhead and broke off the stem of the timer inside the camera. No problem. John will act as timer and I will manually hold the shutter open for the required twenty seconds.

While John and I are discussing this, and are taking care of the hundred and one small items that are necessary preludes to dumping cabin pressure and opening the hatch, suddenly we get a surprise visitor. On the radio comes the gravelly voice of our boss,

Deke Slayton. "John, this is Deke. You guys are doing a commendable job of maintaining radio silence . . . Why don't you do a
little more talking from here on?" This may not sound like much
of a censure, but in our world this is a BIG DEAL. Deke, who is close-
mouthed himself, comes on the radio only in extremis, and he
must have caught a huge ration of abuse around Mission Control
to actually incite us to chatter. The ground thinks somehow that
the 36 percent fuel would miraculously be restored to 60 percent if
only their participation in the rendezvous had been greater; they
think that the upcoming EVA may provide only ultraviolet photographs, but the reporters at the news center in Houston are not
going to be satisfied with a vague promise of scientific results to be
published at some future date: they want hard news, they want
quotes, and they want them right now. The American public has a
RIGHT TO KNOW! Never mind that we are busier than two one-
legged men in a kicking contest. Never mind that we have been
given four days' work to do in three, and have tried our best to
compress it all—systems management, Agena commands, on-board
navigation, myriad experiments, double rendezvous, the whole
bit. Never mind all that—we are not talkative enough, and we have
been commanded to speak. John is pissed. "O.K. What do you
want us to talk about?" Deke backs off. "Well, anything that
seems appropriate. Like EVA." John rubs it in a little bit. "All
right. Mike is talking right now as a matter of fact." Mike is not
only talking, he is singing like a canary, prating endlessly about
things that are better left for a terse sentence or two during the
debriefing after the flight. I describe the Agena, I describe a broken
movie camera and ask advice in fixing it, and Houston responds
with a description of the Houston Astros in a series with the New
York Mets. Jesus Christ! Here I am asshole deep in a 131-step
EVA check list and they want to talk about baseball! One little
boo-boo at this stage of the game and all the oxygen will depart my
suit and I will die, and they will be talking about the color of the
infield grass, and I will have to interrupt them to describe my last
gasp, just in time to make the deadline for the city edition.

Actually, this EVA shouldn't be that tough, but a little melodrama usually doesn't hurt a performance. All I have to make sure of, really, is that I have properly installed extension hoses in the intake and outflow lines, hoses long enough to allow me to stand up in the open hatch while still connected safely to the spacecraft's oxygen supply. Since I will open the hatch at dusk and egress in total darkness, it may be a bit tricky, but hopefully no big surprises are in store. The main problem for John and me will be handling the equipment in our pressurized suits, making sure that the appropriate camera settings, time schedules, and so forth are observed. Promptly at sundown, we dump our cabin pressure by opening a small valve. Like letting water out of a bathtub, it takes a little while, and we don't dare open the hatch until the cabin pressure reads zero, because otherwise it will be blown out of my hands with considerable force and possibly damage itself. When the hatch does open, it swings freely, and with one hand I am able to move it all the way open with only light pressure. This is good news, because hatches have been known to bind, and since a sealed hatch is necessary for a successful re-entry, fooling with the hatch is always accompanied by a bit of nervousness.

It is pitch black now as I cautiously emerge, waist high, turned slightly to my left. The nose of our Gemini is pointed south, aligning the camera (which I have now inserted into its bracket with some difficulty) with the southern constellation Centaurus. John, with his clock, is acting as my timer, and I hold down the shutter mechanism for twenty seconds on each exposure. Something doesn't feel right on my left side. I feel restrained, somehow pulled back into the cockpit. Suddenly it occurs to me that my left shoulder harness is still engaged, a remnant of the precautions taken during the last Agena burn, and now I fiddle with the release mechanism. Unable to see it, I finally locate it among the thicket of wires, tubes, and straps on my chest, give it a jerk, and release it. Now I feel more relaxed, and as the initial shock of finding myself "outside" subsides, I look around and like what I see. I am pointed east-southeast, roughly facing our direction of travel, with the

Agena to my right and the adapter section of the Gemini to my left. My first impression is a feeling of awe at the wide visual field, a sense of release after the narrow restrictions of the tiny Gemini window. My God, the stars are everywhere: above me on all sides, even below me somewhat, down there next to that obscure horizon. The stars are bright and they are *steady*. Of course I know that a star's twinkle is created by the atmosphere, and I have seen twinkle-less stars before in a planetarium, but this is different; this is no simulation, this is the best view of the universe that a human has ever had. Down below the earth is barely discernible, as the moon is not up and the only identifiable light comes from an occasional lightning flash along a row of thunderheads. There is just enough of an eerie bluish-gray glow to allow my eye to differentiate between clouds and water and land, and this in turn allows motion to be measured. We are gliding across the world in total silence, with absolute smoothness; a motion of stately grace which makes me feel God-like as I stand erect in my sideways chariot, cruising the night sky. My only complaint is that the protective coatings on my visor do not allow an even more spectacular look at the stars. I crank my head around to the left, toward the north, and seek out the familiar "seven sisters," the Pleiades. On a good night in the Arizona desert, the naked eye can count eleven or twelve rather than seven, but now I am disappointed to discover that only seven make it through the filter of my visor. On the other hand, Venus appears so bright that I have to convince myself that it really *is* Venus, not by its appearance, but by its position in the sky, at the spot where Venus should be.

The camera work is going well; as dawn approaches we have a full roll of twenty-second exposures of the ultraviolet secrets of the southern sky, and we are ready to proceed with the next task. Again it is photography, but of a totally different variety. With all the grandeur of the universe to record on film, we have—believe it or not—brought along our own object to photograph. It is a titanium plate, about eight inches square, divided into four colors: red, yellow, blue, and gray. It comes with its own bracket and three-foot extension rod. I attach it to my camera, and I am supposed to take pictures of the plate in direct sunlight at various exposure settings. The idea is, by comparing my pictures with earth laboratory pictures of the same plate, to determine emulsions and developing processes which will most accurately reproduce the true colors of space. This is deemed to be of great importance in scientific analysis of the photos which will be brought back from the lunar surface. John and I think there are easier ways of getting this information than photographing this stupid plate, and hence this is not our favorite experiment, but as dawn approaches I am diligently fiddling with camera, film, plate, and rod, getting everything ready.

The sun comes up with its usual fierce burst of piercing, white light, and as it does my eyes begin to water. By the time I have taken a couple of exposures the watering is really getting bad, despite the fact that I have lowered my sun visor and retracted my head, turtle-like, down into my suit to keep my eyes shaded. Why should my eyes be watering? The only thing I can think of that makes Gemini 10 any different from its predecessors is that, following Gene Cernan's visor-fogging problems on 9, a special anti-fog compound was added to our equipment list. It comes in wet wipe form, and I have liberally rubbed one damp pad of it over the inside of my visor just prior to this EVA. My theory is that somehow the sunlight has caused this compound to vaporize and form a gas which the eyes find irritating. It should get better shortly, as the oxygen flowing over my face will diffuse it. Meanwhile, however, I can't see the f-stop markings on my camera, so I pass it back in to John and ask his help. "I've got a problem here, John." "What's the matter, babe?" "Well, as soon as the sun came up, my eyes started watering, and I'm not sure whether it's this compound that's on the inner surface . . . or what it is. But my eyes are really watering like crazy, to the point where it's real difficult to keep them open to see what the heck I'm doing. I'm serious." John thinks it's the sunlight. "Don't look at the sun." "Well, I'm not . . . I've got my eyes closed right now and . . . you know, I can tuck my head in like a turtle and get down inside the suit . . . my head is in the shadow from the sun." Now John slips me *his* news. "Mine have been watering too, babe, the whole time." "Is that right?" "Yes." Grand! That's all we need, two blind men whistling along with the door open, unable to read check lists, or see hatch handles or floating obstructions that impede hatch closing. I sound like a broken record. "My eyes are watering." Maybe if I say it often enough, the problem will disappear. John is cheery. "Mine are too, Mike. I can't see a gosh-darn thing." "O.K., I'm going to heave this bracket. It's gone." John gets the message. "O.K. Come on back in. Let's close the hatch." "All right." If we can. I grab the familiar handle and give a heave. "O.K., here comes the hatch, babe." "Having trouble?" "No." Beautiful, that hatch is beautiful. As I wedge my body down as low as possible, I feel the hatch engage properly, and I now unstow the locking handle and crank it shut. Somehow we find the right valves and soon our little cabin is filling up with oxygen again. John is still having problems. "I can't see anything." "Just close your eyes, John. It goes away after a while if you close your eyes." John is dubious. "Can you see?" "Yes, I can see just fine. You just don't sweat a thing, buddy. Cabin pressure's just lovely. It's coming back up, cabin locks, door locks easily and all that." "Just all of a sudden, Mike. It must be something in the oxygen circuit." "Yes, I'll bet you it's this stuff they rub on the inside of the visor. Or something in the lithium hydroxide, maybe . . ." "I don't know, boy, it really hurts. I'll bet that's what it is, lithium hydroxide." "O.K., pressure's up to 2 psi and the oxygen pressure's holding good." "O.K. I'm sorry, Mike, but I can't see a thing." "You just sit there and don't sweat it." "Can *you* see anything?" "Yes. I can see pretty well now." And I can; it gets better every minute. Now we are finally in contact with a ground station and explain our problem, and they ask us a hundred questions, but none of us can say with any certainty what caused the problem.

Lithium hydroxide is what we use to absorb the carbon dioxide we exhale. The more John and I discuss it, the more convinced we become that it is the culprit, but we cannot find any granules of it as proof. Besides, it is supposed to be contained inside a canister, and if it were leaking, would we not have noticed it before now? Fortunately, there isn't time to worry about it, for our workload of experiments must continue. We can both see well enough now; we must get on with it. Still . . . it's not a very healthy feeling, and I can't help but wonder what would have happened if I had been out on the end of a fifty-foot tether, blinded suddenly, instead of just standing in the hatch. I hope we can work out something before tomorrow.

In the meantime, I am fiddling with the stars again, this time in a navigation experiment which is designed to overcome the difficulties of finding the horizon visually. This time I have in my hand, not a sextant, but a photometer. Instead of trying to measure the angle between star and horizon, I simply aim on a known star, which is disappearing into the murk near the horizon, and allow the photometer to measure the decreasing intensity of the light from the star. It is restful work, and as it progresses the atmosphere inside our cabin improves to the point that our crying has stopped completely and, except for a swollen redness about the eyes, John and I are back to normal. Between commands to the Agena, fuel-cell purges, star sightings, and discussions of eye problems, we sort out tomorrow's activities. It will be a big day, with a long series of maneuvers using Agena's engine and our own, designed to end in a rendezvous with the 8 Agena, after which I will do my EVA and retrieve the micrometeorite package from it. Today has been a mixed bag, not so good from a weepy-eye standpoint, but great from a fuel consumption point of view, as we still have 30 percent remaining. The ground tells us that they are working on a test to be performed in the morning which should clarify the lithium hydroxide matter. Meanwhile, we have just a few housekeeping chores to do, like eating, and then we can sack out for a good eight hours.

Tonight sleep comes easily. In the first place, my knee has stopped hurting, and this little cockpit no longer seems such an outlandish place in which to snooze. I know where to put my head, and my wandering hands be damned. But above all, I am shot, I am pooped, I am bone tired and ready to sleep no matter what. It has been a satisfying day, really, and if we are still short of fuel, and if tomorrow's rendezvous and EVA are still plagued with uncertainty, well at least we seem to be crawling up out of the hole we dug for ourselves yesterday. We have a good chance of getting all tomorrow's work done as scheduled.

The next thing I know, seven hours have passed, and there is a voice in my ear, as our faithful friends on the ground summon us to begin Day 3. While we are eating breakfast, they outline the test to ascertain if the lithium hydroxide, or whatever it is, will reoccur during our planned EVA. Basically, we will simply drop the cabin pressure part way and see if we have any odor or eye irritation. In addition, we will use only one of the two fans in the system, the theory being that the use of dual fans yesterday might have stirred up loose lithium hydroxide, whereas one fan alone will not. Actually, during my upcoming EVA (if there is one), I will be using a different oxygen system—one coming directly from the supply tank via a high-pressure, small-diameter hose routed through my umbilical into my chest pack. In the chest pack, the pressure is reduced, and a separate purification and cooling system is employed. But John . . . John will still be hooked to his same oxygen hoses, and he must be able to see. His delicate job during EVA will be to fly in formation with the 8 Agena, and to act as my guardian and my eyes, to prevent me from getting my umbilical line fouled in places I cannot see. Furthermore, as John maneuvers the Gemini to keep it parked next to the Agena, he will of course be firing combinations of the sixteen thrusters mounted in various places around the Gemini. These little rocket motors of one hundred pounds thrust spit out gases at very high temperature and velocity, and no one knows exactly how close to one I may safely venture in my vulnerable suit. If the Gemini is rising up toward the Agena, John must translate downward to avoid striking it. To do this, he must fire the up-pointing thruster, the one mounted directly behind the cockpit, adjacent to my handrail and my nitrogen valve. But he better not fire it while I am trying to hook my umbilical connector into the nitrogen valve! No, we will need all our wits about us, and all four eyes in good working order, in order to carry off this EVA. Yesterday's simple hatch stand was straightforward enough, but today's stroll will be complicated, demanding, and unforgiving of error.

The lithium hydroxide test comes and goes. We pass with flying colors and are on to the next order of business: finding the 8 Agena we have been chasing for two days. It requires two last small maneuvers using our own Agena; then finally we are rid of it, forty-five hours into the flight. We had originally planned to dump it twenty-three hours ago, but, of course, the ability to substitute its fuel for our own has been a godsend, and we would not be able even to attempt this second rendezvous without it. Even so, we are tight on fuel, and we discuss various cut-off points. We agree with the ground that when we get down to 7 percent fuel remaining, we will discontinue further attempts to rendezvous. In the meantime, we have a couple of very small orbital adjustments to make, based on ground computations, and then we are on our own, with only our eyes to guide us. We know we are aligned precisely with the inert Agena, eight miles below it and slowly overtaking it. We should be able to see it soon in reflected sunlight, and finally, about twenty-five miles out, John does see it, a tiny speck 15 degrees or so above us. Now John concentrates on keeping our nose pointed exactly at it, and I respond by measuring the rate at which our angle above the horizontal is changing. By comparing these actual angles with a chart full of theoretical angles, I am able to tell John when we should make our move to depart our orbit for the Agena's, to cause that eight-mile difference in altitude to go to zero, smoothly and parsimoniously. Closing the gap must also be done on time, for, with no radar to guide us, we must get there before sundown or we will miss the Agena in the darkness. All this has been carefully calculated months before, to make the best use of our fifty-five minutes of daylight, but it's a tight schedule nonetheless, one which calls for us to spot the Agena at dawn, make our move as the sun is directly overhead, and reach the Agena with five minutes to spare before darkness. Things look good as we close on it. I calculate corrections at two points after we make our move to intercept the Agena's trajectory, and they are both comfortingly small. The first tells us to thrust upward four feet per second and the last to thrust down one foot per second. If these numbers are to be believed, we are in good shape. As we get within a couple of miles, the Agena grows from a dot of light into a cylindrical shape, and I pull out my sextant to measure the angle it subtends. By comparing its growth with my clock, I can give John a crude opinion of our range and closing velocity, but beyond that I can only shout encouragement and anti-whifferdill sentiments as we close in on it and he brakes to a halt. Finally all motions are stilled, and we are riding serenely next to it, with 15 percent fuel remaining! Not much compared to our pre-flight estimate of 40 percent at this point, but a hell of a lot better than our 7 percent cut-off threshold. We have arrived.

Normally, one would choose to follow the excitement of a rendezvous with a quiet period, but in this case our workload continues unabated. The sun is going down and we flip on our searchlight. John will spend the coming thirty-seven minutes of darkness working hard, never allowing the Agena to wander out of the tiny circle of artificial light, and I must get on with my EVA preparations, for I am scheduled to pop open the hatch at dawn. I have previously done as much EVA preparation as possible, but the key to it is the chest pack. There is not enough room to attach the chest pack *and* manipulate rendezvous charts and sextant, so this is the first chance I have had to unstow the chest pack and to hook up the fifty-foot umbilical. There is a seventy-step check list to complete, and instead of its being a cooperative effort with John and me crosschecking each other, this time we are compelled to operate independently—each of us preoccupied with our own set of problems. From what I can tell from fleeting glimpses out the window, John's world is in good shape, and fortunately the Agena is not tumbling, but appears rock-steady. I can at least put this six-month worry behind me as I wade through my check list.

As dawn arrives, I am precisely on schedule and ready to go. It never occurs to me that the ground might not agree. "Gemini 10 . . . you have a GO for the rest of the stationkeeping." John sounds a bit nonplussed. "Roger. How about the EVA? You want it?" "That's what we mean exactly." "Glad you said that, because Mike's going outside right now." And so I am. My first job is to get back behind the cockpit next to Thruster 16, where the nitrogen valve is and where a micrometeorite detection plate has been exposed for the past two days. I remove the plate with no difficulty, making sure that John doesn't fire 16 while my hand is practically crammed down its nozzle. "Watch that thruster there, babe. All right, don't translate down. I'm by it." "O.K.," John replies, but not O.K. for long, as it's not more than a few seconds before he blurts, "Well, babe, if I don't translate soon, we're going to run into that buzzard." I have no idea where the Agena is, as I am facing the side of the Gemini with my feet lazily swinging back and forth as I try to make my way along a handrail with the micrometeorite plate in one hand. I guess I'm far enough away from 16 now. Or am I? "Wait! O.K., go ahead." I now manage to reach the cockpit, hand the plate in to John through the open hatch, and then start back to connect my gun to its nitrogen supply. Again I warn John, "Don't translate down," and again I get the same answer, "O.K., we're going to hit this thing if we can't translate down pretty soon." I get out of the way. "Go ahead, translate down now. You're pretty clear." John gets a new idea. "Hey! Can we back out a little?" I don't care about those thrusters, I'm not near them. "Yes, go ahead, you want to go down?" "No, I want to back out." "O.K., back out. O.K. Don't go down until I tell you."

My problem is that there are two handrails, one raised manually by me, and the other, which is supposed to pop up automatically. I have raised the former prior to retrieving the micrometeorite plate, but the latter has only popped up at one end, the end farthest away from the cockpit, and the near end is still almost flush with the skin of the Gemini. My plan calls for winding my nitrogen line under the handrail and then connecting it to the nitrogen valve, but clearly there is not enough room to get the bulky nitrogen connector under the flattened rail, despite a couple of good healthy tugs. Finally, I decide to make the connection without the loop under the rail. I remove the cover plate from the nitrogen valve, get positioned as best I can, using the two handrails to torque my body into place directly above the valve; and then holding a rail in my right hand and the connector in my left, I ram the connector down onto its mate. Missed, goddamn it; the sleeve on the connector has sprung forward and must be recocked (a two-handed operation). In the meantime, the reaction to my shove has caused my body to lurch over to one side, and my legs bang up against the side of the spacecraft. John feels the commotion and so does the Gemini's control system, which resents the unwanted swaying motion I am creating and fires thrusters to restore itself to an even keel. "Boy, Mike, those thrusters are really firing." What can I say? "O.K." John continues, "Take it easy back there, right?" I'm trying, as I am now poised for another stab at the nitrogen connector. When my body is in the right position once again, I quietly release both hands and—floating free for a second—reach down and recock the connector, find the handrail again, and give the connection another shove. Made it! "O.K., I'm hooked into the nitrogen." "O.K.," says John.

Now we have to contend with the floating loop in the nitrogen line, which is the result of my having hooked it up without first detouring it around under the handrail. If the slack is not taken up, it will assuredly drift over on top of good old Thruster 16 and get severed just when I need it. The solution lies in the cockpit. "I'm coming back into the cockpit area for just a second here." John is apparently drifting away from the Agena, because he replies, "O.K., I have to translate up. O.K.?" "All right. Just a second. O.K., go ahead. Translate up, that's all right. See that loose nitrogen line? You're going to have to snub that down some place. Can you do that?" "Where is it?" I dangle it down inside the open right hatch. "See it?" "Yes." "Got it?" "Yes." I don't know what he's going to do with it, sit on it if necessary, but he's in a better position to cope with it than I. Perhaps we can trade jobs for a moment. "Good boy. O.K., I'll watch the Agena while you take care of the nitrogen line. O.K.?" As I hang on to the open right hatch, I look up and slightly to my right at the Agena, which must be twenty feet away. I can't see the earth, only black starless sky behind the Agena, so I guess the earth must be somewhere behind me. I realize with a start that I have not been conscious of the earth one instant since I opened the hatch; I really don't give a damn where it is; all I care about is assembling the claptrap I require to get over to that Agena and retrieve that micrometeorite package. The position of the earth is immaterial. So is our speed, which is nearly eighteen thousand miles per hour; but so what? What counts is our speed relative to the Agena, not relative to the earth. "O.K., I've got it secured." John has solved the nitrogen line problem and is now maneuvering us closer to the Agena. "What I'm trying to do is put you right next to it." Great! Peering out through his tiny window over on the left side, John edges us slowly forward and upward toward the end of the Agena, the end with the micrometeorite equipment on it. But now he comes to his limit. "If I get in closer to it, I won't be able to see you or it." I can see it just fine, not having to peer out through John's paltry window, and I can tell we are in good position. "O.K., I can almost leap right now, but I'd rather not if you can get a little bit closer. I'll give you directions, John . . . John?" It takes John two seconds to decide. "O.K." I take John forward from fifteen feet . . . ten feet . . . six feet —he's blind now, as the Agena is almost directly over my head. "O.K., better stop right there, John." "O.K." I want him to back out a bit now, but he will be doing it without me. "Translate aft. O.K., you're in a good position. I'm going to leap for her, John." In a soft, fatherly voice he lets me go. "Take it easy, babe." "O.K."

Gently, gently, I push away from Gemini, hopefully balancing the pressure of my right hand on the open hatch with that of my left hand on the spacecraft itself. As I float out of the cockpit, upward and slightly forward, I note with relief that I am not snagged on anything but am traveling in a straight line with no tendency to pitch or yaw as I go. It's not more than three or four seconds before I collide with my target, the docking adapter on the end of the Agena. A cone-shaped affair with a smooth edge, it is a lousy spot to land, because there are no ready handholds, but this is the end where the micrometeorite package is located and, after all, that is what I have come so far to retrieve. I grab the slippery lip of the docking cone with both hands and start working my way around it counterclockwise. It takes about 90 degrees of hand-walking in stiff pressurized gloves to reach the package. As I move I dislodge part of the docking apparatus, an electric discharge ring which springs loose, dangling from one attaching point. It looks like a thin scythe with a wicked hook, two feet in diameter. I don't know what will happen if I become ensnarled in it; I suspect it is fragile and will pull loose easily from the Agena, but on the other hand, it *is* made out of metal of some sort . . . Best I stay clear of it. By this time, I have reached the package, and now I must stop. Son-of-a-bitch, I am falling off! I have built up too much momentum, and now the inertia in my torso and legs keeps me moving; first my right hand, and then my left, feel the Agena slither away, despite my desperate clutch. As I slowly cartwheel away from the Agena, I see absolutely nothing but black sky for several seconds, and then the Gemini hoves into view. John has apparently watched all this in silence, but now he croaks, "Where are you, Mike?" "I'm up above. You don't want to sweat it. Only don't go any closer if you can help it. O.K.?" "Yes."

Where I am slowly comes into view and perspective. I am up above the Gemini about fifteen or twenty feet, in front of it and looking down at John's window and my own open hatch. I must be just out of John's view. The Agena is below me to my left, and slightly behind me. A loop of my umbilical is stretched out menacingly toward the fouled front end of the Agena, which is why I don't want John to get any closer, but that is a very ephemeral worry, because I am moving so as to take the slack out of the umbilical, moving up and to my right on the end of my cord. My motion is taking me away from the Agena, but it is tangential relative to the Gemini, which is not pleasant, because the laws of physics tell me that as I get closer to it (as my radius decreases), my velocity will increase and I may splat up against it at a nasty rate of speed. This conservation of angular momentum is what causes twirling ice skaters to spin faster as they pull their arms in closer to their bodies. This isn't what I had in mind. Fortunately, I have my gun, my maneuvering unit, stuck on my hip. It can null, or at least reduce, my tangential velocity so that I can safely make it back to the Gemini. I reach for it. It's gone! I grope until I find the hose leading to it, and discover the gun isn't really gone, it's just trailing out behind me. I reel in the hose, open the arms of the gun, and start hosing. I am squirting nitrogen out through two tiny nozzles pointed in a direction I have selected to (1) reduce my tangential velocity, (2) increase my radial velocity toward the Gemini, and (3) keep me pointed toward the Gemini. By the time I begin this procedure, I have swung up higher above the Gemini and off to its right, so that it is now low down to my left and I am moving off toward the rear of it. The gun is not capable of changing this path entirely, but it does modify it enough so that I come sailing around in a slow arc and straighten out as I fly behind the Gemini, a location I have never intended to explore. "I'm back behind the cockpit, John, so don't fire any thrusters." "O.K., we have to go down, if we want to stay with it." "Don't go down right now. *John, do not go down.*" If he does, it will not only fire thrusters near me, but, worse yet, it will cause my target to sink below me just at the moment I am having difficulty getting down to it. "O.K.," he says. Now things are getting better: I am approaching the cockpit from the rear, and I have only that one thruster next to the nitrogen valve to worry about. "O.K., John, do not fire that one bad thruster, O.K.?" "Which one bad one?" Goddamn, it's no time for numbers games. "You know, the one that squirts up." "Oh, 16."

My approach to the Gemini isn't exactly graceful, and it's still a bit too swift for my liking, but as I reach the open hatch, I snag it with one arm and it slows me practically to a stop. It is now a simple matter to get the umbilical cord pulled in, and to stuff it and myself back inside the open hatch. Time for another try. "O.K., John. Want to give it a new try over there?" "Yes." "O.K., let's try it one more time." "O.K." This time I decide to use my gun to translate over to the Agena, so John doesn't have to fly so close to it that he loses an important part of his visual field. If he stays fifteen feet or so away, he should be able to see most of it and all of me. When John gets us into position, below the Agena, I depart the cockpit by squirting my gun up, pointing it right at the end of the Agena. Up I glide, miraculously it seems, pulling myself up by my bootstraps. As my left bootstrap reaches the top of the instrument panel, my foot snags on something briefly and causes me to start a gentle face-down pitching motion. Just as a diver wants to hit the water headfirst, not flat on his back, so do I wish not to splat into the Agena back first, so I have some quick adjustments to make with the gun. I hold it in the proper position to create an upward pitch, and after a few seconds of squirting in this direction, I have restored my desired orientation. I now discover to my horror, however, that I am gently rising up and that my path is no longer taking me to the end of the Agena but just above it. Fortunately, I have just enough time to make one last frantic correction, and as I cruise by, I am able to reach my left arm down and snag the Agena, just barely. As my body swings around, in response to this new torque, I am able to plunge my right hand down into the recess between the docking adapter and the main body of the Agena, and find some wires to cling to. I'm not going to slip off this time! After all this, I have lost my bearings, and I don't know which way to move to find the meteorite package. Of course, since the end of the Agena is circular, I will wind up at the right spot eventually, but I don't want to discover the Indies by sailing west, especially if that means getting entangled more easily in that dangling metal loop. John is worried too. "See that you don't get tangled up in that fouled thing." I see what he means. "Yes. I see it coming." I continue, hand over hand, past the menacing obstacle. John is still worried. "Don't get tangled up in that thing. It's going on behind you now."

I can't stop now, and I can't see my trailing umbilical anyway. "If it starts to look bad, let me know. I'm going to press on up here." Finally, I make it around to the meteorite package, which is protected by a fairing and held in place on rails. The fairing can be removed by depressing two buttons and then yanking. It is attached to the micrometeorite package itself, a square metal plate some six inches on a side, by two wires ending in pins imbedded in holes in the package. I am supposed to remove the fairing, dump it, and then pull the micrometeorite experiment itself off and hang on to it for dear life. At this stage of the game I wouldn't be at all surprised if the whole thing were welded in place on the side of the Agena. Therefore, what a pleasant surprise when the two buttons stay depressed and the fairing jerks loose with only a moderate pull, bringing with it the package dangling by the two wires! For some reason the precarious nature of my hold on the package never enters my dull skull, and I am delighted that I have the whole thing in one hand with no fuss or bother.

John is still worried about my getting fouled with the Agena, which is starting to move now. I have yanked, pulled, and twisted the end of it on two occasions, and its response has by now become obvious to John. I can't see its motion, but I can feel that when I push against it, it seems less than rock-steady. John registers his alarm. "Come back . . . get out of all that garbage . . . just come on back, babe." I have my package, which has been on this Agena for three months, and although I am supposed to replace the old micrometeorite experiment with a new one, it doesn't seem a wise move. The Agena is tumbling slowly, John is worried, and the loose, hook-shaped wire is uncomfortably close. I am ready to return to Gemini and so inform John. "Don't worry. Don't worry. Here I come. Just go easy." "You want me to turn around to meet you?" John asks. He is offering to maneuver the Gemini around until it faces me directly, an unnecessary frill. "No, don't do a thing." This time, with no tangential velocity to worry about, I am coming home the easy way, hand over hand on my umbilical, but *slowly*, to avoid going fast enough to splat up against the side of it when I get there. I just don't want John firing thrusters in my face. "Don't fire any thrusters if you can help it. I'm getting back that way," meaning I have swung around toward the rear of the Gemini. John is obviously concentrating on *my* position, for suddenly he asks, "You don't see the Agena anywhere, do you?" "No. Yes. I see it. O.K." I find it up above us, and behind the Gemini for the first time, which is why John can't see it. We are well clear of it.

I am now back at the open right hatch, and I make a very sad discovery. I have lost my camera, the 70-mm. still camera which I had stuck into a slot on the side of my chest pack. I have no one to blame but myself, for this unconventional rig is of my own design. A couple of months ago I asked for a special bracket to be made, so that the camera could be attached by wedging a metal finger on the bracket into a keyhole-shaped slot on the chest pack. The system really worked well during training, and I had gotten quite adept at one-handed operation of the camera film advance and shutter mechanisms, even in my unwieldy suit. As a safety measure, I had a lanyard connecting the camera to a ring on the chest pack. Throughout the past half hour I have found that my zero-G wrestling match with the Agena caused the camera continually to work free of the keyhole slot and to dangle aimlessly on the end of its tether, banging and twisting from side to side. This has happened half a dozen times, and each time I have grabbed it, taken a quick picture, and stuffed it back into its slot. In addition, I have taken quite a few other exposures along the way, so that this roll of film must have at least a dozen of the most spectacular pictures ever taken in the space program—wide-angle pictures of Gemini, earth, and Agena—and now it is gone! I can't see the camera anywhere, but the lanyard dangles forlornly from my chest pack, making aimless little pirouettes to draw my attention to it. Each time the camera came loose and banged around, the screw must have backed off a little bit, until finally—*pfft!* Adios, beautiful pictures.

The next task in this EVA is to evaluate the gun properly, and I have mapped out a series of maneuvers which I will do out on the end of my tether, in front of our Gemini, where John can see me. The precision with which I can perform these maneuvers, as measured by John's movie camera and my impression, will allow engineers to evaluate the potential of the gun as a maneuvering device. But first, a couple of other things have to be done. I report my camera loss to John and he sounds sad, but he still has his mind on the Agena. "We're not going to hit the thing, are we?" "No. We're clear. I'm watching it. We're good and clear." Adios, 8 Agena. We've finished our three-month love affair with you, and now you are once again a derelict free to drift, a menace to be avoided. The next thing I must do is talk to the ground, before they chew us out again. All the talking John and I have been doing has been on intercom and the ground has heard none of it. The way it works, if I just talk, John hears me, but the only way the ground can hear is if my microphone button is held down, and it is back in the cockpit where only John can reach it. Thus, I need his help to talk to the ground, and so far both of us have been too busy to fool with it. But now, with the Agena gone, I had better make my report before the ground has a baby.

"How about pushing my mike button. Can you do that?" "Yes. Go ahead." "O.K., Houston, this is Gemini 10. Everything outside is about like we predicted, only it takes more time. The body positioning is indeed a problem, although the nitrogen line got connected without too much of a problem. I—when I translated over to the Agena, I found that the lack of handholds is a big impediment. I would—I could hang on, but I couldn't get around to the other side, which is what I wanted to do. Finally, I did get around to the other side and I did get both the S-10 package and the nose fairing off. John now has them. However, there is a piece of shroud hanging—or part of the nose of the Agena that came loose, and I was afraid I was going to get snarled up in that. So did John, and he told me to come on back. So the new S-10, which I was going to put on the Agena, I didn't, and I just now threw it away. Also, I lost my EVA Hasselblad inadvertently, I'm sorry to say. I'm getting ready now to do some gun evaluations. O.K., John, you can let go." The ground has other ideas. "We don't want you to use any more fuel. No more fuel, over." This means that John won't be able to maneuver the Gemini, to hold it steady while I practice my little routine out on the end of my string. John grumbles, "Well, then he'd better get back in." The ground agrees. "O.K. Get back in." John now serves me formal notice. "Come on back in the house then." I don't wanna, I don't wanna, but there's no basis upon which I can argue the point and I know it. The main thing was to get the micrometeorite S-10 package off the Agena, and besides—if we're out of gas we're out of gas.

John has to remind me to disconnect the nitrogen line, and then I let the ground know what we're up to. "O.K., Houston, Gemini 10. I've disconnected the nitrogen line and I'm standing up in the hatch here. John's not firing the thrusters any more. We're just going to take a little rest here and make sure we both know what we're doing before we press on with the ingress." "Roger. This is Hawaii. Take your time and get all squared away and they'll pick you up over the States shortly." Houston—Hawaii, what difference does it make what piece of real estate we are over? It's all "the ground" in my mind, and besides, they make it sound as if the hatch won't shut properly unless we're over the States. They don't have the remotest idea of my problem, which is that this goddamned umbilical cord is wound around me at least a couple of times and I don't seem to be able to get far enough down to . . . Shit, don't think about it. John sees my predicament and is able to reach over far enough with his stiff pressurized right arm to unloop one coil. Now I have to back out, and John guides my feet and yanks a bit, and I am nearly free, with only one persistent loop around me. The rest of the fifty feet seems somehow to be all down in the cockpit, a nearly opaque mass of white coils that obscures the instrument panel and John and everything else below the level of the hatch frame. Good grief, a full house already, with most of me still outside. I wedge my body down through the coils, forcing my legs deep into the footwell, jackknifing my knees until my torso swings down and inward in that old familiar motion I have practiced a hundred times in the zero-G airplane. I grab the hatch above me and swing it gingerly toward the closed position. Its first contact will be either the hatch frame or my helmet. If the latter, I will have to repeat this ghastly process all over again. Click! Success! Now I have only to unstow the hatch handle, and crank and crank, until—*voilà*—it is actually sealed shut! Somewhere along the line the ground wants to know if John can give them a propellant quantity reading. "Get serious," he growls.

Neither one of us can see much besides links of umbilical cord as we fumble for the valve which will fill our cabin up with oxygen, allowing us to depressurize our suits and get on with the monumental task of restoring some semblance of order to our tiny home. I manage, in my thrashing around, to turn off the radio by accident, which at least gives us a few moments of blissful silence. When we finally re-establish contact, John tries to be funny. "He's down in the seat because there is about thirty feet of hose wrapped around him. We may have difficulty getting him out." I try to be funny. "This place makes the snake house at the zoo look like a Sunday-school picnic." Pretty thin humor, but we are trying to tell them just how glad we are to be back inside again, with the spacecraft pressure up at a comfortable five pounds per square inch.

We have one more EVA to go, the final one being simply a quick hatch opening to dump all unnecessary equipment. We prepare a tremendous duffel bag, into which we manage to cram the finally tamed umbilical, the chest pack, empty food bags, and everything else we don't absolutely need. This time I wedge myself way, way down in the footwell before dumping cabin pressure, so that the hatch will swing open and closed a good six inches above my helmet. All goes according to plan; I pitch out the bag, and finally the cabin pressure is up again, and we have (hopefully) trusted the little old ladies of Worcester and their glue pots for the last time. The next time my hatch opens, the only thing I want to worry about is blue Atlantic sea water gushing in. The cockpit has taken on a new aura of spaciousness, especially over on John's side, as for the first time he is able to stretch his legs full length without bumping into equipment lashed to the floorboards. Somewhere along the line we have lost an experiment package we wanted to keep and our flight-plan book. Finally, we find the flight plan, flat on the floor under John's feet (where he couldn't see it until after the equipment dump), but the experiment apparently really is gone, presumably having floated out the right hatch somewhere along the line.

We have a couple of things to do before this day is over, but not too many, please, as the propellant quantity gauge tells us we have but 7 percent remaining. First, and most important, we want to get our orbit down from 250 miles, and we use up most of the 7 percent in a burn which reduces our perigee to 180 miles. The lower the better, in case we have difficulty with our retrofire burn tomorrow morning. The retrofire energy comes from four separate solid propellant rocket engines, but in case one doesn't fire, or we have difficulty controlling the direction of the thrust, we want to be as low as possible. A perigee of 180 isn't the greatest, but it beats 250. Once we are safely in our new orbit, we make up for some lost experiment time and race through an involved sequence designed to determine whether the positively charged ions whizzing past our craft can be used to determine the direction we are pointing, a very simple scheme compared to the complex and heavy gyroscopes we now employ for this task. It seems to work well, and the needles connected to our ion sensors move in close agreement with the information coming from the gyroscopes.

Once this is behind us, we are on our own for the first time in the fifty-three hours since we left Cape Kennedy, with a ten-hour sleep period ahead of us, with no Agena to keep us rigidly upright, with no navigation or rendezvous or other complicated tasks to consume our every minute. We have only one constraint, and that is, we must not use fuel, as our gauge is nearly on zero, but that is no hardship as we drift carelessly over the world. We congratulate ourselves and receive compliments from our friends on the ground. "We'd like to let you know that we're pretty doggone happy down here . . . it was a great job today. It was fabulous." John agrees. "I tell you it was a tremendous thrill. It was really incredible. I didn't believe part of it myself." I think he's talking about my climbing on the Agena, which God knows *was* unreal. Who would believe it? "Hope those pictures come out" is all I can say.

Meanwhile, we have turned off our attitude-control system, to save gas, and are doing some very un-airplane-like things. In a fighter it is possible to do loops and rolls, even spins, but one never goes sideways or backward—at least not without violent consequence. But we are tumbling now—slowly, smoothly, aimlessly—and as we go, the square snout of our spacecraft traces graceful arcs in the sky, sometimes in front of our direction of travel, sometimes to one side or the other, sometimes behind. It is a three-dimensional roller-coaster ride in slow motion, with no noise, no banging around, no hollow feeling in the pit of the stomach. If this is what the old Edwards test pilots meant about an astronaut just being along for the ride, being a canned man, then I am for it. It is suppertime now and I have missed lunch in the hurly-burly of EVA preparations; I am famished, and have finally gotten a chance to fill up a plastic tube of dehydrated cream of chicken soup with cold water (the only kind we have). After kneading it for a couple of minutes, I snip off the end of the feeding tube with my surgical scissors (which are powerful enough to cut through the fifty-foot umbilical in case of an emergency, like being unable to disconnect from the nitrogen valve. I know, I have tried it). Anyway, I finally get my first swig of soup and it is the best thing I have ever tasted, better than a martini at Sardi's, better than the pressed duck at the Tour d'Argent. And the view out of the window is absolutely breathtaking!

I will try to explain it. First, some arithmetic. At two hundred miles above a sphere whose radius is four thousand, we are just skimming along one twentieth of a radius above the surface. The atmosphere itself is ridiculously thin, thinner than the rind on an orange, and we are just barely above it. The curvature of the earth is apparent, sure, but it is not a dominant impression, not any more than running your eye over a curved dinner plate causes you to think about its concavity rather than the design on it. Nor is the speed impressive. It's certainly not the blinding speed of the Indianapolis 500. The rate at which something appears in the window, crosses it, and vanishes is not much faster than in a commercial airliner. That is because our much faster orbital velocity is balanced by our higher altitude, so that *angular* changes (the most important visual cues of speed) are still within the realm of the commonplace. These, then, are the similarities, to which one must add color, for, although the sky is absolute, unrelieved black instead of blue, the colors below look about the same as they would if seen from an airplane. Let a six-year-old child have a peek and he would be back at his coloring book within a minute. Then what is so impressive, what makes it different? It is the eye of the adult, which balances what he sees with a lifetime of crawling the surface of the planet. Supertourist is up, and what a feeling of power! Those aren't counties going by, those are countries or continents; not lakes, but oceans! Blanche, we can still make Yellowstone today, if we drive six more hours. Forget all that, baby! In six hours we will have circled this globe four times! Look at that, we just passed Hawaii and here comes the California coast, visible from Alaska to Mexico, and my cream of chicken soup not yet finished. San Diego to Miami in nine minutes flat; if you missed it, don't worry, it will be back in ninety minutes. Another difference is that we are above it, and it is uniformly bright. No misty days, no towering thunderheads; all that is below, and superbright in the unfiltered sunlight, which spreads a cheeriness over the whole scene below. No doom, no gloom, only optimism. This is a better world than the one down there. Fantastic!

Up until now we have charted our progress by the clock, not the scenery. The fuel cells must be purged at 51:36; never mind that it's over the delightful island of Ceylon. Now, as we munch on pressed bacon cubes and suck thin grapefruit juice out of plastic bags, we pay attention to the chunks of real estate which grandly parade past our windows as the Gemini slowly cartwheels. The Indian Ocean flashes incredible colors of emerald jade and opal in the shallow water surrounding the Maldive Islands; then on to the Burma coast and nondescript green jungle, followed by mountains, coastline, and Hanoi. We can see fires burning off to the southeast, and we scramble for our one remaining still camera to record them. Now the sun glints in unusual fashion off the ocean near Formosa. There are intersecting surface ripples just south of the island, patterns which are clearly visible and which, I think, must be useful to fishermen who need to know about these currents. The island itself is verdant—glistening green the color and shape of a shiny, well-fertilized gardenia leaf. Then back over the Pacific again in a race for Hawaii and the California coast. I could stay up here forever! Er, amend that, I don't want to stay up here past 70:10, which is retrofire time, and it is now 56:14 and time to get some sleep. Up the girls go on the windows, blotting out the bright shiny world, and leaving John and me with our thoughts in the darkened cabin.

As I drift off to sleep, I can't help but compare this night with the previous two. Let's face it, the first night was miserable, with the twin embarrassments of the Module VI fiasco and the excessive rendezvous fuel, plus my painful knee and my generally keyed-up state, resulting from this strange and apparently hostile environment. The second night was different, better certainly, but still full of unknowns, and my extra tiredness had created gloom, not giddiness. Tonight is something else again. We have carried off that second (and more difficult) rendezvous, and I have actually space-walked from one satellite to another and retrieved a package from it! I can't get it out of my mind. Too bad it got cut short, too bad I didn't have a chance to savor the experience, to let the view soak in a bit. Funny there was no sensation of motion, or of falling, especially considering that I get a bad feeling in the pit of my stomach when I peer over the edge of the roof of a tall building. I really wasn't very conscious of the earth at all, only the Agena and the spacecraft—one at a time, depending on what had to be done next. Work, work, work! A guy should be told to go out on the end of his string and simply gaze around—what guru gets to meditate for a whole earth's worth? I think nirvana must be at an altitude of 250 miles, not down below in the teeming streets of Calcutta or up above in the monotonous black void. I am in the cosmic arena, the place to gain a celestial perspective; it remains only to slow down long enough to capture it, even a teacupful will do, will last a lifetime below. "I found truth in orbit." Wrong, I haven't. "I found God outside my spacecraft." Wrong, I didn't even have time to look for Him. Would that I could, like Mercury of the winged heel, convey some swift message of value, a message of splendor and beauty, of hope and praise, a message which accurately mirrors what I have seen today. John Magee would have known how to do it. Behind my head, stowed away in a small bag with some flags, rings, and other trivia, is a small file card on which my wife Pat has typed his poem "High Flight."

> Oh, I have slipped the surly bonds of earth
> And danced the skies on laughter-silvered wings;
> Sunward I've climbed, and joined the tumbling mirth
> Of sun-split clouds—and done a hundred things
> You have not dreamed of—wheeled and soared and swung
> High in the sunlit silence. Hov'ring there,
> I've chased the shouting wind along, and flung
> My eager craft through footless halls of air.
> Up, up the long, delirious, burning blue
> I've topped the windswept heights with easy grace
> Where never lark, or even eagle flew
> And, while with silent, lifting mind I've trod
> The high untrespassed sanctity of space,
> Put out my hand, and touched the face of God.

All that from the cockpit of a Spitfire. What could he have said after one orbit? I cry that he was killed.

Day 4 begins with the usual bugling from below—"Gemini 10, Canary CAPCOM"—over and over until we answer. We've only been asleep a few hours, but no matter—today will be short, just a few experiments, retrofire, and . . . God, if they'd only give you a few minutes to wake up, but no. "Cryo quantity switch to the O_2 position?" "Roger, will do." Breakfast is the last meal up here, the ground reminds us, and I am enjoying it thoroughly, perhaps too much, as John wryly observes. "Shoot, you should see him; he's eating *my* last meal too." After breakfast I hook a full urine bag to the overboard dump valve and am rewarded with the usual snowstorm of escaping white particles. The constellation "Urion," as Wally Schirra has dubbed it, is formed by the instantaneous freezing of the urine stream as it reaches the vacuum of space and breaks into thousands of individual miniature spheres. Cascading out in an irregular stream, they whiz past the window and tumble off into infinity, glistening virginal white in the sunlight instead of the nasty yellow we know them to be. The fairytale quality of this transformation is typical of this place, an unreal world far above the unseen squalor below.

This morning everything seems to be going well. We find the right stars in the right places at the right times, and we cheerily wade through the remaining experiments until, suddenly, we are on our last circle before retrofire. This is traditionally a time for a little formality, to thank the ground crews who may have spent weeks in Kano, Nigeria, or on a small ship in the middle of the Indian Ocean, in support of our flight. "Gemini 10, Canary CAPCOM." "Go!" "O.K., we have nothing for you. We'll be standing by. Have a good trip home." "Roger, thank you very much. Enjoyed talking to you. It's been a lot of fun . . . want to thank everybody down there for all the hard work." John isn't kidding. Without some very precise and rapid calculations on the part of the troops in Mission Control, we would not have been able to press our Agena into extra duty to compensate for our own fuel shortage. Of course, that cat down there in the Canary Islands didn't do it, but his people in Houston did, and they saved our ass. Over Kano I announce to the world, "Boy, I really hate to come back. This is really something up here." "Take more groceries next time," I am advised. Wise bastards. This should have been a full four-day flight anyway; we have just been too harried to do all of it well in . . . let's see . . . it's now 69:21:05 . . . retrofire is at 70:10:25.

The weather is supposed to be good in our landing area: scattered clouds at two thousand feet, visibility fifteen miles, wind out of the southwest at eight knots, waves two to three feet high, a few showers in the area, the helicopter carrier *Guadalcanal* on station in the Atlantic some four hundred miles north of Puerto Rico. We will be firing our four retrorockets out over the Pacific as we face aft, putting the rocket nozzles forward, where their thrust will reduce our speed below orbital velocity, causing us to start our slow fall back down into the atmosphere. Between now and retrofire time we have our check list to go through, and no check list is performed with the reverence reserved for the pre-retro check list. You don't get a second chance at this one, you have to do it right the first time, and as we run through the steps—oh, so slowly—I check them off one by one with my pencil. Some of the more critical ones we discuss beforehand, even the details of *how* I am going to punch a particular row of buttons. "Push them in the center; push them hard and hold them down for a good, fat one second," John tells me. "Roger." "Have a little separation between them," he adds. "Yes, I will. About two seconds?" "No. Make it one." "O.K." As the time approaches, we chop the fuel and electrical lines leading aft into our adapter section and then jettison the adapter section itself. This exposes the four blunt solid propellant rocket motors which will ripple fire, one after the other, taking just enough energy out of our orbit to cause us to enter a terminal glide. Assuming, that is, that we are pointed in the proper direction; if somehow we should get it backward, we will enter a terminal climb, and I mean *terminal*, as there would be no way to get us down from our higher orbit. We pay a good deal of attention to the direction in which we are pointing as retro time approaches. Canton Island comes on the air with the inevitable countdown . . . $10 - 9 - 8 - 7 - 6 - 5 - 4 - 3 - 2 - 1$ . . . RETROFIRE!

After nearly three days of weightlessness, I have forgotten the measure of acceleration; all four retrorockets shouldn't give us more than one-half G, but as the first one goes off, it feels more like 3 Gs to my sensitized body, as I am shoved back in my seat. But it's not the Gs I am worried about, it is the count: One—two—three—four! "I count four beautiful ones, John babe." "Yes," says John to me, and to the ground, "That was a superfine automatic retrofire: 303 aft; five right; 119 down." He means that the retrorockets have imparted a change in our velocity of 303 feet per second in the aft direction, five feet per second to the right, and 119 down toward the earth. This compares with the perfect retro-fire of 304 aft, zero left-right, and 114 down. Pretty damned close! Now we can jettison the retrorocket package, exposing our heat shield, and John can start steering this moose toward our landing area. As is our general pattern, John will manually fly the machine and I will help out by interrogating our computer and doing some back-up calculations with charts and graphs.

As we enter the upper atmosphere, a sheath of ionized gas will surround our spacecraft and create a barrier which our radio signals cannot penetrate, giving us five minutes of "blackout," or radio silence; but before that we will be over California, where we will get a preliminary look at our trajectory. Yes, here comes Houston, relayed through the California station now. John inquires, "I say, have you got Super-retro down there helping us?" "That's affirmative. He's right here." Super-retro is John Llewelyn, one of the flight-control team specializing in re-entry, and if we make a mistake, Super-retro is going to get so pissed off he will stick his strong Welsh arm up here and yank us out of the sky. We discuss numbers with Super and he certifies our upcoming success. Now we pass an altitude of 400,000 feet, the height generally considered to be the top of the atmosphere and the point at which John has to get busy with his controls, changing our bank angle as directed by our computer. It's sort of like making gliding turns in an airplane, except we are upside down and backward, presenting the blunt heat shield behind us to the increasing friction of the thickening atmosphere. The heat shield, made of a fiberglass, honeycomb structure filled with a silicone material, is designed to dissipate heat by the process of ablation, in which the heat shield will partially erode, literally evaporating and carrying with it the heat of atmospheric friction. It's not long before I can tell that the heat shield is doing its job. We are developing a tail. Tenuous at first, then thicker and more startling, it glows brightly, a red and yellowish gassy trail curving off into the lightening sky. I am impressed. "Man, that's starting to look like something . . . look at that son-of-a-gun burn!" Occasionally a small chunk of heat shield breaks loose and adds sparkle to the halo. Now we start to pick up some Gs, some deceleration. "How many Gs going, John?" He has the only G meter, and it's way over on his side. "About one half." "You're kidding!" To me it feels more like three, and yet it's not even up to one G, the level which I have spent a lifetime considering normal. How could I forget, in just three short days?

Somewhere out of an old classroom comes the theoretical computation of maximum G during re-entry: it is the inverse of the vehicle's lift-to-drag ratio. Our Gemini has an L over D of about one fourth, therefore peak G should be four. I don't know why I go through this exercise at a time like this; I know darned well from the simulator it is four, and besides there is nothing I can do about it anyway. "Fly that thing, John! You're doing a beautiful job!" We're right at 4 Gs now, and then I feel the load lighten as we finish our blackout period and the ground starts yelling at us again. We are down to 2 Gs as we pass one hundred thousand feet, and everything looks good. At thirty-eight thousand feet we let fly our drogue parachute. Only six feet in diameter, this little rascal is designed to stabilize us in a vertical position and slow us down enough to enable us to deploy our main chute (fifty-eight-foot diameter) at ten thousand feet. Wild! Far from being stabilized, we begin to sway back and forth, dangling in a crazy arc which takes us 25 degrees or so to either side of our supposedly vertical descent. "Shoot," says John. For some obscure reason I, who have never seen a drogue chute before in my life, decide John is worrying needlessly, despite the obvious fact he has this one to compare with the chute on Gemini 3. "Quiet, John, it's all right." Just wishful thinking, I guess. By now it is quieter, and we get ready to deploy the main parachute a bit early. "There goes the chute!" John tells Houston, and indeed it does, completely filling our windows with red-and-white reassurance. We are coming down on our backs now, at thirty feet per second, and have only one thing left to do: release the chute harness from our nose, so that our spacecraft will swing out horizontally, and when we splat, it will be heads up, with our hatches above the water line. This maneuver—switching from a single-point vertical suspension to a two-point horizontal system—has caused Gus Grissom on Gemini 3 to whack into the instrument panel, cracking his helmet visor. John remembers this well and warns me. "You ready? Brace your arm, Ace!" We each put a soft pressure suit arm between our heads and the panel, and are a bit disappointed as we do no more than swing gently forward. We are at fifty-five hundred feet now and looking good. We can't see the boat out the window, but we must be close to our aim point, for Houston reports they can see us on TV. Now we notice a strange thing: the clouds are whizzing by our windows sideways! Not only are we coming down, but apparently we are spinning on the end of our parachute line. The motion gently slows, and finally stops; then reverses itself and starts to pick up speed in the opposite direction. I don't ever remember hearing about this, and I don't like it. I figure it's going to cause us to descend more swiftly. "Boy, we're going to hit like a ton of bricks!" Amazingly we don't, but plop gently into the Atlantic. We must have caught the lip of a descending wave.

It is a soft and gentle day and the eight-knot wind seems to have disappeared; the sea is mercifully quiet. My ears are blocked from the sudden descent, and there is an unaccustomed aroma of burned chemicals in the air. Outside my window the pair of nose-mounted thrusters hisses and smokes, emitting an occasional tendril of weak flame. I realize with a start how dry and cool it has been for the past three days, for now we are back, and the un-filtered ocean air is moist and fetid and hot—mostly hot. My pressure suit with its thick EVA cover layer has fulfilled its purpose. No longer needed, it is sheer encumbrance now, a stifling, suffocating blanket that I am rapidly filling up with my own sweat. Outside there is immediate motion, as a helicopter flashes past our prow, to be followed by swimmers who surround us with a stabilizing, inflated rubber collar. Hot! We establish radio contact with the swimmers. Gemini 10 flies no more; we are but awkward trespassers, bobbing at the mercy of a new set of experts. They are EVA now, thrashing around out there somewhere. "Hey, boys, take your time! We're not in any hurry. We don't want anybody getting hurt out there." It is over.

Anders, Borman, and Collins in front of their monster

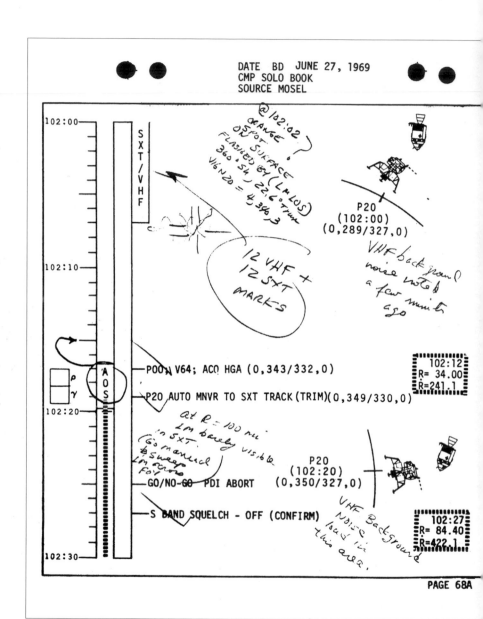

A page from my rendezvous book

Our LM in cramped quarters atop the now useless Saturn

Eagle emerges from its nest and spreads its legs

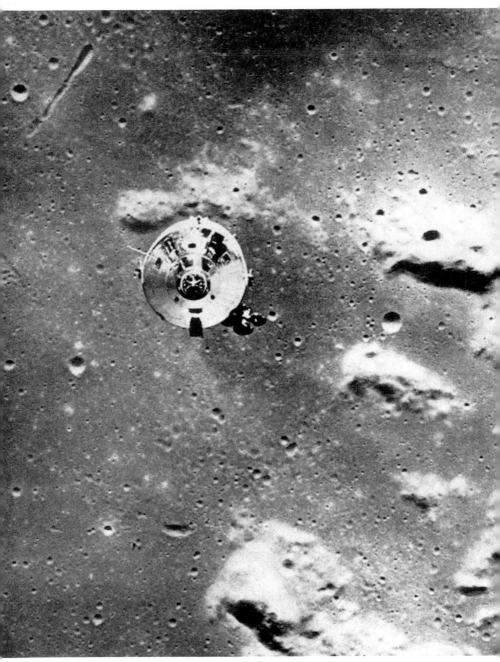

Columbia, my happy home, as seen from *Eagle*

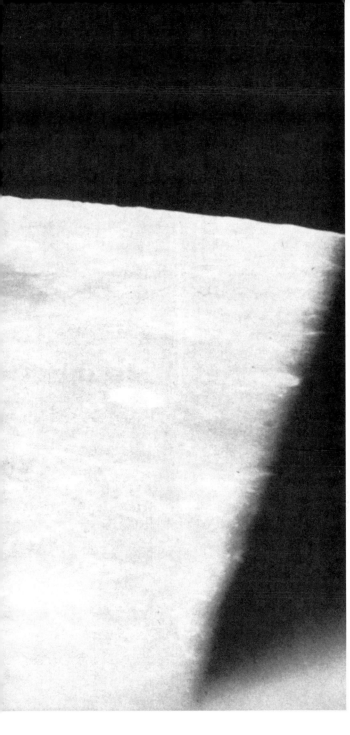

The best sight of
my life: Neil and
Buzz returning!

Hearing it from the Chief

9

When it is dark enough, men see the stars.

—Ralph Waldo Emerson

Being the seventeenth American to fly in space is a small distinction, and in a small way it changed my life. My name didn't become a household word ("Let's see, aren't you Scott Cooper?"), nor did people keep my checks for the signature instead of cashing them, as they reportedly did John Glenn's. But I was a "real" astronaut now, and I began a lifetime of explaining "what it was like up there." My first formal opportunity was at our post-flight press conference, where John had the impossible task of clarifying, in layman's terms, the reasons behind our out-of-plane whifferdill and its exorbitant fuel consumption. My part was much simpler: to describe floating in space and trying to do useful work without benefit of handholds or footrests. People were kind to us. Gilruth allowed as how it was the most complicated mission ever undertaken, while Bob Seamans, Webb's deputy, said that the way was now paved for our lunar program. The press pushed gently for some admission of guilt in regard to the rendezvous fuel, and in addition suggested that losing equipment amid piles of umbilical cord was a bit gauche.

But no one claimed that the flight was a PR disaster because of our silence. Perhaps it would have been better had they done so, because at least we could have gotten our side of it out on the table, namely that we had been too damned busy to get everything done properly *without* the necessity of explanations in quotable English each step along the way. Furthermore, what was important *was* duly recorded and properly delivered to the appropriate experts after the flight. As test pilots, that was what we had been trained to do, not to emote over the radio. In a test pilot's world, boring is good, because it means that you haven't been surprised, that your planning has been precise and your expectations matched. Conversely, excitement means surprise, and that generally is bad.

We did have one narrow escape at our press conference. We found out after the flight that the two young lovelies whose photos adorned our window shades were friends of a McDonnell engineer, and were employed as Bunnies in the St. Louis Playboy Club. He had told the girls that their pictures had flown with us, and now the pair had decided to come to Houston for the press conference. Wow! John and I shuddered at the scene we conjured up, the two of them suddenly standing up in the crowded auditorium, their spectacular appearance commanding instant silence, and then tee-heeing something to the effect of "What was it like going around the world forty-five times with us?" We were saved by a speedy phone call from Deke Slayton to some high McDonnell official, threatening I know not what form of ruin and damnation for all concerned. The girls never arrived.

After the press conference and the technical debriefings, John and I were no longer a team, but were again individuals in the growing pool of astronauts available for future flights. I put in my travel voucher for the Gemini 10 flight (three days at $8.00 per diem = $24.00) *, took two weeks at the beach to get reacquainted with my family, and returned to my desk to find a mountain of mail awaiting me. One of the great disappointments of the flight was the fact that there were no photos of my space walk over to the Agena. What with camera malfunctions and John's understandable preoccupation with keeping us from hitting the Agena (without firing that bad thruster, whatever its number was), all we had was the film from one movie camera, which was pointed straight ahead and which automatically recorded an uninterrupted sequence of black sky a few feet below the Agena. In the second half of the twentieth century, an event must be seen to be believed, and a description of a trip without pictures seemed anachronistic at best. I was really feeling sorry for myself, unable to produce graphic documentation for my grandchildren of my brief sally as a human satellite, when I discovered among my mail something that restored a bit of my perspective. It was an account, by Captain J. C. Frémont of the Topographical Engineers to the Senate, describing his trip to Oregon and northern California in 1843-4.

> Unhappily, much of what we had collected was lost by accidents of serious import to ourselves, as well as to our animals and collections. In the gorges and ridges of the Sierra Nevada, of the Alta California, we lost fourteen horses and mules, falling from rocks or precipices into chasms or rivers, bottomless to us and to them, and one of them loaded with bales of plants collected on a line of two thousand miles of travel; and, when almost home, our camp on the banks of the Kansas was deluged by the great flood which, lower down, spread terror and desolation on the borders of the Missouri and Mississippi, and by which great damage was done to our remaining perishable specimens, all wet and saturated with water, and which we had no time to dry.

Whoever sent me that account, thank you—what is one camera lost in the bottomless chasm compared to fourteen horses and

---

* I should have claimed seven cents a mile, which would have come out to nearly $80,000, except someone (Schirra?) had already tried that during Mercury and had received instead a bill for a couple of million dollars—the price of the Atlas booster.

mules, one loaded with two thousand miles' worth of meticulous collecting?

I also received a basketful of "Dear Astronaut Collins, You are my favorite astronaut. I wonder if you would . . ." but I had long since learned about these. Certain fourth-grade teachers assign astronauts just as they used to assign states. Instead of Arizona, little Susie had drawn Collins, and a report was expected within fourteen days. Favorite, indeed, but anyway I sent each a ton of material on the space program that rivaled anything the Arizona Chamber of Commerce might produce—and it was already written in fourth-grade language.

I was spared at least one post-flight ritual: as a result of having been born in Rome and having kept moving thereafter, I had no home town; hence no home-town parade or other homecoming was possible. John was off to Orlando, under duress, but I was free now and could relax for the first time in two years. I had a space flight under my belt, a creditable one, and I was now available for assignment to the Apollo program at what seemed to be just the right time. Gus Grissom was talking about getting Apollo 1 airborne before the end of the year, possibly in a dual flight with Gemini 12, and Seamans had indicated that the lunar landing might be possible as early as 1968. Therefore, I was delighted when Frank Borman told me I had been assigned to his crew. The lineup would be Borman, commander; Tom Stafford, command module pilot; and me, lunar module pilot. Our job was back-up to the second manned Apollo flight, whose crew would be Schirra, Eisele, and Cunningham. This in itself wasn't particularly good (Schirra's flight didn't even have a lunar module for me to pilot), but the important thing was that obviously we would go on to other assignments, and perhaps even become the first lunar landing crew. I thought we had a good crack at that because we were the most experienced trio to have been assembled so far, the only Apollo crew with no rookies. Borman had as much time in space (Gemini 7 for fourteen days) as anyone, and Stafford more rendezvous experience (on Geminis 6 and 9) than anyone. And I would be lunar module pilot, which suited me just fine, as I had always been fascinated by Grumman's outlandish-looking landing bug. Besides, I might just get to land it on the moon and put some of my extravehicular experience to work.

At any rate, August 1966 was a time of great optimism for our program as well as for me personally. Not only was Gemini winding up in highly successful fashion, not only was Apollo about to start flying, but the space age seemed truly to have arrived. The moon was getting closer and closer, for both the United States and Russia, as lunar orbiter and Luna XI photographs and measurements poured in. Nor did the moon seem to be the only goal, as leaders such as General John McConnell, the Air Force Chief of Staff, spoke of future Air Force pilots flying space machines to landings on conventional runways. The NASA leaders were looking past Apollo to a program called Apollo Applications, and thoughtful editorial writers were urging less haste in moon racing and more foresight in planning for the long haul.

My small piece of all this action was the care and feeding of one Apollo command module undergoing final assembly and testing at the North American plant in Downey, California. It would be the second such machine to fly with men in it, and was known simply as 014, its serial number. In the white room at Downey (a much larger version of McDonnell's St. Louis white room, where I had spent so many hours nursing Geminis 7 and 10), the spacecraft's test crew had hung a sign on it, a great leaping green frog. This was a not-so-kind reference to the fact that Gus Grissom's spacecraft, 012, the first manned vehicle, was having more than its share of difficulties, and if Gus and his crew didn't watch it, old 014 was going to leap right on past and be the first one ready for flight. Both 012 and 014 were called Block I spacecraft, and they were prototypes of a sort, incomplete in some areas and rendered obsolete by later designs in other areas. For example, they were unable to dock with a lunar module (LM), having no docking apparatus, and the heart of their guidance system, the inertial measuring unit, was pointed off in a strange direction; all its measurements had to be converted into the spacecraft body axis co-ordinate system. Working with Block I machines was a bit frustrating for Borman, Stafford, and me, as there were only two of them scheduled for manned flights and we would assuredly move on to a Block II spacecraft and also get an LM of our own.

On the other hand, it was great to have a specific crew assignment, which could only serve to expedite our learning the basic Apollo system. In fact, I felt thrown into it before I really had had a chance to assimilate sufficient background information. The prime crew of Schirra, Eisele, and Cunningham had been at it for months, and Borman had not been far behind them. More recent Gemini graduates, Stafford and I were the newcomers, but Tom seemed to be a swifter study than I. He had been number one in his class in electrical engineering at the Naval Academy, and he sopped up wiring diagrams with the same ease with which some teenage girls memorize the words of popular songs. It came harder for me, but as September wore on, I found the interior of the great wedge-shaped chunk in the white room starting to make a little sense. Gemini had seemed complicated at first, but it was simplicity itself compared to this beast; there were over three hundred of one type of switch alone, not to mention the scores of pipes, valves, levers, brackets, knobs, dials, handles, etc., etc. In addition to the controls necessary to operate the machine itself, there were a multitude of experiments, both medical and scientific. These filled every nook and cranny not already crammed with lockers and boxes containing fourteen days' worth of food and household supplies—the interior was a mess. Worse yet, it was a dynamic mess, never static, as workmen came and went with alarming regularity, rerouting wires, replacing black boxes, and rearranging equipment. It was a bit disconcerting because no one I talked to seemed to know exactly what was, or was not, in the spacecraft at any one time, but then I figured someone somewhere must be keeping book on it. My job was to be present during the interminable test sequences, running around the clock, which were necessary to measure 014's readiness for flight. It soon became apparent that North American had a long way to go before it could match the smooth professionalism of the McDonnell test teams, veterans of Mercury and Gemini. Match them North American did, but not until two years later, and at this stage of development, it was Amateur Hour, not helped any by this amateur, struggling to learn the new Apollo nomenclature and systems.

The NASA engineers assigned to Apollo were not particularly helpful, either. Gemini had always been billed as a testing ground for Apollo, a means of gaining practical experience that could be applied to the lunar effort. But strangely enough, when Gemini methods were suggested to the Apollo engineers, there was no eagerness to accept or even to listen. At the first mention of the word Gemini, a veil came down, the eyes became slightly glassy, and one was informed—generally in a cool and faintly supercilious tone—that it was simply not done *that way* on Apollo. To disagree was simply to admit that one had not yet mastered the superior intricacies of the Apollo system.

If the test procedures left a lot to be desired, the ancillary arrangements did not. With a twenty-four-hour-a-day schedule, the spacecraft test crew might need an astronaut in the cockpit on short notice at any strange hour, so we were provided with a plush lounge, complete with bunks and color TV. We also received frequent and special invitations to the executive dining room (the Golden Trough), where we got to know the North American managers and began to appreciate the enormous engineering complexity of the Apollo program. We were afforded every courtesy, and our weekly visits to Downey were made as pleasant as possible.

Still, I found the place unpleasant and unsettling. Southern California rococo-a-go-go! I stayed at an ersatz thatched-roof village, a garish plastic replica of Tahiti, where go-go dancers sweated far into the night, bobbing and weaving in bored syncopation with overamplified electric guitars. I ate breakfast one plate-glass window away from the bus stop, and watched twelve-year-old girls in tight skirts and blackened eyes take one last drag from their cigarettes before boarding the bus with a shrug of ennui and resentment. At work, craftsmen seemed more interested in talking about getting their campers up into the High Sierra on weekends than they did in explaining why the last test sequence had been goofed. Somehow go-go land didn't seem the place for the meticulous attention and the very square discipline required to assemble these machines of unparalleled complexity. Perhaps underlying all this was the fact that I didn't really want to fly 014. If *any* space flight could be considered boring, then this one would be. Grissom-White-Chaffee in 012 would have the distinction of being the first ones up, for two weeks if the plumbing held together; Schirra-Eisele-Cunningham in 014 was just a repeat performance, another data point on a chart which said that more complex later missions could be safely planned. Furthermore, 014 was chock-full of medical experiments, such as a collapsible bicycle device called an ergometer, which the crew (if the cramped space allowed) could ride periodically to give the medics some measure of cardiovascular deconditioning. Even a frog was to be carried on board, not a leaping frog, but one whose inner ear was wired and who was encased in a miniature centrifuge. After enjoying weightlessness for prescribed intervals, the frog was to be periodically spun up on the centrifuge, and his otolith function duly recorded. The crew couldn't even warn the poor suffering little bastard, but merely flipped a FROG switch to ON, and away he went. Imagine that, a FROG ON-OFF switch, hardly work for test pilots. No sir, no 014 for me, please; let's just get this thing tested and out of go-go land, so we can get started with Borman, Stafford, Collins to the moon.

Between stints in Downey, I was working fairly hard in Houston, trying to get caught up with Apollo work, reading a bit about the LM when I could, and trying to give my family at least twice as much time as I had during the hectic first half of 1966. Unfortunately, twice a small number is still a small number, and with four trips to Downey in September and four in October, I was more than a bit resentful of Wally's crew and their leaping frog.

One pleasant interlude was helicopter flying in Houston. NASA had sent me to helicopter school more than a year earlier, and I loved flying the two agile little Bell machines parked on the ramp at nearby Ellington Field. Flying helicopters is not easy, especially the early models which didn't have automatic control of rotor blade rpm; both hands are busy doing strange un-airplane-like things. It's not unlike rubbing your stomach with one hand while patting the top of your head with the other. The left hand controls the up-down, plus the throttle. If you want to go up, pull the left stick (called the collective) up, and the rotor blade will twist so as to take a bigger bite of air, producing more lift. Unfortunately, the rotor blade rpm will probably decrease as a result of this, so you must also twist your left wrist as you pull up, adding fuel to the carburetor and keeping the rpm constant. Old helicopter pilots learn to judge rotor rpm by sound alone, but I never got the hang of that and had to keep one eye practically glued to the rpm gauge. In the meantime, the right hand cannot let go of the right stick (called the cyclic) or the unstable little rascal will lurch out of control, left-right or up-down. The lower torso doesn't feel left out, because both feet stay busy with the rudder pedals. The reason NASA trained us in helicopters was that the lunar descent and landing in the LM were similar in some ways to a helicopter approach, especially the last few feet prior to touchdown. Furthermore, part of an LM pilot's training consisted of several flights in a weird-looking, flying-bedstead contraption called the lunar landing training vehicle, and it was deemed necessary to have two hundred hours of helicopter time prior to checking out in the LLTV. NASA had also built a simulated lunar surface not far from Ellington, an acre or two of gray slag, laboriously arranged to duplicate crater patterns in imitation of lunar orbiter and Ranger photographs. We used this "rock pile" for simulated lunar approaches, chopping throttle five hundred feet or so above it, and gliding down in the helicopter at various angles and speeds, under varying lighting conditions, etc., to familiarize ourselves with the problems of gauging height and sink rates accurately. This was good fun and good therapy between Downey trips.

Things were going all right in Downey with 014, but not with Apollo in general. The feeling of optimism prevalent during the summer seemed to collapse as winter approached. Grissom's 012 was not about to fly with Gemini 12 in November, but was daily slipping farther into 1967. There was no single cause for the slippage, but rather a long list of minor ailments and imperfections requiring replacement and retesting of components. The environmental control unit was especially troublesome. A plumber's nightmare at best, the unit installed in 012 seemed forever to be springing leaks that required time-consuming removal and repairs. Nor was Apollo's trouble limited to spacecraft 012. On October 25, one of spacecraft 017's service module tanks ruptured in a pressure test at Downey, sending a wave of consternation throughout the space community. The propellant tanks in the service module were vital to the lunar mission, for if one failed in lunar orbit, it would be impossible to get home. Even worse was the rumored sad state of the second stage of the giant Saturn V rocket, whose first and third stages were coming along nicely. Inspectors kept finding minute cracks in the aluminum skin and welds of the second stage, and those who understood such things inferred that North American had picked the wrong alloy of aluminum. While light and strong, this particular alloy was too brittle, so the story went, and the second stage was a long way from being a healthy vehicle to ride. We astronauts were very tuned in to these technical difficulties, but were much less prone to concern ourselves with recurring press reports of rising costs and questionable management. Costs were not our concern, and management seemed O.K., even if some of the Apollo people who had never worked on Gemini did seem a bit naïve and overconfident, perhaps even arrogant. But we Gemini people may have been a bit arrogant too, in which case we were seeing our own attitude mirrored back at us. At least we could say that we had been through a tedious spell of midwifery once before, even if it was only for a puny earth-orbiting two-seater.

The Gemini program was still alive and enjoying a robust old age. Pete Conrad and Dick Gordon had gotten off in Gemini 11 in mid-September and had flown a very nice flight indeed. Just as John Young and I had easily broken the Voskhod II altitude record of 167 miles, raising it to 475, so it was effortlessly taken away from us by Gemini 11, which soared up to 850 miles. From this altitude the curvature of the earth is very pronounced, and Gemini 11 returned with spectacular pictures of the Persian Gulf and surrounding territory. The technical aspects of the flight were equally impressive, and Pete Conrad came through as a cool and competent commander without losing any of his cheerful bubbly exuberance. Dick Gordon's performance was also near perfect, his only problem coming during EVA, when he quickly became overheated and tired and had to be recalled to the cockpit by Conrad. Dick's task was to connect a tether between his Gemini and the Agena to which it was docked, so that the two vehicles could later undock and experiment with using the tether as a fuel-saving device, keeping two vehicles together by spinning them slowly and keeping the tether taut. Dick had been able to make the necessary attachments without undue effort in training, but in flight he found it an extremely tough task, and his body flailed about as he tried to maintain a proper position astride the nose of the Gemini while groping with the balky connector. "Ride 'em, cowboy" was the only advice Conrad could offer to the sweating Gordon, who returned to the cockpit shortly thereafter, having finally won the wrestling match at the expense of the remainder of his EVA experiments, which had to be canceled.

In view of the difficulties encountered during EVA by Gene Cernan on Gemini 9 and Dick Gordon on 11, including high heart rates and obvious physical distress, plus Ed White's and my comments, the program managers decided to use the last flight, Gemini 12, to test a variety of devices to make the space walker's task easier. Buzz Aldrin was the reluctant guinea pig, reluctant because he had hoped to be able to propel himself about, Buck Rogers style, using an elaborate back-mounted maneuvering device, the AMU. Instead he had the less dramatic but equally useful task of evaluating a group of restraining devices, handholds, footrests, and so forth. All in all, Buzz spent five and a half hours outside, working in a highly programmed fashion, resting at regular intervals, and he had no difficulty whatsoever.

Gemini 12, which flew from November 11 to 15, 1966, also accomplished a successful rendezvous and docking and a long list of experiments. It was a very well-executed flight and a good one to end a proud series. Still, it irked me more than a little to read various articles to the effect that NASA had finally been able to solve the problems which had exhausted all the previous space walkers. Ed White hadn't had any problems, and neither had I, except for the obvious fact that the Agena had never been designed to be space-walked upon. Certainly Buzz's tinkering with a board full of restraining devices was child's play compared to removing an experiment package from a second (inert and unprepared) satellite. I was also puzzled that Gemini 12 had been given a fourth day to complete its assigned tasks, although its list of experiments was no longer than Gemini 10's. Oh, how John and I could have used a fourth day!

The following information, in tabular form, recounts some of the highlights of the Gemini program. It does not include the main points, which are (1) Borman and Lovell on Gemini 7 proved that man could stay weightless for fourteen days without serious physical problems, and fourteen days was nearly twice as long as a lunar landing mission; (2) rendezvous in various forms, using a variety of techniques, was proved to be practical on Geminis 6, 8, 9, 10, 11, and 12—and if we could do it in earth orbit, so could we around the moon; (3) an intricate, competent testing, planning, and flight control team was assembled—capable of minimizing the very real risks in venturing a quarter of a million miles from earth and returning safely, despite the defects which inevitably occur when fallible man assembles complex machinery. Now some of the details:

PEAK HEART RATES DURING LAUNCH AND RE-ENTRY

Mission	Crewman	Peak rates during launch beats/minute	Peak rates during re-entry beats/minute
3	Grissom	152	165
	Young	120	130
4	McDivitt	148	140
	White	128	125
5	Cooper	148	170
	Conrad	155	178
6	Schirra	125	125
	Stafford	150	140
7	Borman	152	180
	Lovell	125	134
8	Armstrong	138	130
	Scott	120	90
9	Stafford	142	160
	Cernan	120	126
10	Young	120	110
	Collins	125	90
11	Conrad	166	120
	Gordon	154	117
12	Lovell	136	142
	Aldrin	110	137

I don't know what to make of these data. "If you can keep your head when all about you . . ." perhaps you don't understand the situation. NASA has conducted some interesting tests involving two men in high-speed aircraft which show that anxiety (or at least a high heart rate) shows up as a result of who has the *responsibility* for the flight, rather than who is doing the actual piloting. If applied to the above table, this logic would tend to drive up the heart rates of the commanders of each mission.

RE-ENTRY SUMMARY

Mission	Distance of miss, nautical miles from target
3	60
4	44
5	91
6	7
7	6.4
8	1.4
9	.38
10	3.4
11	2.65
12	2.6

For a time, John and I were the record holders for this event, despite the fact that 9 landed much closer to the ship than we did. The reason? Simply that the Fédération Aéronautique Internationale established a new class record after the flight of 9. As the first flight to go after the new category was established, John and I were automatic winners. I suppose 11 took the record away from us, but I don't know, because the FAI only notifies—and sends certificates to—the new champs, not the dethroned ones.

RADIATION DOSES ON GEMINI MISSIONS

Mission	Duration Day:hour:minute	Mean cumulative dose in millirads Command pilot	Pilot
3	0:04:52	20	42
4	4:00:56	42	50
5	7:22:56	182	170
6	1:01:53	25	23
7	13:18:35	155	170
8	0:10:41	10	10
9	3:01:04	17	22
10	2:22:46	670	765
11	2:23:17	29	26
12	3:22:37	20	20

The reason John and I got so much was simply that our high-altitude orbits were grazing the inner Van Allen radiation belt, especially in the critical area of the South Atlantic (the "anomaly," where the belt dips down to lower altitudes). Even 765 millirads, however, is considered well below the threshold of danger to health.

CREW WEIGHT LOSS (POUNDS)

Mission	Command pilot	Pilot
3	3	3.5
4	4.5	8.5
5	7.5	8.5
6	2.5	8
7	10	6
8	Unknown	
9	5.5	13.5
10	3.0	3.0
11	2.5	0
12	6.5	7

Generally speaking, the longer the flight the greater the weight loss to be expected. I guess the 13.5-pound loss experienced by space walker Gene Cernan during the three days of Gemini 9 was caused by dehydration resulting from his tough EVA struggles with bulky equipment.

EXPERIMENT PERFORMANCE

Mission	Percentage of mission time planned for experiments	Number of experiments	Number accomplished
3	5%	3	2
4	16%	11	11
5	17%	17	16
6	12%	3	3
7	22%	20	17
8	21%	10	1
9	21%	7	6
10	37%	15	12
11	29%	11	10
12	30%	14	12

Percentage-wise, John and I carried the heaviest load; only the two long-duration flights (5 and 7) carried a greater number of experiments than we did.

HEART RATE DURING SPACE WALKS
(not available for Ed White)

Mission	Average heart rate	Peak heart rate
9 (Cernan)	150	180
10 (Collins)	118	165
11 (Gordon)	140	170
12 (Aldrin)	105	155

Aldrin's workload was carefully controlled and interspersed with rest periods. The rest of us did whatever was necessary at whatever frantic pace was required to get the job done. Despite this, the only part of me that felt tired was my fingers, from working inside the cumbersome pressure suit gloves.

PROPELLANT USAGE (POUNDS) DURING RENDEZVOUS MANEUVERS*

Mission	Actual fuel used	Minimum theoretically possible	Ratio of actual to minimum
6	130	81	1.60
8	160	79	2.02
9 Rendezvous 1	113	68	1.66
Rendezvous 2	61	20	3.05
Rendezvous 3	137	39	3.51
10 Rendezvous 1	360	84	4.28
Rendezvous 2	180	73	2.46
11 Rendezvous 1	290	191	1.52
Rendezvous 2	87	31	2.81
12	112	55	2.04

* This table and its six predecessors are all taken from NASA SP–138, "Gemini Summary Conference," February 1–2, 1967.

No doubt about it, we were the champion fuel guzzlers on 10.

With Gemini safely and successfully completed, NASA management homed in on the faltering Apollo program and made a series of changes that radically affected the lives of some of us crew members. First, the leaping frog leaped into oblivion—the 014 flight was canceled, the reasoning being that it was an unnecessary and unjustified repeat of the 012 flight. If Grissom-White-Chaffee were to spend fourteen days in earth orbit in a Block I command module, then why did Schirra-Eisele-Cunningham have to do it all over again? So 014 was canceled, and the Schirra crew became back-ups to the Grissom crew, releasing McDivitt-Scott-Schweickart, who were promptly made prime crew of the second manned flight—a complex earth-orbital test of the first Block II command module and the first manned LM. Aha, this was significant indeed! Schirra could easily have been given this plum instead of McDivitt, but someone obviously thought McDivitt's crew was better suited to absorb the horrendous complexities of this mission, involving the simultaneous flight of two manned spacecraft. The LM was especially tricky, because once McDivitt and Schweickart got in it and separated from Scott in the command module, they *had* to get back, having no heat shield on their frail craft, and therefore being unable to return to earth separately without burning up. Schirra's assignment, on the other hand, looked to us insiders to be a dead end for the Mercury and Gemini veteran, an assessment that was shortly—and tragically—to be proven incorrect.

Of course, the cancellation of 014 also freed Borman-Stafford-Collins for reassignment, and reassigned we were, but not as a unit. Tom Stafford moved up a notch and acquired his own highly experienced crew, John Young and Gene Cernan; they became McDivitt's back-up. Score one for Tom. Borman and Collins got promoted to prime crew of the third manned flight, picking up Bill Anders as our third member. In the process, Collins also got "promoted" from lunar module pilot to command module pilot, and lost right then and there his first chance to walk the surface of the moon. The reason I had to move up was that Deke at that time had a firm rule that the command module pilot on all flights involving an LM must have flown before in space, the idea being that he didn't want any rookie in the CM by himself. Since Bill Anders had not flown, I was it. Slowly it sunk in. No LM for me, no EVA, no fancy flying, no need to practice in helicopters any more. Instead I was the navigator, the guidance and control expert, the base-camp operator, the owner of the leaky plumbing—all the things I was least interested in doing. Years later, I have answered a thousand times the question "How did you and Armstrong and Aldrin decide who was going to stay in the command module and who was going to walk on the moon?" I have answered it a hundred ways, none of them completely honest, but then it's so hard to say, "Listen, lady, when they canceled 014 I lost my chance," even if that is 99 percent of it. From late 1966 on, I became a command module specialist, and though I changed crews, I never changed specialties again.

All was not gloom, however, as there was a great deal of satisfaction in switching from Cunningham's back-up to a prime crew slot on the third manned flight. Our new flight was going to be a fascinating one. It would be the first manned flight aboard the giant Saturn V moon rocket, and while we would not leave earth orbit, I would regain my former altitude record and then some—soaring to a four-thousand-mile apogee from which the entire earth could be seen, from pole to pole. Then Borman and Anders would give the LM a workout, while I remained in the CM. This was heady stuff, and it was especially pleasant to be setting our own pace, instead of being tied to Schirra-Eisele-Cunningham, who could be exasperating co-workers. Wally was late every morning, never apologized, and never tried to catch up with the schedule, but instead wasted another forty-five minutes on guffaws, coffee, and war stories before finally settling down to work. Cunningham bitched constantly, at Wally and the world, and Eisele served as a good-natured referee who didn't quite understand what was going on half the time. Now we were free of them, free to move out as only Borman could move, free to plan and organize and drill for 1967. First we had to get Grissom airborne, perhaps in February, then McDivitt, and then—hopefully by late summer—it would be our turn. As we discussed plans with our families over the Christmas holidays, none of us had, so far as I know, the slightest premonition that disaster was about to overtake Apollo. No crew would fly for the next twenty months, and that crew would be none other than Schirra-Eisele-Cunningham.

On Friday, January 27, 1967, the astronaut office was very quiet, practically deserted, in fact. Al Shepard, who ran the place, was off somewhere, and so were all the other old heads. But someone had to go to the Friday staff meeting, Al's secretary pointed out, and I was the senior astronaut present, so off I headed to Slayton's office, note pad in hand, to jot down another week's worth of administrative trivia. Deke wasn't there either, and in his absence, Don Gregory, his assistant, presided. We had just barely gotten started when the red crash phone on Deke's desk rang. Don snatched it up and listened impassively. The rest of us said nothing. Red phones were a part of my life, and when they rang, it was usually a communications test or a warning of an aircraft accident or a plane aloft in trouble. After what seemed like a long time, Don finally hung up and said very quietly, "Fire in the spacecraft." That's all he had to say. There was no doubt about which spacecraft (012) or who was in it (Grissom-White-Chaffee) or where (Pad 34, Cape Kennedy) or why (a final systems test) or what (death, the quicker the better). All I could think of was, My God, such an obvious thing and yet we hadn't considered it. We worried about engines that wouldn't start or wouldn't stop; we worried about leaks; we even worried about how a flame front might propagate in weightlessness and how cabin pressure might be reduced to stop a fire in space. But right here on the ground, when we should have been most alert, we put three guys inside an untried spacecraft, strapped them into couches, locked two cumbersome hatches behind them, and left them no way of escaping a fire. Oh yes, if the booster caught fire, down below, there were elaborate, if impractical, plans for escaping the holocaust by sliding down a wire, but fire inside the spacecraft itself simply couldn't happen. Yet it had happened, and why not? After all, the 100 percent oxygen environment we used in space was at least at a reduced pressure of five pounds per square inch, but on the launch pad the pressure was slightly above atmospheric, or nearly 16 psi. Light a cigarette in pure oxygen at 16 psi and you will get the surprise of your life as you watch it turn to ash in about two seconds. With all those oxygen molecules packed in there at that pressure, any material generally considered "combustible" would instead be almost explosive. And combustible materials—books, clothing, supplies—there were aplenty, also plenty of ignition sources. There was supposed to be none of the latter, but let's face it, the inside of a Block I spacecraft was a forest of wires, a jungle which had been invaded over and over again by workmen changing, and snipping, and adding, and splicing, until the whole thing was simply one big potential short circuit.

As we sat there stunned, the red phone rang again and delivered additional details—rescue crew on the spot but unable to enter because of excessive heat . . . damage confined to command module alone—no word from the crew or sign of activity within. Hell no, nor would there ever be—the only question was: How quickly, how quickly? Were they baked, roasted, cremated, or asphyxiated? Five seconds or five minutes? How about their families? They had to be told something somehow, and swiftly. We had learned *that* the hard way in the case of Ted Freeman, whose death had been announced to his new widow, Faith, by a newspaper reporter at her door seeking additional details. I called the astronaut office and got Al Bean, level-headed, reliable Al Bean, who said he would organize the notification of the wives while I stayed near the best source of information, the red telephone. Within a few minutes, Al had found astronauts and wives to go quickly to the Grissom and White households, but he had not been able to find anyone to notify Martha Chaffee. It couldn't be just anyone; it should be a close friend who was an astronaut himself, who could somehow say the unspeakable and ease her into thinking the unthinkable. Gene Cernan, her next-door neighbor, would have been the one, but he was off somewhere. With an awful sinking feeling in my stomach, I realized I was next best—Al Bean could do more good staying on the office telephone coordinating details—so I told Al I would tell Martha, and drove very slowly the mile and a half to her house, three doors down from ours. Al had sent his wife Sue and some neighborhood wives over to screen calls and visitors, so Martha knew something bad had happened, but *this?*

They were all waiting for me in the family room, big-eyed and silent. Martha stood out in the small group—she radiated concern, awareness, and quiet resignation. Usually Martha stood out because of her beauty. She had everything: the healthy cheeriness of a college cheerleader, wide cheekbones, a perfect chin, a model's poise, an athlete's trim body. Of all the astronaut wives, Martha Chaffee stood out in a group, her bronzed blondness shining like the Cape of Good Hope beacon. Besides all that, she had a good head—optimistic, bright, helpful to Roger in his zany new career—Christ, she even studied lunar topography with him. Now I had to come along and do this. "Martha, I'd like to talk to you alone." "Yes," she said, and headed down the narrow hallway with me a step or two behind. I hadn't expected Apollo to mean this, risk certainly, but not the trauma of telling beautiful women that their husbands had fried.

Arlington Cemetery was no better. For the third time in less than three years, we were assembled to bury dead comrades. At least we knew the routine by now: take the unfamiliar uniform out of the cedar closet; load widow, children, one close friend, and one preacher aboard the Gulfstream; and meet them in Washington that night at the Georgetown Inn. Up the next morning, assemble glumly in the lobby and peer around at the strange mix of unfamiliar uniforms ("I wonder how he happened to get that DFC?"); then into limousines for the short trip to Arlington. Arlington at the end of January was at its soggy, misty, gloomy worst. At least we didn't get rained on as we, the honor guard, walked along in mournful cadence on either side of Gus and Roger's horsedrawn caskets. Finally meaningless words were spoken, rifles were fired, and the two men were moved for the last time. Ed White was being buried the same day at West Point, and I had wanted to be there, as much as anyone can want to be at a funeral. Ed and I had been West Point classmates a hundred years before, and we had worked together as the Gemini 7 back-up crew. But Roger was a classmate too, in a very real sense; we had been hired the same day, with twelve others, and during the past three years the fourteen of us had developed a strong bond of friendship and support. Furthermore, although there wasn't a thing in the world I could do to alleviate Martha Chaffee's grief, I wanted to try, to follow through somehow with help and solace, for after all I had been with her from the ghastly beginning. I had explained this as best I could to Pat White, and she understood.

Martha was left with a folded flag and two children to rear, and the other wives were left with the knowledge that space flight, just like airplane flight, *did* kill. Always before it had been a theoretical possibility, a paper exercise in percentages, but it had never happened, not all through the Mercury and Gemini years. Now this Apollo had destroyed three without even flying one; what was the pattern? Would one disaster follow another, just as airplane accidents seemed to occur in clusters, or could the program absorb one catastrophe and still recover quickly? How could NASA get going again? How many astronauts would decide they hadn't signed up to be incinerated and quit? How many wives would quit if hubby didn't? The answer of course is that no one quit, not husband or wife, and that is a record to be proud of, I think; but how many close calls there might have been, no one can say. I know I never discussed it with Pat in those terms. We didn't discuss it in any but the most superficial and peripheral way. I suppose I was afraid to measure the depth of Pat's resentment and hostility toward this Apollo which held us both captive.

In the dismal early months of 1967, it became increasingly obvious that the fire in 012 was not simply a one-time freakish occurrence but indicated a generic weakness in the command module family. The first mistake was the environment—pure oxygen at 16 psi. The second was the abundance of combustible material exposed to that highly inflammable environment. Third, and most insidious, was the lack of an iron-clad system for controlling last-minute changes to the spacecraft; it seemed that too many changes had been approved and those too sloppily executed. The investigating board that spent months examining the charred carcass of 012 never did determine what triggered the fatal spark, but they discovered something even worse—that there were scores of possible sources, and that the mound of paper work which accompanied the spacecraft was not a completely accurate representation of the true condition of the vehicle. It appeared that some work had been done that never showed up in the paper work; on the other hand, some jobs had been recorded but not properly finished. There is a facetious saying in the airplane business to the effect that not until the weight of the paper work equals the weight of the airplane will it be cleared for takeoff. In the space business, paper is *the* most important material. Without paper, chaos results, and no one knows which jobs have been talked about but not performed, and which performed but not talked about. This is particularly true in the harried, hurry-up, three-shifts-a-day environment that inevitably precedes a manned space flight, especially the first one in a series. The midnight shift, finding itself with an exposed wire bundle fifty-six wires thick, may just splice into the wrong one unless the accompanying paper work has been kept to perfection by the previous shift.

In my judgment, NASA and North American each dug into the problem with a great deal of professionalism. Initially, there was some caterwauling and finger pointing,* but soon each side realized that the main point was not to assign blame but to take concerted action to get the program moving again—and safely this time. The most difficult job was replacing combustible material with nonflammables, especially for garments, towels, food bags, and other personal gear. Practically any substance will burn if ex-

---

* Partisans could take either side of this argument. NASA: North American has been incredibly sloppy and doesn't even know what was inside 012 at the time of the fire. North American: We were only responding to NASA pressure to hurry up, and besides, NASA was supposedly supervising and approving our work at each step along the way.

posed long enough to a sufficiently hot flame in the presence of abundant oxygen. Even stainless steel burns readily in pure oxygen. The exterior layers of the Gemini and early Apollo pressure suits were made of Nomex, a high-temperature nylon fabric which had to be heated to over 700 degrees F to burn, and then it burned slowly; but it was replaced as a result of the fire with Beta cloth (woven glass filaments). Glass underwear can be a bit scratchy, and Beta cloth as an outer garment wears out very rapidly, resulting in minute glass particles floating free throughout the cabin, where they can be inhaled into the lungs. So Beta cloth should be coated with something, such as Teflon. Thus, what seems like a simple problem escalates. Each new material considered had to be subjected to exhaustive testing, and of course, it was not only suit coverings which were to be replaced, but virtually any exposed component inside the command module. It quickly became apparent that although solutions were fairly straightforward, it was going to take one hell of a long time to implement them.

In addition to new materials, new mechanisms were required. The side hatch, for example, needed redesign to permit swift egress: there were actually two hatches, and the inner one had to be laboriously removed using a torque wrench, exposing the outer one, which could be removed only after the heavy inner hatch had been set aside. The two were combined into one, and the latching mechanism was vastly simplified. All these things took time, and our hopes for getting three manned flights off in 1967 quickly evaporated. A year's delay at least, said the smart money, but it ended up being closer to two years.

Meanwhile, work throughout the Apollo program continued without letup, and the astronaut office was no exception. We spent more time in Houston and less in Downey, but there were still a million details to be ironed out before anyone in his right mind would remotely consider setting sail for the moon. The fire, horrible as it was, gave the other parts of the program much needed breathing room, while North American struggled with its redesign problems. Saturn V problems, LM problems, ground radar tracking and computing problems—everywhere one looked, people were struggling to catch up. I don't think the fire delayed the first lunar landing one day, because it took until mid-1969 to get all the problems solved in areas completely unrelated to the fire. One small example was the on-board computer which both the CM and the LM carried. This compact little unit had a thirty-eight-thousand-word vocabulary, which wasn't much when one considered the complex rendezvous and other problems it would be called upon to solve. Clearly, each of the thirty-eight thousand words had to be picked and arranged for maximum advantage to the crew—and the language had to be efficient, direct, and simple—which it was not. Nor were complaints limited to the computer and its software. Many items of hardware were suspect as well.

The Borman-Collins-Anders crew was tied up with its own problems. Borman was a member of the board investigating the fire and spent virtually all of his time during early 1967 at the Cape sorting out what had happened inside 012, or at Downey, supervising the implementation of the changes the board had recommended. One of these was that the Block I CM should be shelved entirely, which meant that I could begin learning the Block II systems. In fact, I inherited my own Block II spacecraft to look after, serial 104. At the same time, Bill Anders was absorbed in LM problems, and I became more and more involved with the planning of our flight, which was known by a variety of names and numbers but which was generally called 503.*

The first unusual thing about our flight was that the behemoth Saturn V rocket was to be flown manned after only *two*

---

* It takes bookkeeping more precise than mine to explain the various systems of nomenclature in their entirety, but the highlights are: the Grissom-White-Chaffee flight would have been called Apollo 1. It was also called 204, because it was to be carried aloft by the fourth booster of the second Saturn series, the Saturn 1B. After the fire, the numbers were changed, and the Schirra flight became known as Apollo 7 because it had been preceded by six unmanned test flights. It was also still 204, since it used Grissom's booster. The Borman-Collins-Anders flight was called 503 because it was the third flight of the Saturn V, following the unmanned test flights 501 and 502. Sandwiched between Schirra and Borman was the McDivitt flight, which had so many different numbers at one time or another that I won't even attempt to list them.

unmanned tests. The Atlas booster of the Mercury program had had about fifty unmanned tests before John Glenn rode one, and nearly as many unmanned flights of the Titan II had preceded Grissom and Young's Gemini 3. The second strange thing about 503 was that the third stage of the Saturn V would be reignited, just as if it were a lunar mission. However, ours would be shut down early, causing us to stay in earth orbit, and to soar up to an altitude of four thousand miles. This little detail created all sorts of planning problems, because one could only escape from this lopsided orbit at certain prescribed intervals, and if one had troubles and was forced to return to earth prematurely, it was entirely possible to end up landing in Red China. Following our high-altitude stint, we would give our LM a workout, conducting an elaborate rendezvous sequence.

The matter of rendezvous was no small item; in fact, if there was a hot potato around the astronaut office, it was the question of how many preliminary rendezvous would be necessary to validate the Apollo equipment prior to attempting the lunar landing. Rendezvous is such an arcane art that I find it difficult to explain to the layman, but in general terms, the problem was that there were a host of important variables. For example, the LM could approach the CSM (command/service module) from below, as in the normal case, or from above, as was necessary in certain cases where normal timing was not possible. The LM had two guidance systems, a primary and an alternate, called the abort guidance system; the latter should be allowed to prove its worth by successfully demonstrating at least one rendezvous. Next, the LM rendezvous could be conducted with just its upper half (the ascent stage), pretending the descent stage had been left on the lunar surface, or it could be practiced with ascent and descent stages still joined. There were a lot of significant differences between the two configurations. Finally, the overtaking LM could close from a small differential altitude (Δh) at slow velocities, or it could come whizzing in at a great closing rate from a large Δh. These, then, were the four main variables:

1. LM above or below
2. Primary or abort guidance system
3. Ascent stage alone, or descent stage attached
4. Small or large Δh

When people think of astronauts in training, they seem to relate the training to the roadwork of the prize fighter or the meditations of a solitary man seeking his karma. Nobody ever mentions Δh, but that is what was important. The rendezvous heavyweights —Tom Stafford, Buzz Aldrin, Pete Conrad, Neil Armstrong, Dave Scott—may have run or meditated, or they may not have, but they sure as hell spent hours worrying, and talking, and planning rendezvous with all its complexities. They were not always right, but they were always influential, and the fact that Apollo 11 was first to land on the moon, instead of Apollo 10 or Apollo 14, was due in large part to their deliberations. As I said, they were not always right, and I can remember one watershed meeting on April 26, 1967, at which the astronaut office took the firm position, among other things, that it was *mandatory* to demonstrate one rendezvous with the LM approaching from above and another with the abort guidance system in control. Neither of these rendezvous was ever performed. However, by and large these rendezvous planning sessions, and to a lesser extent their counterparts concerned with EVA and other phases of the upcoming flights, were not only useful to us as crew members but also helped solidify NASA's planning. The name of the game was to get to the surface of the moon as swiftly as possible, while at the same time investigating as many potential problems as possible in as few preliminary flights as possible. This philosophy made 503 and the other early flights as complicated as the flight and ground crews could handle, and planning for 503 certainly kept me hopping during the rest of 1967.

Fortunately, there were a number of breaks in my schedule, to rescue me from Houston meetings and midnight vigils with spacecraft 104 in Downey. In April a group of us was sent off to Key

West, to attend a one-week Navy scuba-diving school. The idea was not to enthrall us with the beauties of the deep, but simply to teach us the bare fundamentals of underwater operations, since ballasted underwater training was becoming a popular substitute for the short-duration parabolas of the zero-G airplane. A large water tank in Houston (and a huge one at NASA Huntsville) was built during this time as zero-G simulators. In addition, we kept up our geology training, which, although not nearly as interesting as scuba diving, seemed to be a necessary evil. I think all of us felt we had more compelling, more immediate things to do, but none of us was willing to abandon the possibility of lunar prospecting which geology expertise seemed to enhance.

Another welcome diversion came at the end of May 1967, when NASA sent Dave Scott and me to the Paris Air Show, at which a couple of cosmonauts were supposed to appear. I had never been to the Paris Air Show—excuse enough for me to want to make the trip. Also, it was one of those rare occasions when our wives were invited, and I knew Pat would enjoy another look at Paris—even if she did have to suffer through an air show or two. Sure enough, two cosmonauts showed up, and Dave and I were fascinated by our first glimpse of the competition. How should Dave and I treat Colonel Pavel Belyaev and Mr. Konstantin Feoktistov, products of a hostile system, yet comrades in a sense Marx never envisioned? We decided to be as friendly and open as possible. I expected that they would learn more from us than we from them, but so what; the Russian experts could glean everything they wanted to know about Apollo from unclassified literature anyway. On the other hand, the Russian program was hidden from view, secret and mysterious, and if our side knew what was going on, the information never trickled out of the CIA files down to us working troops in Houston.

Our first meeting with the Russians took place in their pavilion, out on the open floor, and a huge melee immediately ensued as the two of them tried to show Dave and me around. Photographers, autograph seekers, security men, and confused tourists, all back-to-back and belly-to-belly, we swirled around in an aimless eddy, moving in whatever direction the pushing and shoving took us. In the midst of all this, our wives joined the fray, helping the situation not one bit. Finally, one of the Russians had a brilliant idea and invited us on board their jetliner, a TU-134 which was parked just outside. At the airplane the Russian security people were able to prevent the uninvited from climbing the boarding ramp (except for a reporter or two who slipped through), and soon the four of us were seated around a table, chatting amiably through an interpreter and chugalugging vodka. They inquired as to the health of the Grissom-White-Chaffee wives, and we responded with similar concern for the widow Komarov, whose husband had recently been killed when the parachute on his Soyuz I became fouled. We drank toasts to no more space accidents, we drank toasts to increased cooperation between our two nations, and we drank toasts to a couple of other things that slip my mind. We found Belyaev and Feoktistov good fellows indeed, although Feo the Fink, as we dubbed him, was adding soda water to his vodka, while the three of us were tossing it down neat as each toast was proposed. Feoktistov, Scott, and I were in civilian clothes, but Belyaev was resplendent in his uniform, which was festooned with all kinds of interesting-looking doodads. We heard later that Feoktistov, in addition to serving as a crew member of the three-man Voskhod flight, was one of their top spacecraft design people, but at the time this was not apparent. He kept very much on the periphery of the conversation, and with his glasses, gray hair, and serious mien, he seemed an unlikely cosmonaut. The autograph seekers, who usually have a pretty good eye for such things, earlier had jostled Feoktistov aside to get at the glittering, bemedaled Belyaev.

In Belyaev we found a kindred spirit. I liked him, and I would have flown with him. He exuded not only good humor but an air of quiet competence. He asked the right questions and seemed to grasp the answers immediately, although on the more technical questions both sides got hung up in the translation. We had a dear, sweet lady from the American Embassy who spoke fluent Russian, but she was not a technical person, and we spent a lot of time trying to explain things to her so she could explain them to the Russians, and vice versa. We found that the Russians had a group of cosmonauts training on helicopters, and that Belyaev himself expected to make a circumlunar flight in the not-too-distant future.* If the Russians weren't interested in a manned lunar landing, if—as they subsequently said—they were not racing us to the moon, then why on earth (no pun intended) were they training cosmonauts to fly helicopters in 1967?

Later the two cosmonauts paid a return visit to the United States pavilion, and this time we continued the conversation over American coffee instead of Russian vodka. Even with the lesser stimulant, the session was extremely friendly, almost boisterous, and we discovered a long list of common interests and complaints, such as distaste for the medics and enthusiastic endorsement of the attractive little "poochies," the girls outfitted in colorful miniskirts by Emilio Pucci who swirled around us serving coffee. As we parted, I wondered just how meaningful our budding friendship could be. It would be the ultimate in naïveté to imagine that space pilots' mutual concerns could measurably modify the disparate national interests of their respective countries, but still, if close rapport could be so quickly and easily established in one field, why couldn't it spread to others and eventually bring the two countries closer together? At least it was nice to know that somewhere another group like ours was struggling with similar problems, even if they were flying MIG-21s instead of T-38s. Dave and I left Paris feeling like diplomats carrying a signed treaty.

From Paris, Pat and I detoured through eastern France and visited Chambley, the village near Metz where we had been married ten years before. Actually, we had been married there twice, once in a civil ceremony and once in church. Despite this, we really weren't all that keen to renew our acquaintance with the nonde-

* Poor Belyaev never flew in space again, but died of complications following surgery for stomach ulcers in January 1970.

script city hall, the cluster of bleak farmhouses, the steamy community manure pile—but everyone insisted. NASA insisted, our embassy in Paris insisted, our French friends insisted—so we surrendered and went. When we got there, the reason immediately became clear: we had been mousetrapped into getting married a third time. A huge entourage (the entire village plus a couple of hundred outsiders) greeted us. A brass band blared, and a squadron of schoolchildren, released from classes especially for the occasion, escorted us with great whoops and whistles to the town hall. There waited the mayor, resplendent in his Sunday suit, engirdled with a crimson sash of office and festooned with ribbons and medals. With a great flourish we were ushered inside and immediately got down to the serious business of wedding number three.

God, I remembered wedding number one all right, the day of the shakes. When we arrived at city hall at the appointed hour, we found that the mayor was nowhere to be seen, but his substitute (the assistant mayor?), an octogenarian in obviously frail health, was prepared to do the honors. Whether from nervousness or palsy, the poor man got the ceremony off to a shaky start, literally, as the twitches of the book in his hand were matched only by the quavers in his voice. Pretty soon Pat began to quiver and shake, and then me. The two official witnesses, farmers in blue berets and hip boots, grinned toothlessly from the corner. By now the three of us were synchronized, standing there jiggling in unified if unsteady obeisance to the solemn marital laws of France. At certain intervals it was necessary for us to signal assent, at which times the tremulous voice would stop without warning and the farmers would stage-whisper, "*Dites oui!*" and we would croak, "*Oui.*" God, I remembered wedding number one. Wedding number two was by comparison a smooth and flossy ceremony in the nearby air-base chapel, with more Chillocothe, Ohio, in it than Alsace-Lorraine.

Now we were back for thirds, back at the very spot of number one, and I started to feel cold and clammy, even amid all the grinning faces. Would someone think to say, "*Dites oui*"? My bride of ten years, however, was a steadying influence, as usual, and toward the end I even began enjoying the shakeless ceremony. I most surely enjoyed the champagne that followed. Champagne, I am firmly convinced, is most pleasurable when one is surprised by it, when it appears unexpectedly and out of context, like in a Norwegian sauna at ten in the morning. An unexpected third marriage certainly rates champagne, and plenty of it. Pat and I finally escaped amid a hailstorm of rice, feeling very married indeed. The children would be pleased, and we would be pleased to see them and Houston again.

After the welcome break this week in France provided, it was back to Apollo with full force. As 1967 drew to a close, the hardware started to become more familiar and the 503 flight plan began to fall into place. With familiarity comes confidence, and I began to feel comfortable in the Apollo shoes which had squeezed and squeaked so badly the year before. Gemini became a memory, rather than a competitor, and the moon, which had almost disappeared in the aftermath of the fire, once again shone and monthly seemed larger and more inviting. But we still had a hell of a long way to go; we weren't going to get Wally and crew airborne until summer of 1968 at the earliest, and they were going on a puny Saturn IB rocket. How about the Saturn V, the moon monster needed to propel 100,000 pounds' worth of CSM and LM into a lunar trajectory? If the Saturn V crumped, it was a whole new ball game, and the moon might be impossibly far away, at least as far as our American program was concerned.* In addition to worrying about the future of our space program, I had a more personal reason for sweating out the Saturn V, namely that it was going to be my pink fanny riding the third one. Consequently, it

---

* The Russians were rumored to be building their own moon rocket. While the Saturn V was to produce 7.5 million pounds of thrust, far more than anything the world had seen so far, the Soviet behemoth was considered by Western "experts" to be in the 11–14-million-pound class. I don't know whether it ever existed or not, but press reports over the years have it undergoing a series of trials and tribulations; and once in the summer of 1969, so the story goes, it exploded violently on its launch pad, killing a blockhouseful of high-ranking space officials and military officers.

was with a great deal of trepidation that, on November 9, I stood on a causeway three or four miles from Launch Pad 39A at Cape Kennedy* and watched 501 go.

The monster squats there, all 365 feet of it, and steams daintily in the thin November sunshine. The first stage burns liquid oxygen and high-grade kerosene, which will produce a highly visible flame, unlike the almost transparent exhaust of the Titan, so it should be a spectacular sight as the five first-stage engines ignite, *each one* gulping over *three tons* of propellant per second. It will be even more spectacular if the son-of-a-bitch blows up, in which case four miles away may not seem too much. The public-address system is blaring away, describing pressures and valves and blarney for the few who are listening. I should be interested in the details, but I'm really not. I just want a yes or a no, a thumbs up or down. This monstrosity determines my immediate future either way, and 503 is so sensitive to the vagaries of 501 that I feel utterly dominated by this white pencil across the lagoon.

Von Braun and his Huntsville cohorts have been really clever: they have not only built the machine, they have taught it to reveal itself, to report temperatures and pressures from every nook and cranny to a regiment of skeptical engineers, some of whom have been launching birds since Peenemünde days. Even after the engines ignite, *especially* after the engines ignite, 501 will frantically report its well-being, and the regimental commander will not order its steel vise grip released until satisfied that all is in working order. That is reassuring. The tempo of noise from the loudspeaker picks up in the terminal count (what a ghastly term), the small crowd hushes, and all faces turn north. Slowly a trickle of flame appears, directly under the rocket, seems to burp once or twice, and then cascades out to either side and circles back skyward. This is

---

* Actually the Saturn V launches don't take place from Cape Kennedy, which is the Air Force side of the launch complex, but rather from the NASA side, known as MILA or Merritt Island Launch Area. The causeway on which I stood separates MILA from Cape Kennedy. However, it's easier to bow to popular nomenclature and call the whole region "the Cape," which irritates my wife, who insists that everyone knows "*the* Cape" means Cape Cod.

normal, the work of the flame deflector, a gigantic steel and concrete double scoop below the rocket, designed to sweep aside the exhaust gases which burn all in their path. The flame pattern resembles a gigantic pair of hands, one on either side of the rocket, cradling it gently. The fire at the beginning was a conventional red-orange, but now it changes to incandescent white at the source and dirty brown on the outer edges. Is it just going to sit there and be consumed by fire, or is it going to go? Just when I begin to despair, the motion begins, imperceptibly at first, then a stately, dignified climb up the side of the launch tower. "Lift-off, we have a lift-off," crows the loudspeaker, but "ooze-off" would be a better word. Certainly not the dramatic "blast-off" the television announcers are so fond of. It almost seems to prove that motion is impossible, as one of Zeno's paradoxes would have us believe. Over two millennia before 501, the Greek philosopher had it figured thus: motion can never occur, because to reach a certain place one must first go half the distance. But to go half the distance one must first go a quarter of the distance, and to go a quarter, one must first go an eighth, and so on. That first yard off the pad is the tough one.

So far the launch looks good, but somehow it lacks the intimacy of a Titan II launch. Big as this Saturn V is, the safety people have pushed us so far back that its scale is completely lost, and besides, it's climbing into the sky without a whisper. Of course, that's it! We are so far away the sound hasn't reached us yet, but when it does, it is a surprise, a jolt, a shock—even for one who thought it overdue. God, it's not a noise, it's a presence. From tip of toes to top of head, this machine suddenly reaches out and grabs you, and shakes, and as it crackles and roars, suddenly you realize the meaning of 7.5 million pounds of thrust—it can make the Cape Kennedy sand vibrate under your feet at a distance of four miles. Supposedly, the acoustic energy kills birds who fly by too closely; what must it be like to ride one?

On the trip back to Houston I have plenty of time to ponder this, as results from 501 keep pouring in. A nearly perfect flight, it was not only the maiden voyage of the Saturn V but was also a test of the CSM, which reached an apogee of 9,700 miles and a maximum speed of 36,500 feet per second, roughly equivalent to the atmospheric entry speed of a spacecraft returning from the moon. The temperature at the surface of the CSM's heat shield had hit 5,200 degrees F, about as expected, and apparently all was comfortable within the cockpit despite this fantastic velocity and temperature. November 9, 1967, was a good day for the program in general and for 503 in particular.

If the Saturn V was an institutional worry, then my own personal hairshirt was the on-board computer and its associated hardware. Together with a telescope, a sextant, and a three-gimbal inertial platform, the Apollo command module guidance and control system was an intricate maze, one which I didn't follow particularly well. I had made several trips to the Massachusetts Institute of Technology near Boston, and had tried my level best to suffer through a couple of weeks of "simple" explanations of the system by their experts, but I always came away shaking my head. They didn't speak my language and I didn't speak theirs.

The basic idea behind the Apollo guidance and navigation system was simple enough. It all began with the stars, whose position in inertial space was well known and *unchanging*. The spacecraft platform, with its three gyroscopes isolated from spacecraft motions, could then be aligned in relation to the stars, providing a fixed frame of reference. The study of the stars themselves was interesting, I thought, one of the most interesting parts of our training, despite its being pure rote. There is something fascinating about the stars. Even today, when I fly in the night sky, particularly in the pure desert air of the Southwest, I look up and experience an almost physical wave of nostalgia. There are my old friends, the stars, ready to guide me back to the moon or past it into the black velvet where only they are visible. They are so far away, of course, that they appear the same whether one is on earth or a mere lunar distance away. They remind us of the puny distances we men travel; the nearest one, Alpha Centauri, is four light-years away, far beyond our capability to visit today or in our lifetime, yet beckoning, mocking us and our dreams. Old friends they are, with friendly yet mysterious names of predominantly Arabic origin. We learned them in the Morehead Planetarium in Chapel Hill, North Carolina, and damned carefully at that—a star sighting using the wrong star would be embarrassing at best, and could easily be disastrous. In our training, we clutched at crutches for memory aids: from the handle of the Dipper we "Arc to Arcturus, Speed on to Spica." The banality of the words was lost in the awe of coming to know these symbols of infinity: we became masters of the sky without leaving the planetarium chamber. Also, one learned to show off. Of course that's Zuben el Genubi, just on the other side of Zubeneschamali. What else could it be, dumdum? The strange names rolled off the tongue with a certain grace and majesty, even with an overlay of North Carolina mountain twang. Altair, Deneb, Vega. Beautiful—I brought one star home with me and named our new pup "Dubhe," leaving a trail of bewildered vets behind. "Do-what?"

The guidance wizards at M.I.T. have read the chicken entrails and decreed that the astronauts' computer shall be given the celestial coordinates of thirty-seven selected stars.* The astronaut, peering out through either his telescope or his sextant, finds one of the chosen few, superimposes a + on it, and pushes a button at the instant of perfect alignment. He then tells the computer which star it was, by number. Repeating this process on a second star allows the computer and the platform to determine which way the spacecraft

---

* For my own amusement I list the thirty-seven, giving each the octal number by which we knew it. (The computer couldn't count to ten, but used a base of eight.)

1. Alpheratz	13. Capella	25. Acrux	37. Nunki
2. Diphda	14. Canopus	26. Spica	40. Altair
3. Navi	15. Sirius	27. Alkaid	41. Dabih
4. Achernar	16. Procyon	30. Menkent	42. Peacock
5. Polaris	17. Regor	31. Arcturus	43. Deneb
6. Acamar	20. Dnoces	32. Alphecca	44. Enif
7. Menkar	21. Alphard	33. Antares	45. Formalhaut
10. Mirfak	22. Regulus	34. Atria	
11. Aldebaran	23. Denebola	35. Rasalhague	
12. Rigel	24. Gienah	36. Vega	

is pointing. So we now know which way is up? Well, not exactly, because "up" is a rather fragile concept meaning away from the center of the earth, a direction opposite the gravity vector used to clutch us tightly. But suppose we cannot even see the earth in our window, suppose we are floating free of earth's gravity. What now, M.I.T.? Back to our friendly stars. We simply define a new up-down and left-right, using the stars in place of the earth. All will be well as long as we all play the game by the same rules, as long as the ground controllers send us instructions (I almost said send us "up" instructions) using the same stellar frame of reference. Now we are free of all terrestrial conventions and can correct our course to and from the moon by pointing our rocket engine in the proper direction relative to the stars.

In 1964, I had no more idea of all this than "a pig on roller skates," a phrase Congressman Wayne Hays used a few years later to describe my entry into the State Department. By 1967, it was beginning to come through, and by 1968, I was even writing memos myself in M.I.T.-ese. ". . . The following procedure is recommended for translunar and transearth burns: MSFN computes external AV maneuver. MSFN updates state vector, using LM state vector locations. Crew transfers the MSFN vector from the LM to CSM locations in the computer. This is called the 'unzap' procedure. Crew executes burn, using MSFN vector. Crew transfers post-burn vector from CSM to LM, called 'zap.' Crew continues on-board navigation, updating CSM state vector but not changing LM state vector. This keeps the W-matrix happy . . ."

The moon was looking a lot closer to Borman-Collins-Anders as 1968 wore on, because people around Houston were reviewing the goals of the first three manned flights, and quiet support was building for changing the third flight (ours) from a four-thousand-mile apogee to a 230,000-mile apogee—that is, to fly around the moon. A modification to this idea proposed actually going into orbit around the moon, not to land of course, but to get within sixty miles of it, although I thought it was rather far-out to entertain such notions before the first manned flight got airborne.

At about the same time, another concern started intruding more and more forcibly into my consciousness, and soon had Apollo nearly completely blotted out. There was something wrong with me, something insidious, something getting worse, something obviously serious. It began with an awareness during handball games that my legs didn't seem to be functioning normally, but I discounted this warning. I had heard of boxers whose legs suddenly gave out, and at age thirty-seven I figured my time had come—instant middle age, with a loss of speed and agility. Soon, however, I began having other difficulties: occasionally, walking downstairs, my left knee would buckle and I would nearly fall. Furthermore, my left leg felt strange, tingling in some places and numb in others. Hot and cold water each produced abnormal nerve reactions: cold hurt, while hot water on my left calf caused no sensation, even when it was uncomfortably hot to surrounding areas. Worse yet, the abnormal area was spreading, working its way up past my left thigh into my side.

Finally, with great reluctance, I turned myself in to the NASA flight surgeon. The flight surgeon is supposed to be the pilot's friend, but every pilot knows that if he walks into the doctor's office on flying status, there are only two ways he can walk out: on flying status or grounded. Since his status can only change for the worse, why risk it? Maybe it will get better . . . but this one I knew wasn't getting better.

The NASA doctor didn't know what it was and finally said so. He readily agreed that I should go into Houston and see a specialist, a neurologist. When I did, I had my diagnosis within an hour and an X-ray to prove it: a bony growth between my fifth and sixth cervical (neck) vertebrae was pushing against my spinal cord, and the pressure had to be relieved by surgery; the sooner the better. This was on Friday, July 12, 1968, and I blithely made an appointment with the doctor for early the next week to have the operation, and then returned to NASA, whereupon all hell broke loose. No, they said, it isn't that simple, you need corroborating opinions, you need to see the best people in the country, you are an Air Force officer and your body belongs to the Air Force, etc., etc. So I hastily retrieved my X-rays from downtown, apologized to the local neurologist, and hustled over to San Antonio to see what the Air Force had to say. More X-rays but same conclusion, followed later in the week by a Harvard Medical School man. Same. The only difference was that Harvard and Air Force proposed one type of operation, while Houston preferred a different approach. Harvard/Air Force said come in through the front, slitting my throat so to speak, removing the spur and some adjoining bone, and then fusing the two vertebrae together with a small dowel of bone removed from my hip. Houston said via the rear, don't fuse anything, just remove a section of bone to relieve the pressure from the spinal cord. In addition to winning 2–1, the frontal approach had one overwhelming advantage: simply that a bone fusion was required by the Air Force to permit my return to flying status, the argument being that the other operation would leave my spine too weakened to withstand the jolt of an ejection seat. So an anterior cervical fusion it would be, at the Air Force's Wilford Hall Hospital in San Antonio. On Sunday, July 21, Susan Borman came over to stay in our house with Kate, Ann, and Michael, and Pat and I drove the two hundred miles to San Antonio to have dinner and check me into the hospital.

Along the way I had plenty of time to think, too much in fact, as the monotonous highway stretched westward. If one followed it far enough, he ended up in Downey, California, where iron men were assembling the fabulous Apollo command module. If one stopped short, he ended up with his throat cut and a very cloudy future. What a small thing forcing the poorer choice, a bone spur—how could it have happened? Apparently it had grown as the disk adjoining it had deteriorated and become thinner, but then what had caused the disk to degenerate? A sudden jolt, such as ejecting from my F-86 a dozen years before, or hitting the ground too hard during Gemini parachute training, or was it a disease process unrelated to any such activity? What would happen to me afterward? I had already been replaced on my crew by Jim Lovell, but did that leave a slot for me somewhere down the line? Somehow it didn't seem too important; Apollo seemed to have lost its urgency, compared with the necessity of getting this aging carcass glued back together. Stop the creepy crawlies in my side, that's all that mattered, and hope that health would be followed by a return to flying and Apollo.

Pat and I enjoyed a leisurely dinner at a delightful Mexican restaurant in the heart of old San Antonio, on the banks of the beautifully reconstructed canal which gives the place a quiet tropical charm. It was an anniversary of a sort—two years to the day since Gemini 10 had splashed down, but a poor two years they had been, in the main, not at all what I had expected of life as a "real" astronaut. I certainly had no reason to feel bitter, but to end up *this* way, after the past five years' work? I paid the bill, left an extra-large tip for luck, and, with a heavy, resigned feeling, headed for the hospital and the knife.

10

The earth is the cradle of humanity, but mankind will not stay in the cradle forever.

—Konstantin Tsiolkovsky

Ward T-2 (the neurosurgery ward) at the Wilford Hall Hospital was a scary place. Part of it was devoted to a couple of intensive-care rooms, where the really bad ones were, and I never went into them during my ten days there. The rest of ward was enough. Neurosurgery can, I suppose, be performed on nearly any part of the human anatomy, but backs and brains seemed to be the parts most frequently requiring the attention of the surgeons in T-2. There were old patients, youngsters, accident victims, those too weakened by disease to undergo surgery, those about to leave under their own steam, and the others who never would. The most popular catastrophe seemed to be ejecting from a speeding jet without making sure the spine was properly positioned prior to pulling the handle. The second-place favorite was bashing one's skull in an automobile or motorcycle accident. I found I was doubly lucky, first lucky in that I was not incapacitated, and second lucky in that my case was going to be swiftly attended to by the chief of neurosurgery. Pat had had her misgivings about my being in a military hospital, based upon various encounters over the years with military doctors who were either very inexperienced or who were practicing outside their specialty area, but we both knew that this was not the case at Wilford Hall, as this was a major installation to which difficult cases were referred from various parts of the United States, as well as overseas. Furthermore, any lingering doubts Pat might have had were dispelled by her first look at Colonel Paul Myers. The man was all authority and quiet competence. Busy as he was, he nonetheless slowed down long enough to give us a forthright, businesslike briefing, patiently answering our questions as he went. It was to be a complex operation ("This is no tonsillectomy, you know"), but when he left, Pat and I both had the feeling I had come to the right place.

The first step in the procedure was a myelogram, for which I was wheeled lightly sedated into an X-ray room and strapped stomach down on a tilt-top table. Then a globule of dye was injected into the lower end of my spinal column. The dye had two important characteristics: it was visible in a fluoroscope, and it was more dense than the spinal fluid. Therefore as long as my head was kept higher than my backside, the glob stayed put, but as the table was carefully tilted downward, it could be seen on the fluoroscope slowly descending the inside of my spine, between the cord and the surrounding bony structure. The whole idea of the procedure was to see whether the glob moved freely, or whether it was obstructed along the way. Sure enough, just past cervical vertebra 6 it slowed nearly to a halt, and the mere trickle that oozed past indicated the severity of the blockage pressing upon my spinal cord at that point. Thus the myelogram corroborated the X-rays and confirmed the need for, and defined the location of, the next day's surgery.

The next morning, very early, I had an enema and a long shower and contemplated my future. The only thing I could be sure of was the cleanliness of my inner and outer body. Beyond that, a veil came down and I could not really evaluate my predicament. I only knew that I had never been so eager to make the minutes fly, to peel off the pages of the calendar, to find out one way or another whether I was an astronaut detained briefly on his way to the moon, or a hopeless cripple who would spend his remaining days boring his VA hospital compatriots with stories of his former prowess ("EVA, man, I'm not sure what it is, but I think it's edema of the veins and arteries"). When the nurse came with her needle, I was eager for it, and grateful for the cloudy, carefree plateau it placed me on. Slice away, baby, I really didn't care, and as they wheeled me into the operating room, I was only vaguely aware of very bright lights and some fool who wanted me to count backward from one hundred. I didn't get very far . . .

The next thing I remember is telling my wife a lot of very interesting things, but somehow her beautiful, sensitive face remained unaltered, impassive, as though she had not heard me. Finally she drifted away, silently, and returned a long time later with Dr. Myers; now the three of us really spoke, and Myers talked of the disk which had worked its way completely loose from its vertebrae and had fallen down into the spinal tunnel and made such a mess of my myelogram and my life. It was gone now, it was in a bottle, and I must rest.

The next time I saw Pat I was really awake, and I could feel as well as hear. It felt O.K. I had a plastic ring around my neck and chin, preventing any motion, and there was a dull pain, as promised, on the point of my right hipbone, from which a circular chunk had been removed. In addition, I had a hell of a time swallowing, but that was all. I wasn't paralyzed or in great pain, and the operation had apparently gone well. Perhaps I was home free. My leg didn't feel any different, but maybe that would come later. I slept and was hungry, and ate, and returned to the world. I sat up, shaved, and limped up and down the halls to the amusement of the nurses, who were very adept at piling good-natured abuse on the heads of we fusionites who were favoring our aching hips. Good therapy! I went to X-ray and to physical training, lifted weights and regained my strength, and before long I was hungry for life outside the hospital.

After a week of this, they let me go home, flying in a commercial airliner back to Houston and a month's convalescent leave. After ten days at home I was really frisky, and Pat and I decided to throw Dubhe into the kennel and the kids into the car and head for Padre Island, our Shangri-La on the Mexican border. We made a two-day trip of it, with Pat doing the driving for a change. I had to keep my collar on twenty-four hours a day for three months, which wiped me out behind the wheel, as I really couldn't see well to left or right. Besides, it was a rare pleasure to sit in the rear seat like a king in his sedan chair and issue navigational commands to my for once quiet and subservient wife. It was a great trip. After a week of sunshine we headed on back to Houston, again making a leisurely two-day jaunt of it.

We spent the night at Rockport, near Corpus Christi, at the Sand Dollar Motel, and when I got up the next morning (August 19, 1968), the local paper delivered a delightful surprise. Apollo 8, as my old flight was now called, was probably going to fly around the moon! All that far-out planning had apparently made sense to someone, although in the final analysis the decision could not be made until after Wally Schirra and crew had flown Apollo 7 and all the flight test data had been analyzed. But the plan was to get 7 off in October and perhaps 8 by the end of the year. Of course, there was no hope for me to make the flight. I had been dropped from the crew as soon as the necessity for my operation had become apparent. Dropped like a hot potato, I thought; Slayton and Borman figured that even if I were back on flying status in time, I would have lost too much valuable training while I was laid up. So Jim Lovell, my back-up, would be flying around the moon, and I didn't have the foggiest notion of what I would be doing.

In fact, my future now depended on a couple of X-rays which would reveal the progress of the fusion supposed to be welding cervical vertebrae 5 and 6 together in an ever stronger bond. The plug from my hip, which was wedged between the two, could not do the job alone—it merely served as a bridgework for the growth of new bone, whose density would be measured by the X-rays. When the first one, thirty days after the operation, revealed good growth, I was much relieved, because I knew what would have happened next if I had flunked the X-ray test: back to Wilford Hall and into a Minerva cast. This contraption, I suppose, received its name from the fact that the goddess Minerva is usually depicted with a helmet covering her forehead, and that is where the cast began—right above the eyebrows. However, it didn't end at the nape of the neck, like Minerva's helmet, but continued down to the waist, preventing the slightest movement of the head and upper torso, to give that fusion as undisturbed a chance for growth as possible. My Minerva acquaintances at Wilford Hall didn't like it one bit, and I was therefore enormously relieved to be going back to work instead.

The first thing I did was find out about the proposed circumlunar flight. The key to it was that the LM development was now lagging behind the CSM by about five months. Because of major difficulties with the rendezvous radar and a long list of minor ailments, it appeared that Borman's LM (3) would not be ready until spring 1969. McDivitt's (LM 2), which was supposed to precede Borman's, was in even worse condition, and was overweight to boot. Meanwhile, Wally was to get the first CSM airborne early in October. If he did, the momentum of this first manned Apollo flight should be kept going, and a second flight should not have to wait for five or six months (or longer?) until all the LM problems were sorted out. Therefore, it made sense to fly a second CSM in the interim. However, just as the leaping frog of earlier days had been deleted because it was too nearly a carbon copy of its predecessor, it now seemed sterile and wasteful simply to put up another CSM to duplicate Wally's achievements (assuming Wally got all his planned work done). What could a second CSM do by itself? Go longer, perhaps, or higher—higher certainly. Once this line of reasoning opened up, it led inevitably to the moon, with either a lunar fly-around or an entry into lunar orbit.

The influence of the Russian program, while difficult to measure, certainly provided an interesting climate for this type of decision-making. NASA officials were publicly speculating about a Russian booster twice as powerful as the Saturn V, and the Soviet interest in circumlunar flight was well documented. In fact, in September Zond 5 flew unmanned into lunar orbit and returned successfully* to earth.

Aside from wanting to get men to the moon ahead of the Russians, there were solid reasons for investigating it at close quarters before landing upon it. The main thing would be to give the entire navigational system a thorough workout, testing ground and flight procedures and equipment under conditions of stark realism, yet with a simpler trajectory, one providing a greater tolerance for error than was possible in a landing. The disadvantage was that the moon was a long way from home in case of trouble; with only one previous manned flight, it was placing tremendous confidence in the reliability of the CSM and all its components. In particular, if one braked into lunar orbit, there was only one engine hanging out the back side of the CSM, and it had to work perfectly or the crew would never return to earth.

Circumlunar thinking had personal implications as meaningful as the technical ones, at least for the crews involved. Second in line after Schirra-Eisele-Cunningham were McDivitt-Scott-Schweickart, and McDivitt was offered the circumlunar flight. In retrospect, he may seem a martyr or a fool for not taking it, but at the time there were solid arguments backing up his decision to stick with his LM. First, Wally's flight was more than a month in the future, and yet it would have to be near perfect, with near perfect performance from fuel cells, gimbal motors, computer, parachutes, etc., etc., to prevent the second flight from becoming

* Critics said that Zond 5 never got close enough to the moon (twelve hundred miles) to provide useful photography, and that its entry into the earth's atmosphere was so screwed up that its excessively steep angle would have killed a human crew from excessive G loads and temperatures. Nonetheless, Zond 5 represented a hell of a capability and made a lot of NASA people edgy. Would Russian men follow shortly?

merely a repeat performance in earth orbit. True, the CSM seemed to have matured nicely in the eighteen months since the fire, but still—no problems on its first manned flight? It was just too much to believe. Further, Jim McDivitt had spent recent months living with the LM, and he wanted to stick with it, to see it through to a successful flight test, even if it meant that his crew dropped from second to third in line. In the final analysis, which would be more important to someone selecting the first lunar landing crew: experience with lunar navigation or with the landing craft itself? It was a tough question.

There were also other implications, details perhaps, but details important to those involved. Dave Scott, for example, agreed with McDivitt's decision but was upset because it meant that he would have to trade his baby, CSM 103, for Borman's 104. While the two were practically identical, Dave had nursed 103 along through countless tests at Downey and the Cape, and had developed supreme confidence in it. I had spent nearly as much time testing 104, and tried to reassure Dave, but it was no good, it just wasn't the same. Nor was all happiness on Borman's crew. Bill Anders was despondent over the loss of his LM. Here he was, an LM pilot without an LM, and flying in lunar orbit wasn't fair compensation for the loss of his machine. All of us knew that decisions were being made which in all likelihood would affect us for the rest of our lives,* yet there was not sufficient evidence to make comfortable decisions, only guesses based on future dice rolls. In such situations, we humans cling tightly to what we *do* know, which I suppose is why a highly intelligent man like Dave Scott was so concerned with the serial number on the side of his CSM. Borman

* If anybody got screwed in all this, it was Pete Conrad. Conrad was McDivitt's back-up. If McDivitt had flown Apollo 8, Conrad would have been commander of Apollo 11 and would have been the first human to walk on another planet, instead of Neil Armstrong. But the Borman-McDivitt switch put Jim on Apollo 9 and slipped Conrad to Apollo 12. Who was the third man to walk on the moon? . . . Uh, I think it was Conrad. Who was the thirteenth? . . . Uh, I don't think there were that many . . . How many men flew the Atlantic solo after Lindbergh? . . . Uh, a lot, but I don't recall their names.

had no such hang-ups. Once he got a whiff of that moon, he gave it the same merciless attention a pointer gives a covey of quail.

Mike Collins watched all this as an intrigued bystander. My overriding concern was, of course, to regain full skeletal strength and thus return to flying duties, but in the meantime I had to do *something*—everyone must be somewhere. Tom Paine was just taking over as NASA administrator from the retiring Jim Webb, and Tom with characteristic kindness offered me a job in Washington, to work on the Apollo Applications program. It would be just "temporary," he said, but I detected something in his voice on the telephone which indicated that he thought he was probably offering a job to an ex-astronaut. At any rate, as I explained to him, if I was to return to an Apollo crew, Houston was the place for me, because in Washington I would inexorably descend the ladder of those who waited to be formed into future crews. Tom (Dr. Paine, I called him) immediately agreed, bless him, and didn't push it. It would have been a lot more difficult to say no to Jim (Mr. Webb, I called him).

Having decided to stay in Houston, and now having the administrator's approval to do so, I decided that it made most sense to stick with my old flight, or Apollo 8 as it was now being called. I could begin by briefing my replacement, Jim Lovell, on a number of details. More important, I had been the flight-plan specialist on the crew, and now the circumlunar aspirations meant that a million crew duties had to be defined anew, and planned and organized in minute detail. The crew badly needed someone in Houston to speak for them, to represent their interests at the endless series of planning meetings, while they spent their time in the simulators or testing the spacecraft. The time spent in lunar orbit, for example, had to be scrutinized minute by minute, because the crew's chores were dependent on a number of variables. At any given time, the spacecraft might be in direct sunlight, in earthshine, or in total darkness. It might be in radio contact with the earth or it might not. If it turned to point its antenna at the earth for communications, the result might be to point the space-craft windows at black space instead of at a potential landing site which needed to be photographed. If the spacecraft was turned to point the sextant at the stars to align the inertial platform, the crew might find that a couple of minutes later it could neither talk to earth nor photograph moon. Each change of direction cost time and fuel, and both were precious commodities in lunar orbit, so the chain of interlocking events had to be cleverly arranged for maximum efficiency. This was tedious, nit-picking work, but at least it caused me to feel useful and focused my thoughts on Apollo rather than on the plastic ring around my neck.

My X-rays continued to improve: by the end of October I was allowed to throw the ring away and was cleared to fly transport-type airplanes without ejection seats; and by the end of November I was back in the T-38s, on full flying status. The X-rays now showed a solid hunk of bone replacing the flexible disk which used to be between vertebrae 5 and 6. Most of my spinal cord damage was apparently permanent, but my coordination did seem slightly improved, and if the hot/cold awareness in one leg was still unsatisfactory, so what, that was something I could live with easily. The main thing was that Dr. Myers's handiwork had been so good that I was now as qualified to fly in space as if it had been only that tonsillectomy he had mentioned so disparagingly. Delighted as I was to be certified whole once again, this period was the most frustrating time in my life. When I had first read of the circum-lunar flight in that motel in Rockport, it had been a world event involving other people, and I had felt undiluted pleasure at the good news. But now the sullen green imp of righteous self-interest surfaced, and my thoughts were ugly. With the flight still a month or two away, why was Lovell going in my place if I was whole? True, he had been sopping up simulator time like crazy, but I had started off way ahead of him, and I *still* knew more about the flight than he did. I wasn't mad at Shaky—it wasn't his fault, and for all I knew he might have preferred not to replace me but to stay where he was, with Neil on our back-up crew. No, it was Borman and

Slayton who were the bastards, but clearly the damage had been done and there was no way to repair it at this late date, especially not by mounting a campaign to replace Shaky. It was just that those bastards Borman and Slayton should have known that it takes 125 days to recover from an anterior cervical fusion!

Wally's flight, October 11–22, was in a very real sense an ending rather than a beginning. It marked the end of the slippage in the program, the end of a series of target dates which were discussed but never met, an end to the fire-damaged talking phase of Apollo. Shortly before the flight, at the end of September, I gave a speech at the Society of Experimental Test Pilots annual symposium in which I outlined our accomplishments of the past year and predicted what the next year would bring. In researching that speech, I dug out NASA's elaborate planning charts, which indicated the lunar landing would take place using CSM 107 and LM 5 in July 1969. I wasn't about to endorse these specifics, and instead waffled, explaining to the test pilots that a lunar landing was theoretically possible before their next symposium in September 1969. But if I had been a betting man, I would have given ten to one against anyone stepping out of LM 5 in July 1969 while CSM 107 orbited overhead.*

The flight itself was kind of a bore compared to the three-day Gemini rendezvous and EVA flights. Wally, Donn, and Walt chatted and fussed their way around the world 163 times. They all caught head colds, which tended to create an overlay of irritation, and they seemed to make mountains out of molehills, but they stayed up there eleven days and did their job, completing all planned tests. They gave their big rocket engine (the one Borman needed to come home from the moon) a good workout, firing it eight times. There were some minor electrical problems, but in the main it was a supersuccessful first flight and cleared Apollo 8 for its

---

* Which happened precisely as advertised, on July 20, 1969.

lunar venture. Of course, NASA didn't say so right away, but fretted and fumed for another month to make absolutely sure there were no contraindications. As the December lunar launch window approached, the moon beckoned to the Russians as well as to us, and there was a lot of speculation concerning who would fly men around the moon first. A review of the two programs up to the fall of 1968 shows some remarkable similarities. Early in 1967, within three months of each other, Grissom-White-Chaffee were killed by the first Apollo and Komarov by the first Soyuz. It took until late 1968, within two weeks of each other, for the two countries to recover from these tragedies, with Schirra-Eisele-Cunningham on Apollo 7 and Beregovoy on Soyuz III. Each side had also traveled to the moon with unmanned spacecraft. Now what?

As December neared, I don't know what the Russians were doing, but things around Houston were building to a crescendo of planning, practicing, checking, and rechecking. I saw less and less of the crew, and finally resorted to communicating with them by memo, since they were mostly in the Cape simulators and I was in Houston meetings. I did occasionally talk to them on the radio, as I was going to be one of the CAPCOMs for the flight, and we practiced with Houston Mission Control and with the crew in the Cape simulator and with the whole network of earth-tracking stations, simulating a flight to the moon chock-full of imaginary problems. In between simulations and meetings, I wrote memos to the crew on such far-out topics as:

1. What happens if the Saturn V goes apeshit and starts putting the spacecraft on an incorrect trajectory? How far do you let it stray before shutting it down?

2. Should you try out the service module engine once before you need it to brake into lunar orbit, just to make sure the goddamn thing is working properly? If the answer to this is yes, which way do you point it so it won't knock you off your heretofore perfect trajectory?

3. How do you want to spend the twenty hours in lunar orbit?

On lunar science, or the technological problems of keeping the spacecraft flying, or on assuring adequate crew rest, or what? Be specific.

4. Which way is up? How do you want your artificial horizon to define up and down? By reference to the earth's horizon, or to the moon's up and down, or to some other vector which will remain constant for the entire flight, or what?

5. Mission Control is not able to keep you on as precise a trajectory as Borman would like (within three feet per second of a free return trajectory), simply because their knowledge of that perfect path in the sky is not that clearly defined. They say they are accurate enough, and three feet per second is unreasonable.

6. Every experimenter in the Western Hemisphere has suddenly discovered that there is a burning need for him to obtain photographs of certain lunar topographical features. Please buck them to me and I will try to filter out the idiots and refer the others to the proper people.

7. For certain launch days in December, the position of the earth and moon relative to the sun is such that the visibility may not be good enough to do *any* on-board navigation at all. Will you still be willing to launch, knowing that you are completely dependent upon (a) the accuracy of the ground's computations and (b) their ability to radio them up to you?

8. When you come roaring back to earth, if a violent storm develops in your splashdown area, would you prefer to (a) land in the typhoon and be sure of getting down, or (b) use an untried procedure to skip back out of the atmosphere and come down somewhere else, hopefully?

9. Under some conditions, we can get you home from the moon quicker if you are willing to accept a higher entry velocity into the atmosphere. Let's discuss.

Of all the work that went into Apollo 8, probably the most important and least understood was the trajectory analysis, wherein computers were used to produce numbers which were kept for possible emergency use. For example, if trouble developed—let's say an oxygen leak—on the way to the moon, one didn't simply turn around and come home. The trajectory, with all the gravitational pulls of earth, moon, and even sun, must first be carefully analyzed. It might turn out that there was simply not enough fuel on board to overpower the moon's gravitational pull. Then one would have to allow the moon to win the tug of war, and swing the spacecraft out and around behind the moon, only then firing the engine to come home. Other situations could develop where one had a choice of a fast return at great fuel cost or a slow economical trip home, depending on whether one was running short of life-support systems or of propellants. All these questions had to be probed by the computers, and a library of answers compiled. The appropriate volume could then be snatched off the shelf in time of need, and a month's work could then be applied in a matter of minutes. Borman touched on this at a pre-flight press conference. ". . . We have regular abort times along the course to the moon . . . then we finally reach a point where it would be swifter to just go on around the moon than it would be to try to abort . . ."

As it turned out, Borman didn't need this kind of information, but his navigator, Jim Lovell, did a couple of years later. In fact, when a tank exploded and Lovell, Swigert, and Haise lost most of their oxygen on Apollo 13, they clearly owed their lives to the Houston computer and trajectory gnomes. Borman at the same press conference hit upon another vital point, namely that Apollo 8 was acting as a catalyst, a forcing function, to clear all closets of skeletons, to force people to state their doubts about going so far from home, and to cause the moon to become an arena of actual operations, not just some far-off place it was theoretically possible to visit. ". . . We designed Apollo, we said we were going to the moon, and . . . finally when we get down to examining the details and saying we are really going, people start getting a little queasy about it." Then Borman went on: "But I have no hesitancy about the hardware." I don't know what blend of guts, brains, and four-leaf clovers it takes to eradicate hesitancy about hardware, but my own feelings were more in keeping with those expressed in a speech by Jerry Lederer, NASA's safety chief, three days before the flight. While the flight posed fewer unknowns than had Columbus's voyage, Jerry said, the mission would "involve risks of great magnitude and probably risks that have not been foreseen. Apollo 8 has 5,600,000 parts and one and one half million systems, subsystems, and assemblies. Even if all functioned with 99.9 percent reliability, we could expect fifty-six hundred defects . . ."

As CAPCOM, my job was to act as spokesman for Mission Control when talking to the crew, and spokesman for the crew within Mission Control. With fifty-six hundred things about to break, we would have plenty to talk about. President Johnson apparently had equipment failure on his mind as well, for he sent the crew a message stating: "I am confident that the world's finest equipment will strive to match the courage of our astronauts. If it does that, a successful mission is assured." *If* is right. Actually, things weren't *that* bad, as Wally's flight had demonstrated a remarkable reliability of equipment design. Again the fire, in its ghastly way, had made its contribution, as about five thousand engineering changes had been made to the CSM in the post-fire redesign. The Saturn V was more of an unknown, in my view. Although 501 had flown a near-perfect flight, 502 had had more than its share of problems, and had barely limped into earth orbit. The first stage had developed a severe oscillation, two out of five second-stage engines had shut down, and the guidance system had overcompensated and put the vehicle into an orbit whose apogee was a hundred miles too high. Yet here we were, on December 21, with 503 about to propel CSM 103 to the moon.

There were three of us CAPCOMs who would alternate in eight-hour shifts around the clock. Along with the engineers who manned the Mission Control consoles, all under the command of a flight director, we were divided up into teams: Jerry Carr would be Black, Ken Mattingly Maroon, and I would be Green (with envy?). Green had the responsibility for the launch phase, and I had been practicing for weeks with the crew and with the Mission

Control Green Team to make sure that we all were able to respond as quickly as possible to any failures. Generally speaking, space flight does not require knee-jerk reactions; usually there is time for questioning and discussion, but not during the boost phase. If CAPCOM says ABORT, the crew had better do it, and right now!

Therefore, I was really nervous at seven in the morning on the twenty-first, when 503 roared into life. I was hunched over my console in Mission Control, watching an electronic blip cross a screen and listening to the experts around me tersely reporting vital signs. Pressure good, temperature good, azimuth good . . . everything good . . . up and away . . . clear of the launch tower . . . 100,000 feet. Mark! Mode 1 Charlie . . . two and one half minutes . . . staging . . . that's one less first stage to worry about . . . escape tower jettison. Mode 2 . . . good guidance . . . five minutes; if the second stage craps out now, we can make it into orbit using the third . . . staging, adios second stage . . . good light on the S–IVB . . . a hundred miles now, building up speed . . . stand by for cutoff . . . SECO! . . . looking good, Apollo 8, right down the alley. Whew! I could relax at my console now, as the spacecraft was in a good orbit where it would coast for nearly three hours.

The next big event was reigniting the third-stage Saturn V engine to set sail for the moon. Known as TLI (translunar injection), this burn had to take place precisely at 10:40 Eastern Standard Time, which meant that before then the crew had to check everything on a long list of equipment, each item of which had been deemed vital to making the trip. If something was broken, we should know about it now, not after TLI, when trajectories became very complicated. Fortunately, the checks went smoothly, and spacecraft 103, Dave Scott's pampered baby, seemed to be purring along flawlessly. Now the big moment came. As we counted down to S–IVB ignition for TLI, a hush fell over Mission Control. TLI was what made this flight different from the six Mercury, ten Gemini, and one Apollo flights that had preceded it, different from any trip man had ever made in any vehicle. For the first time in history, man was going to propel himself past escape velocity, breaking the clutch of our earth's gravitational field and coasting into outer space as he had never done before. After TLI there would be three men in the solar system who would have to be counted apart from all the other billions, three who were in a different place, whose motion obeyed different rules, and whose habitat had to be considered a separate planet. The three could examine the earth and the earth could examine them, and each would see the other for the first time. This the people in Mission Control knew; yet there were no immortal words on the wall proclaiming the fact, only a thin green line, representing Apollo 8 climbing, speeding, vanishing—leaving us stranded behind on this planet, awed by the fact that we humans had finally had an option to stay or to leave—and had chosen to leave.

Outward bound, the wonder of Apollo 8 ceased for a little while, and Frank Borman threw up. He didn't say so, but came on the air with a report of a potentially serious medical problem and requested a private conversation with the medics. Chuck Berry was in hog heaven. Here he had been waiting nearly a decade for someone in flight to solicit his advice, and by gum, the first humans to leave the cradle had called for their pediatrician. The timing was perfect. The Mission Control Center had two identical floors with duplicate facilities, the original idea being that two flights could be controlled simultaneously, or at least one could be in preparation while a second flew. We were occupying the third floor, while below us the second-floor control room was empty, so we gathered up an "elite" group of a half dozen or so and descended to hear Borman in privacy. He really didn't have much to say, merely that he was feeling miserable and had been sick to his stomach. Virus, motion sickness, the onset of a serious illness—who knew? Rest and fluids were prescribed, and only time would tell. Meanwhile, Apollo 8 was steaming away from the earth at an alarming clip, but even though each minute meant several hundred miles farther separation from medical help, what else could anyone do, with only these symptoms to go on? I trudged back upstairs confused and upset.

Motion sickness just didn't make any sense. The Russian literature had long hinted at the fact that weightlessness created inner-ear disturbances in the form of sloshing fluids, which in turn sent messages of malaise to the stomach, which in some individuals resulted in their throwing up. We Americans, however, had tended to pooh-pooh these reports, and when we considered them at all, we based our own asymptomatic flight experience upon the fact that we were flying only experienced test pilots, whose inner ears were accustomed to such buffets and bumps. Now, however, *something* was making Borman sick. Could it be the Russian disease? Anders and Lovell didn't seem too chipper either, although they were not actively ill. They must have felt queasy about sharing the small sick room, and undoubtedly they were depressed by the shadow now cast over their flight. But as time went by, Borman began feeling better, and we all breathed a sigh of relief in Mission Control. It had been estimated that one half of all humanity heard or read about the Apollo 8 launch (there were more newspaper reporters at the Cape than for any flight since John Glenn's) and we didn't want to have to tell half the world we were returning Apollo 8 to earth for unknown medical reasons. Borman thought his problem might be a reaction to a sleeping pill, but hindsight tells us it was probably just the first case of motion sickness to occur in our program. The Mercury and Gemini spacecraft had been too small to allow the astronaut to get unbuckled and float about, but from Apollo 8 on, it would become increasingly obvious that inside the commodious Apollo command module (and later in Skylab) some people got sick for several days and others didn't, just as some sailors are affected by rough weather and others are not. The Russian craft had apparently been enough bigger than ours to reveal this reaction to body motion. We had experienced it only in the zero-G airplane, and had been damned lucky it hadn't occurred during one of the Gemini space walks, where, of course, the inner ear got quite a workout. If a Gemini space walker had ever thrown up while outside, it would probably have meant a messy death, as he would be blinded by the vomit trapped inside his helmet, and possibly asphyxiated by its clogging his oxygen supply circuit. The same, of course, could apply to moon walkers, except that they had the moon's weak gravity as a stabilizing influence.

At any rate, when Borman and company came on television late in the second day, they seemed healthy enough, and in Mission Control we got back to more mechanical worries. Our main concerns were (1) the accuracy of our navigation, which must cause Apollo 8 to miss the moon by a mere eighty miles at a distance of 230,000 miles; (2) the operation of the big service propulsion system engine (the SPS), which must get Apollo 8 into, and *out of,* lunar orbit; and (3) our general concern over being so far from home for the first time (with fifty-six hundred possible failures). The business of navigation was going very well, and we had not even needed our first planned mid-course correction, but as Apollo 8 got closer to the moon, the mathematics of the situation changed slightly. The gravitational pull of the earth was becoming less and less a factor, while the moon's tug was becoming dominant. Our computers acknowledged this fact and arbitrarily had picked one point in the sky at which the mathematical equations shifted from an earth-centered system to a lunar-centered one. We called this leaving the earth's "sphere of influence" and entering the lunar "sphere." At the crossover point, there was a little hiccup in the arithmetic, and our computers' assessment of the space-craft's position in the sky shifted by a few miles. Phil Shaffer, the Green Team's resident expert in such matters, made the gross mistake of mentioning this fact at that night's press conference, as the Green Team came off duty. Never has the gulf between the non-technical journalist and the non-journalistic technician been more apparent. The harder Phil tried to dispel the notion, the more he convinced some of the reporters that the spacecraft actually would jiggle or jump as it passed into the lunar sphere. Big as a professional football player, red-faced and sweating, Phil delicately re-examined his tidy equations and patiently explained their logic. No sale. Wouldn't the crew feel a bump as they passed the barrier and become alarmed? How could the spacecraft instantaneously go from one point in the sky to another without the crew feeling it? The rest of us smirked and tittered as poor Phil puffed and labored, and thereafter we tried to discuss the lunar sphere of influence with Phil as often as we could, especially when outsiders were present.

At 200,000 miles out, on the third day, we got some more television, and the crew shared with us their view of the baseball-sized earth. Noted British astronomer Fred Hoyle had said as early as 1948 that the first whole-earth photographs would unloose a whole stream of new ideas as powerful as any in history. Before Apollo 8, we had seen close-up whole-earth pictures from unmanned satellites, but at 200,000 miles the minuscule egg framed in the spacecraft window seemed eerie, even surrounded by the familiar plastic rim of the television set. Was that *us*? In London, not far from Fred Hoyle's home town, the International Flat Earth Society would shortly profess not to have altered its basic tenet because, although the pictures showed a circular earth, there was no proof that it was a globe. But if they weren't shaken, I was, not only then, but later when Lovell described it as "a grand oasis in the great vastness of space."

The next hurdle for Apollo 8 had the bizarre name of LOI$_1$, and the crew really would feel this one. It stood for lunar orbit insertion 1, and was the burn of the SPS engine which would slow Apollo 8 enough so that it could be captured by the moon's gravitational field. It would be followed by LOI$_2$, a smaller burn to trim the lunar orbit to a sixty-nautical-mile circle. Then, after ten revolutions (taking twenty hours) in lunar orbit, we would have TEI (transearth injection), the burn in which the big engine would set Apollo 8 upon a collision course with the earth's thin atmosphere. Thus did we trust the SPS engine, which had dual plumbing but only one thrust chamber and one exhaust nozzle, a stubby cone sticking out the back end of the service module.

On the fourth day, which was Christmas Eve, Apollo 8 abruptly disappeared around the left edge of the moon, armed with our latest instructions and advice concerning LOI_1, which had to be performed while out of radio contact with the earth. It took four minutes of burn time to reduce the speed by the necessary two thousand miles an hour, and another ten seconds for LOI_2 to circularize the orbit. If LOI_1 had gone awry, Apollo 8 could have arced off into solar orbit or crashed into the moon, but it had not, and now the crew was relatively safe until TEI.

As they circled, they took hundreds of photographs, made scientific observations, and in general acted like three rubber-necked tourists. They saw no colors—only black, white, and inter-mediate shades of gray. The moon was peppered with craters, it was desolate and forbidding, it looked like plaster of Paris or dirty beach sand. Anders thought it "a very dark and unappetizing place." The crew also celebrated Christmas by reading the Bible, each of the three taking a turn at the first chapter of Genesis. It was impressive, I thought, a stroke of genius to relate their pri-mordial setting to the origin of the earth, and to couch it in the beautiful seventeenth-century prose poetry of King James I's scholars. Borman, Lovell, and Anders deserved to make it home for that reason alone, for having thought to bring the rest of us to their moon in humility and reverence. It was a graceful touch.

TEI, however, was the touch we had to have, for all the holy words in the universe would not budge Apollo 8 from its orbit; its only salvation lay in the chemical energy locked within the SPS system which, if released at the proper time in the proper direction, would wing Borman and Bible back to earth. In the Borman kitchen, Pat sat and strained to keep calm as Susan Borman and Valerie Anders waited to hear about TEI, which had taken place out of earshot. When Apollo 8 finally rounded the horn, over on the right perimeter of the moon, and reported a good burn, the gals let out a whoop and were finally able to replace their brittle, forced gaiety with the honest joy of temporary relief. TEI was a good Christmas present. The next test would come soon enough, as the spacecraft, picking up speed, hurtled toward the razor-thin edge of the earth's atmosphere. If too shallow, it missed, slipped by, and was gone forever; if too steep, it burned up. Time enough tomorrow to worry about *that* problem; in the meantime, break out the TEI champagne, let the good times roll, and swallow the frights one at a time as they appear in the flight-plan booklet.

The Gemini had been small, comparatively simple, and could almost be flown like a fighter. The Apollo, on the other hand, was too big, too heavy, and too complex to make sprightly maneuvers. Flying it was, as John Young described it, like driving an aircraft carrier with an outboard motor. It also required a lot of savvy and foresight, and a devotion to the path of righteousness as described by the check list. As the Apollo 8 crew sped toward their splashdown in the Pacific, they established a pattern which would endure throughout the lunar flight series. The trip home was the first chance the crew had to unwind and to consider their situation, and it was painfully obvious to them, just as it would be to the crews of Apollos 10, 11, 12, 13, 14, 15, 16, and 17, that they had one chance, and one chance only, to make a successful entry into the earth's atmosphere. The check list would get them there: it was a fourth crew member whose views had to be considered. Only one midcourse correction was required on the way home to keep Apollo 8 in the center of the entry corridor, and then it was up to the crew to make a good entry, to make no fatal mistakes. The check list, worked out after years of trial and error at Downey, and in the Houston and Cape simulators, described what to do and when to do it, switch by switch. Along the way, we CAPCOMs in Mission Control radioed up supplementary information, details of weather and last-minute checks of velocity, time, and position. The spacecraft would hit a peak velocity of twenty-five thousand miles an hour as it screamed back into the atmosphere. As the air piled up in front of its blunt heat shield, a shock wave would be formed and temperatures at its boundary would soar to 5,000 degrees. The crew would roll the spacecraft, as required, to point the lift vector in the direction indicated by the guidance system, so as to land on target. All this presumed that the check list had been followed to a T in preparing the spacecraft and its systems for this moment, and with no chance for a repeat performance, the crew spent the last day of their flight painstakingly reviewing each item on the long list.

When Apollo 8 finally plopped gently into the Pacific three miles from the U.S.S. *Yorktown* shortly before dawn on the seventh day, Mission Control came unglued and pandemonium filled the normally staid room. People waved miniature American flags and slapped one another on the back; the traditional cigars were broken out and everybody lighted up. None of our worries had materialized, all our fears had been groundless, our planning had been sound, our simulations accurate, our estimates precise. We could have destroyed Apollo 8 in a thousand different ways, but we had instead nurtured and guided it through the most far-ranging week in the history of man. For me personally, the moment was a conglomeration of emotions and memories. I was a basket case, emotionally wrung out. I had seen this flight evolve in the white room at Downey, in the interminable series of meetings at Houston, evolve into an epic voyage. I had helped it grow. I had two years invested in it—it was my flight. Yet it was not my flight; I was but one of a hundred packed into a noisy room. I could wave my flag and smoke my cigar and finger the scar on my throat, but that was about all. For some reason I felt like crying, but I couldn't do that in Mission Control, so I clapped a few good working troops on the back and left.

Apollo 8 splashed down on December 27, and the names of the Apollo 11 crew were announced on January 9. The Apollo 8 back-up crew had been Neil Armstrong, Buzz Aldrin, and Fred Haise (Fred replacing Lovell, who had replaced me). Under Deke's system, the back-up crew of 8 would expect to skip 9 and 10 and pick up 11. However, since Collins had also been an Apollo 8 crew member at one time, clearly a case could be made for his bumping Haise on Apollo 11; I hoped that Deke would make that case. After all, I was a Fourteen, Fred was a Nineteen, and I had been waiting my turn for two years longer than he. Further, I had Gemini flight experience while he did not. On the other hand, I certainly wasn't complacent, because I knew Fred to be an extremely competent chap, and, like Neil, a long-time NASA employee. Therefore, I was much relieved to find my name had been submitted, along with those of Armstrong and Aldrin, to Washington for approval as the prime crew of Apollo 11. The back-up crew would be Lovell, Anders, and Haise.* I cannot for the life of me remember how I acquired this information, from Neil or Deke most likely, but I feel like an ass checking the point ("Er . . . say, Neil, did you ever tell me I was supposed to be on your crew, you remember that one that went to the moon?"). The most important message of my life, and I can't remember the medium of its delivery (what would McLuhan think about that?). Let's just say the news permeated through that membrane of disbelief I had developed on the subject of assignments to, and removal from, Apollo flight crews. Of course, first the three names had to be approved by NASA headquarters in Washington, but as far as we could tell, Washington had always rubber-stamped Houston's recommendations. The first indication we had that they had done so was when Pat started getting phone calls at home asking her to confirm the story and to describe her feelings about it. Poor girl, she knew, but didn't feel she could do anything but "no comment" it all. The stories were coming out of Washington, where the Apollo 8 crew was conducting a post-flight press conference, and it wasn't long before we got the official word and I breathed a huge sigh of relief.

---

* This made sense. Borman had decided he had had enough, Shaky deserved the promotion to crew commander, and Haise was the best-trained guy of those available to add to the crew.

Armstrong and Aldrin. Not only were they near the top of the alphabet,* but they were highly thought of within our tight little astronaut group. Smart as hell, both of them, competent and experienced, each in his own way. Neil was far and away the most experienced test pilot among the astronauts, and Buzz the most learned, especially when it came to rendezvous. But that didn't mean that Neil was all pragmatist or Buzz all theoretician. Among the dozen test pilots who had flown the X-15 rocket ship, Neil had been considered one of the weaker stick-and-rudder men, but the very best when it came to understanding the machine's design and how it operated. Buzz, far from being an academic recluse, was an outstanding athlete, a pole vaulter, and a Korean War MIG killer. As a pair, they brought a formidable array of talent with them, as they approached the problems of landing Apollo 11 safely on the moon. I considered myself damned fortunate to be joining them.

The flight itself was not nearly as straightforward. Apollo 11 was carried on the books as the first lunar landing flight, but that assumed an awful lot. The LM hadn't even flown men yet, and if it acted up on Apollo 9, the landing would probably slip to Apollo 12. Apollo 10 was to be a dress rehearsal for 11, and again it could develop problems which would require delaying the landing. On the other hand, one could point to the aggressive planning which had put 8 around the moon and argue for a landing rather than a dress rehearsal for 10. Therefore, it was far from clear, early in January 1969, that Armstrong-Collins-Aldrin were going to be the first lunar landing crew. If I had been trying to establish odds at the time, I think I would have given Apollo 10 a 10 percent chance, 11 a 50 percent, and 12 or subsequent a 40 percent chance to *attempt* the landing. Be that as it may, I was champing at the bit to get going with our training, as we were scheduled for launch in midsummer. I had just one little unpleasant task to get out of the way, namely to apologize to Fred Haise for bumping him off the crew. I was sensitive to people getting bumped.

* Is that why they were successful? For years, as little kids in school and big kids in the service, they had been first to queue up, first to get the information. Had this kept them a step in front of their contemporaries?

11

The United States this week will commit its national pride, eight years of work and $24 billion of its fortune to showing the world it can still fulfill a dream.

It will send three young men on a human adventure of mythological proportions with the whole of the civilized world invited to watch—for better or worse.

—Rudy Abramson, *Los Angeles Times*, July 13, 1969

The special nature of the Apollo 11 flight was swiftly brought home to me, for as soon as our three names were announced, there was a great clamor around MSC to get our opinions of *this*, or our evaluation of *that*. Joe Kerwin grabbed me first off to review the design of the BIG, one of his areas of responsibility for the astronaut office. The BIG tided one over until he was inside the MQF, which led to the LRL, of course. Behind these meaningless letters lay a vital concern—the possible contamination of this planet with extraterrestrial bacteria or other hazardous life forms. The BIG was the biological isolation garment, which acted as the barrier between our three possibly contaminated bodies and those of three billion other humans. It was vaguely like a pressure suit in design, except much simpler and lighter, without provisions for ventilation and communications. The idea was that, after the command module landed in the Pacific, we would cautiously open the side hatch and the swimmers would throw three BIGs in to us. We would don them before emerging, closing the command module hatch behind us. Then we would be transferred from a life raft into a helicopter and onto an aircraft carrier, where we would walk below decks (still enclosed in our BIGs) into the MQF. The MQF (mobile quarantine facility) was a glorified house trailer, modified with filters and other accouterments to ensure a biological seal. Once locked inside the MQF (along with a physician and an engineer), we could remove our BIGs. When the carrier reached port at Honolulu, the MQF would be hoisted off by a crane and loaded onto a flatbed truck for the short trip to Hickam Field. There it would be loaded aboard a giant C-141 jet transport for the flight back to Houston, where it would again be transferred by flatbed from Ellington AFB to the LRL. The LRL was the lunar receiving laboratory, equipped with elaborate laboratory facilities for examining moon rocks, and us, for possible extraterrestrial life. The principal indicator in this examination was the health of a white-mouse colony, whose inhabitants were to be exposed in various ways to us, our equipment, and our treasures. If the mice started dying, baby, we were in trouble! If the mice stayed healthy, we got released from quarantine three weeks to the day after the lunar landing.

But in January 1969 the BIG–MQF–LRL route seemed so remote, and we had so many other concerns more pressing than healthy mice, that trying on a prototype BIG had an unreal quality about it. The garment had no ventilation of any sort, and one heated up rapidly. There were other flaws in the overall scheme. Some scientists were clamoring to have biological filters added to the CM's post-landing ventilation system. Without such filters, they said, germs could escape through the snorkel-like device that provided us with air to breathe while we awaited the swimmers and their BIGs. However, it seemed to me that filters alone were not the answer, because there was the matter of the open hatch (from which germs could escape), plus the possibility of germs coming out with us on the outside of our BIGs, even though we would soap each other down with disinfectant while in the life raft. It seemed a primitive scheme for protecting the planet, yet neither I, nor apparently the NASA managers, could think up a suitable substitute without either massive redesign of the Apollo equipment or a change in procedure which would keep us inside the CM while it was hoisted on board. We didn't like either of these ideas. Anyway, how real was this threat to the planet? No one knew, and in the absence of concrete data, one could ignore the problem, claim it didn't exist, or predict doom and horror. It seemed to me that if you multiplied a very, very small number (the probability of our bringing back harmful life) by a very, very large number (the implications of turning a hostile organism loose on the earth), you obtained a finite product of sufficient magnitude to cause concern. But it was not my responsibility to measure this entity. A layman who worries about germs is by definition a hypochondriac, and I had my own list of real ailments to concern me.

Rendezvous, for example. How much fuel would it take? John Young and I both knew it could take a lot more than expected. If the LM lifted off from the lunar surface at precisely the correct instant, and got into precisely the proper orbit, it seemed straightforward enough. But if the LM were delayed, for instance, or if it never touched down, a complicated series of events—deviating dramatically from the normal rendezvous sequence—could ensue. The later the LM became, the farther it trailed behind the CM. The remedy, to increase its catch-up rate, was for it to fly lower and lower, to the point where it would be just skimming the lunar mountaintops. The CM could help a bit by going higher, and therefore slower. But there was a point where this logic broke down, and had to be reversed. The LM, instead of the hunter, became the hunted. Now it went high and slow, while the CM descended as low as it dared and tried to make an extra circle as fast as it could to catch the LM from behind. From my point of view in the CM, and to a lesser extent Neil and Buzz's in the LM, these variations required an array of techniques and procedures varying most significantly from the perfect ("nominal") case. The procedures for all these cases (we finally decided, somewhat arbitrarily, that there were eighteen distinctly different situations that could arise and called these "cases") had to be defined and written down in detailed check-list form, to include every computer manipulation and each cockpit switch change. Then these check lists had to be validated somehow; the simulator was our tool for doing that. Hooked up to a huge computer which kept track of the paths of an imaginary LM and CM around an imaginary moon, the simulator would in theory reveal how practical our various rescue schemes were, and how much time and fuel they would consume.

Since the LM had never even flown with a crew on board, the information we were supplying the computer concerning its flight characteristics was a bit sketchy, and was based on *other* simulations and assumptions provided by Grumman. Also, my CM (107) would be equipped with a VHF* ranging device (to help find the LM) that had never flown in space. Therefore, the answers being supplied to us by our simulators were suspect: the first rule of computers is GIGO, or "garbage in, garbage out," and we needed better than garbage as food for our assumed trip. McDivitt's crew on Apollo 9 and Stafford's on 10 would hopefully supply all the hard facts we needed, except for those pertaining to making the landing itself.

The landing site had been picked, in the southwestern corner of the Sea of Tranquility. Perhaps it would be an area of great geologic interest, but its selection was based less on that and more on the fact that it was a flat plain with, hopefully, not enough craters and boulders to prevent Neil from finding a spot smooth enough to set the LM down safely. Another good feature of the

* VHF = very high frequency

landing site was that its central location was convenient in relation to the equatorial orbital plane of the CM overhead, and excellent from the standpoint of communicating with earth. As you look at the moon, Tranquility Base, as it was later named, appears so:

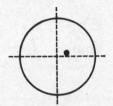

1 degree above the lunar equator
24 degrees right of center line

The next point to consider is the location of the sun. Since Neil and Buzz would be approaching from the east, and since they didn't want the sun's glare in their eyes as they were trying to measure crater depths, etc., the sun at the time of landing should be behind them, thus:

SUN

At what angle behind them? Not too high and not too low. If too high, i.e. overhead, the craters and boulders would not cast shadows, and depth perception and obstacle avoidance prior to touchdown would be a real problem. Too high also meant the surface would be too hot. Too low, and the shadows could get so elongated that they would obscure other useful details and again make a visibility problem for the crew. A sun angle of about 10 degrees was deemed perfect and seemed to work well in the simulator.

Perhaps a review of the classical phases of the moon will clarify the situation.

THE SYNODIC LUNAR MONTH
29 days, 12 hours, 44 minutes

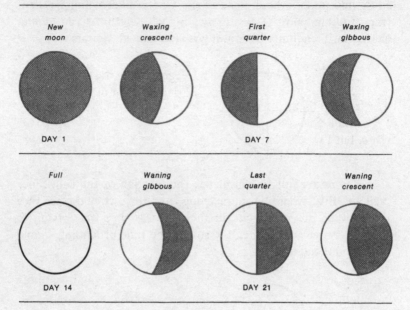

New moon	Waxing crescent	First quarter	Waxing gibbous
DAY 1		DAY 7	

Full	Waning gibbous	Last quarter	Waning crescent
DAY 14		DAY 21	

Since it takes roughly thirty days for the moon to go through this cycle once (i.e. 360 degrees), it turns out (dividing 360 by 30) that at any one spot on the moon the sun angle changes by 12 degrees per day. Therefore, if you want the sun to be approximately 10 degrees above the horizon, behind you as you touch down at a particular spot, there is only one day in the month which is good for landing, and on that day the moon and Tranquility Base will look about like this:

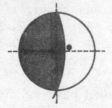

SUN

During July, the month in which we wanted to land, Tranquility Base would reach a sun angle of 10 degrees on the twentieth; therefore, to allow time for the three-day trip, plus time to get all squared away in lunar orbit, we would have to launch July 16. *Voilà!*

In the six months remaining before July 16, I was busier than I had ever been in my life. I could easily have spent eight hours a day in the simulators, learning the huge Apollo CSM and its idiosyncrasies, but I was not allowed to do so as there were a whole host of other events requiring the presence of, or the approval of, the crew. The NASA Center at Langley, Virginia, had conducted an extensive study of the problems of docking LM and CM; they had full-scale replicas of the two vehicles hanging from wires in their huge hangar, and I had to fly up to Langley in a T-38, inspect the rig, and "fly" it, and draw my own conclusions as to possible docking problems when the two real machines met for the first time. The centrifuge in Houston clamored for our attention, because coming back from the moon we could experience a force of 10 or more Gs, eyeballs in, far more taxing than the acceleration produced during any of the earth orbital flights. New pressure suits had to be fitted, at Dover, Delaware, requiring a long day in a T-38, one refueling stop on the way up, and one more coming home. The list seemed endless, spanning the country and the clock. I have never kept a diary, but for some reason the events of April 14 seemed worth recording and I jotted down some notes.

The day began in Houston with the centrifuge, never pleasant, but especially bad when imitating lunar returns. At 10 Gs my chest caves in and my vision narrows, and when I finally reeled out of the torture chamber, I dared not turn my head left or right, lest I tumble my gyros and fall over in an undignified heap. But leave I had to, and quickly, for I was scheduled for the CM simulator, where over and over again I practiced entering the earth's atmosphere, just as I had been doing in the centrifuge. Only this time there was only the familiar one G, rather than the brutal 10 of the centrifuge. Instead, the hazards were more mental than physical, and I was called upon to solve a series of mysteries involving obscure equipment failures. Some I won, but even more frequently I lost, and more than once I plummeted into the sea with parachutes still unfurled, destroying myself, along with Neil and Buzz (assuming I hadn't already left them for all time on the lunar surface). Then I hopped out of the simulator and into a hot and bulky pressure suit, not for more training but for PR, as we had to have a crew picture taken in front of a five-foot faintly blue moon. It was now time to go to the airport, to plan a T-38 flight to the Cape, but first I had to call home and make sure I hadn't forgotten to let Pat know about my absence for a day or two. At home all was chaos—our dog had bitten a neighbor's child, not seriously, but rabies tests required, etc., etc. I was no help, and could only make a few clucking noises while I filled out the flight plan. Then hurry and leap off into the night sky, promptly at seven, with no dinner and no promise of getting to the Cape before everything closed. Please, God, no emergencies tonight, I am just too tired to pump the adrenalin required for coping . . . just let me be, let me sit here and guide this missile through the darkness without any hassles, that is about all I am capable of doing.

Another night, late, I was coming home from Dover and, naturally, passed over my old home town of Washington. God, it looked good to see familiar surroundings! There lay Washington and Baltimore, shimmering like twin jewels in the night mist. I identified the dark band of the Potomac River flowing by Alexandria, where my parents lived for many years after World War II, and my mind wandered off into a string of teenage memories. From forty-three thousand feet I let my eye roam across the river, past the White House, and up Massachusetts Avenue to the Washington Cathedral. Next came a shock—the cathedral had been moved! Could I have been disoriented? Of course! I had been looking at the city from the north, not from the south. Worse yet, if that were true, then I had not been looking at Washington at all but at that alien city of Baltimore. What I had first picked as my old Alexandria neighborhood was no doubt a cluster of honky-tonks on route 40 north of Baltimore. With a jolt I realized it had been a long day and I was making mistakes no alert air cadet would; this guy who couldn't tell Washington from Baltimore was within a few months of navigating to the moon and back.

There were times when I struck out on my own, and other times when Neil, Buzz, and I trained as a unit. They were good workers. Harried and hassled with their LM and lunar surface training, they were at least as busy as I, yet they kept their cool and got their work done. Neil could be slightly disorganized, such as the first morning when we were scheduled for a 6 A.M. takeoff and I had to call an embarrassed Jan and roust them out of bed, but this was a rare occurrence; by and large Neil was clicking along like a well-oiled machine. His air was still nonchalant, but his actions belied this attitude, and he was putting in long hours and was very much on top of a complex and rapidly changing situation. Buzz, of course, was all business as usual. Generally quiet and incapable of small talk, Buzz could nonetheless get wound up on any of a number of technical pet projects of his, and when he did, he could talk the handle off a pisspot, far into the night, with or without a bottle of Scotch for lubrication. Having the constitution of a horse, Buzz would be up again in a couple of hours, none the worse for wear, eager to dive into the simulator once more. They were a tough and knowledgeable pair, both of them far better engineers than I.

As February drew to a close, the time came for the long-awaited flight of the LM. As described by Sam Phillips, the Apollo program director, Apollo 9 would be less spectacular but more complex than its daring predecessor, Apollo 8. It was a flight difficult for the public to understand, for if we had already gone to the moon, why were we now tooling around in earth orbit? But the flight of the "tissue paper spacecraft,"* as Jim McDivitt described his LM, was vital to our program, and there was no point in flying all the way to the moon on its maiden check-out flight. McDivitt, Dave Scott, and Rusty Schweickart had been working on the flight and its problems for a long, long time, charging hard the whole time, and as the deadline approached, their activities became almost frenzied. In the process they became run-down physically and got upper-respiratory infections, which made it necessary that the launch be slipped three days, until March 3. No problem to postpone a couple of days for an earth orbital flight, but a good lesson for the Apollo 11 crew—as we would have had to slip a whole month, to keep our 10-degree sun angle, or else switch to a different landing site, one closer to the western edge of the moon. Best we not get sick.

When Apollo 9 did get airborne, everything worked like a dream, and we were all surprised and relieved. One after another, the major impediments to a lunar landing were removed. First, transposition and docking: the CSM was mounted on the very tip of the Saturn V, with the LM nestled behind it, inside a protective sheath. In order to get into the LM, it was first necessary to separate the CSM from the Saturn, turn around 180 degrees, and then drive the CSM back and dock with the LM. Once the two vehicles were joined in this fashion, nose to nose, the CSM then pulled the LM free from the Saturn, which was no longer needed. On a lunar flight, this transposition and docking could take place only *after* TLI, after we had committed ourselves to leave earth orbit and set course for the moon, and that was no time to discover that the scheme didn't work for some unknown reason. Therefore, an important responsibility of Apollo 9 was to prove the feasibility of

---

* Only a slight exaggeration. In an effort to save weight, the aluminum skin of the LM had been made so thin that if a workman dropped a screwdriver inside it, it could —and did more than once—puncture the pressure bulkhead under his feet.

the transposition and docking concept.* The next and by far the most important milestone was rendezvous. After carefully checking the LM (*Spider*) and testing its descent engine once, Jim and Rusty undocked *Spider* from Dave in *Gumdrop* and performed an intricate series of maneuvers which took them over one hundred miles away from Dave. Then they found their way back again in as exact a duplication as possible of the techniques that would be required of a crew ascending from the lunar surface. In the CSM, Dave tracked them with his sextant and computed his own maneuvers required to rescue the LM if it should crap out. Generally speaking, the CSM can make the exact "mirror image" of a planned LM maneuver, if it is made precisely on time, and keep the same rendezvous strategy going. During the six hours of its excursion, the LM, including the all-important rendezvous radar, performed brilliantly. Interspersed during the ten-day flight were other features such as a space walk by Rusty (a test of the lunar suit and back pack) and another manifestation of the reliability of the CSM† and its systems, but the main things were the rendezvous and docking.

When *Gumdrop* finally plopped into the Atlantic next to the U.S.S. *Guadalcanal*,‡ the NASA planners were presented with a fascinating question: If the CSM and LM had passed their flight checks, and if NASA had flown once to the moon, why not allow

---

* When I first came to work for NASA, it made people nervous (me included) to plan to have the CSM completely free of the LM and Saturn during the turn-around maneuver. Various schemes involving telescoping poles and other devices were proposed to keep the CSM physically attached to the Saturn during the separation and turn-around maneuver. Today such Rube Goldberg schemes seem silly, but at the time (before the first Gemini dockings), there was psychological resistance to the idea of having something float free instead of being tethered. Another example of this, one that has continued to this day, is the requirement to tether space walkers, even though in many ways it might be easier and safer to have them operate without life lines.

† The CSM was old 104, the one I had spent so much time with at Downey. Dave Scott hadn't been at all happy about flying it in lieu of 103, which Borman took to the moon, but after its great performance on Apollo 9, all was forgotten.

‡ The same ship that picked up John Young and me on Gemini 10. A small aircraft carrier, used to carry helicopters, the *Guadalcanal* is a great recovery ship because it is air-conditioned.

Apollo 10, the next flight, to attempt the lunar landing? Apollo 10 was scheduled for May, and featured Tom Stafford and Gene Cernan separating their LM and descending to fifty thousand feet before returning to John Young in the CSM. If they were going to venture a quarter of a million miles from home, so the argument went, it was folly to let them get within fifty thousand feet of the national goal, in a qualified landing craft, without actually landing. I could well understand this logic, and secretly agreed with it, but there were powerful arguments on the other side as well. For one thing, the moon was an entirely different environment in which to attempt rendezvous, and our earth orbital experience might not be transferable. The lighting conditions were different, the orbital velocities were considerably different, the ground tracking capability was completely different. Under these circumstances, it was prudent to conduct as simple a rendezvous as possible, and that meant keeping the LM in precisely the same orbital plane as the CSM and keeping the velocities of the two vehicles fairly evenly matched. Under these conditions, getting back to the CSM should be fairly simple, and the CSM could be pressed into service as a rescuer at any point along the way. Not so if the LM landed. Then it would be a whole new ball game, with the LM on the surface at zero velocity and with the requirement to take off precisely on time and to reach precisely the same orbital plane as occupied by the CSM whizzing by overhead.

The rendezvous situation was further complicated by the fact that the gravitational pull of the moon was not evenly distributed; there were local heavy spots, called mascons,* which caused an orbiting spacecraft to dip slightly as it passed over them. Apparently the rocks underlying some of the *maria*, or lunar seas, were heavier on the average than those in other parts of the moon. In the spring of 1969, these mascons were not well understood, and the tracking people had not been able to feed an adequate mathematical explanation of them into their computers, which meant

* Mascon = mass concentration

(GIGO!) that the computers could not predict exactly when and how they might perturb* the orbit of an LM or CSM. One thing for sure, though. If both LM and CSM were in nearly the same orbit, the effect on one should be nearly the same as the effect on the other, but if one were sitting on the ground and the second were orbiting all that time, the cumulative error would be much greater. Tom Stafford, a rendezvous expert if there ever was one, was very hip to these arguments. Since he had always been on the conservative side in our astronaut office discussions, and had insisted that a variety of rendezvous situations be demonstrated in flight before committing to a lunar landing, Tom wouldn't, or couldn't, reverse himself now and display great enthusiasm for an Apollo 10 landing, even if it meant he (or Gene Cernan) would be first to walk on the moon.

The final argument, and probably the clincher, was that Stafford's LM (4) was overweight. A few extra pounds didn't matter much as long as the LM stayed in lunar orbit, but they were terribly important if they had to be lifted off the surface. The chances of off-loading anything were slim, so Tom would probably have had to wait for a later LM, and that opened Pandora's box from a managerial and scheduling point of view. So the original plan was adhered to, and Stafford's flight would be as full a dress rehearsal as possible but not the real McCoy. Stafford and crew seemed happy enough about it . . . still, had it been my decision, I think I would have delayed Apollo 10 a couple of months, given the crew some more training and our LM (5), and let them land. Man, to go to within fifty thousand feet and then take a wave-off, that's just too much!

The flight of Apollo 9 was not only a great engineering triumph, but it also was a milestone of a different sort in the small, closed world of Apollo crews. It moved Neil, Buzz, and me one rung up the ladder, and our new pecking order gave us a great deal more clout in reserving airplanes for trips and, more important, in

---

* The language of orbital mechanics is peculiar. Any disturbance to an orbit is referred to as a "perturbation."

scheduling simulator time. The system was such that the next crew to fly, Apollo 10, in this case, had the world at its feet and received lavish support from all components of NASA. Beyond that, help trailed off steeply; it made a great difference to be number two instead of number three. There were two CSM simulators at the Cape and one at Houston, but the one at Houston was devoted more to research and development than to crew training. Having McDivitt out of our hair meant that we owned one of the Cape simulators now, except for the times that Stafford's was broken and his crew pre-empted us. At the risk of repeating some of the things I have said earlier about simulators, I would like to emphasize that they were absolutely vital in preparing the crew. The crew needed simulator time, above all other activities, to master the intricacies of the very complex Apollo flights, which required split-second timing and perfect execution of hundreds of different tasks. There would be no instructor to slap our wrists in flight, no second chance if we erred.

Consider someone who has never ridden in a car before. Without actually letting him drive, prepare him for a trip across Los Angeles, from Downey down the Long Beach Freeway, then up the San Diego Freeway to the airport. Along the way, let him have a high-speed blowout, change the tire, and proceed. Explain to him the rules of the road, the operation of the machinery, the physical coordination of clutch, brake, and gas, the feel of the wheel. Condition him to red lights, flashing ambers, the glare of oncoming brights. Trust him to make that trip the first time, by himself, without dinging a fender. Far-fetched perhaps, but that is how we did it on Apollo. To car or cockpit, one hooks up a huge computer, to change speedometer readings in response to gas-pedal pressure, or to change state vector in response to star-to-horizon measurements. The computer makes every gauge, pedal, handle, and switch perform as realistically as possible. The gas tank empties as the miles click by; the oxygen pressure increases as the heaters are turned on, etc., etc. Outside the window the real world is more difficult to duplicate, but in the most important respects it can be done. Filmed cars can be made to whiz by in adjacent lanes, star fields can be precisely duplicated out the spacecraft windows. Navigation is easy: the signs pointing to the San Diego Freeway can be duplicated precisely, the lunar horizon less accurately, but accurately enough to measure the angle between it and a selected star. Malfunctions can be inserted. The steering wheel can be made to shimmy and shake for a few seconds before pulling violently to one side as a front tire blows. The stabilization and control system can suddenly develop a short circuit, causing one thruster to fire continuously and to topple the spacecraft end over end. For trip by auto or by Apollo, one can duplicate the view and the hazards, and practice normal and emergency procedures as often as required to get the hang of it.

The reason one does not do it for an auto trip is that there are easier, cheaper, more reliable ways of making the trip (let your wife drive), but for space flight a good simulator is an absolute must. The first time John Young saw the CSM simulator, he dubbed it the Great Train Wreck, in tribute to its physical bulk and geometric complexity. It was a huge and gaudy affair, with a carpeted staircase leading up to the cockpit entrance, some fifteen feet above floor level. The cockpit itself was standard size, naturally, since it was an exact replica of an Apollo CM cockpit, but surrounding the cabin on all sides was a jumble of huge boxes, stuck on at odd angles, containing the various visual effects we saw out our windows. The computer itself stood aloof from all the confusion, isolated behind glass in an adjacent air-conditioned palace, where it was attended to night and day by a huge crew. It was the queen bee, licked by drones and fed by workers, who brought it punched cards containing the latest flight information, or trajectory, or rendezvous sequence, or lunar gravitational model, or what have you. The whole thing cost millions of dollars and took hundreds of people to maintain, working three shifts, seven days a week. Naturally, it wasn't always in perfect operating order, and especially early in the Apollo program, it seemed easier to get the actual spacecraft to fly than to get the simulator to duplicate it. A favorite trick of the computer was to allow a sequence of operations, such as in a rendezvous, to proceed up to a point and then to stop abruptly. While technicians swarmed over the queen, the crew fidgeted in the cockpit. If order was restored, it usually meant that the problem had to be started all over from the beginning, rather than picking up again at the point of the malfunction. This meant, in the case of rendezvous, which took hours, that the crew had worked right up to the moment of truth for nought and had spent hours lying on their backs without ever knowing whether the rendezvous would have been successful. It was even worse when the LM crew got in the LM simulator, and the people in Mission Control got behind their consoles, and the three parties, each with their own computers, all tried to hook up together and simulate the same problem at the same time. That was tough sledding, and rarely could the problem be carried all the way through without some component "bombing out," as we called it. In the astronaut office, one of the most common greetings was an accusatory "Aren't you supposed to be in the simulator?" and the usual answer was "Yeah, but the frapping thing bombed out again."

During my abortive Apollo 8 training, I had managed to accumulate 150 hours of simulator time, which helped me to understand the complex guidance and navigation system, at least in a rudimentary sort of way. Then, during the early months of Apollo 11 training, in Houston and at the Cape, I had sniveled whatever simulator time I could, but it was not until after McDivitt's flight, in March, that things started to get rolling, and none too soon, if we were to meet our July deadline. In fact, we had a long list of mandatory simulations to be successfully completed in conjunction with Mission Control before we would officially be declared sufficiently trained to attempt the flight; for a while it looked as if we simply weren't going to make July.

At least, flying the simulator was a programmed activity, and I could lie there and either converse silently with the machine or out loud with one of the instructors stationed at a nearby console.

Outside the simulator, however, the real world was not so orderly, and despite the fact that we had a small staff of schedulers who organized our time, we still found a host of people clamoring for our attention. Most were engineers who just wanted to make sure we *really* understood their own particular subsystem, or who wanted to share with us some newly discovered problem or its solution. These were conscientious people who could not be turned off, even if frequently they rambled on about things over which we had absolutely no control. Still, occasionally one garnered useful information. Then there were people with whom we had worked for years, and who merely wanted to chat and wish us well, or who wanted us to carry some little trinket of theirs to the moon. If the person was a NASA or contractor employee, and if the object was small enough, I generally consented, as each of us was allotted a PPK (personal preference kit) which could be jammed with personal mementos belonging to us or to others. On Apollo 11 I carried prayers, poems, medallions, coins, flags, envelopes, brooches, tie pins, insignia, cuff links, rings, and even one diaper pin. Some things belonged to me and some to others. The only criterion was that the object had to be small, but none matched the ingenuity of one gent for whom I carried a small hollow bean, less than a quarter of an inch long. Inside it were fifty *elephants*, carved from slivers of ivory, which he planned to distribute to his co-workers after the flight. If you don't believe that story, I'll show you the bag I carried the elephants in. Then there were the people who wrote us enclosing trinkets to carry, or who offered advice, or who simply wished us bon voyage. As we got busier and busier, we paid less and less attention to the mail, and I have since met more than one friend who clearly felt that I had gotten so big for my britches that I had chosen to ignore his heartfelt message. Most of the messages were pretty straightforward, but there were a few weirdos in the bunch. One poor man, obviously deranged, wrote from Israel, enclosing reams of data on the gigantic ants that infested the moon. It would be disaster, he pointed out, for the LM to come down on or near one of these anthills, and he would be happy to provide us with his detailed maps showing the location of each hill—for a price, of course.

There were also a variety of non-technical chores, such as thinking up names for our spacecraft and designing a mission emblem. We felt Apollo 11 was no ordinary flight, and we wanted no ordinary design, yet we were not professional designers. NASA offered us no help along these lines (wisely, I think). On Gemini 10, which in my view has the best-looking insignia of the Gemini series, artistic Barbara Young had developed one of John's ideas and come up with a graceful design, an aerodynamic X devoid of names and machines. This was the approach we wanted to take on Apollo 11. We wanted to keep our three names off it because we wanted the design to be representative of *everyone* who had worked toward a lunar landing, and there were thousands who could take a proprietary interest in it, yet who would never see their names woven into the fabric of a patch. Further, we wanted the design to be symbolic rather than explicit. On Apollo 7, Wally's patch showed the earth and an orbiting CSM trailing fire. On 9, McDivitt produced a Saturn V, a CSM, *and* an LM. Apollo 10's was even busier. Apollo 8's was closer to our way of thinking, showing a figure eight looping around earth and moon, on a command-module-shaped patch, but it had, like all the rest, three names printed on it. We needed something simpler, yet something which unmistakably said peaceful lunar landing by the United States. Jim Lovell, Neil's back-up, introduced an American eagle into the conversation. Of course! What better symbol—eagles landed, didn't they? At home I skimmed through my library and finally found what I wanted in a *National Geographic* book on birds: a bald eagle, landing gear extended, wings partially folded, coming in for a landing. I traced it on a piece of tissue paper and sketched in an oblique view of a pockmarked lunar surface.

Thus the Apollo 11 patch was born, although it had a long way to go before final approval. I added a small earth in the background and drew the sunshine coming from the wrong direction, so that to this day our official insignia shows the earth over the lunar horizon like ◐ when it should really look like ◯. I also penciled an APOLLO around the top of my circular design and ELEVEN around the bottom. Neil didn't like the ELEVEN because it wouldn't be understandable to foreigners, so after trying XI and 11, we settled on the latter and put APOLLO 11 around the top. One day outside the simulator I was describing my efforts to Jim Lovell, and he and I both agreed that the eagle alone really didn't convey the entire message we wanted. The Americans were about to land, but so what? Tom Wilson, our computer expert and simulator instructor, overheard us and piped up, Why not an olive branch as a symbol of our peaceful expedition? Beautiful! Where do eagles carry olive branches? In their beaks, naturally. So I sketched one in, and after a few discussions with Neil and Buzz over color schemes, we were ready to go to press. The sky would be black, not blue, but absolute black, as in the real case. The eagle would be eagle-colored, the moon moon-colored, as described by Apollo 8, and the earth also, so all we had left to play with, really, were the colors of the border and the lettering. We picked blue and gold, and then we had an illustrator at MSC do the artwork for us. We photographed his finished design and sent a copy through channels to Washington for approval. Washington usually rubber-stamped everything. Only this time they didn't, and our design came back disapproved. The reason? The eagle's landing gear, powerful talons extended stiffly below him, was unacceptable. It was too hostile, too warlike; it made the eagle appear to be swooping down on the moon in a very menacing fashion, according to Bob Gilruth. What to do? A gear-up approach was unthinkable to this pilot who had woken up more than once with cold sweats over a dreamed wheels-up landing. Perhaps the talons could be relaxed and softened a bit, made limp like a receiving-line handshake. Then someone had a brainstorm: just transfer the olive branch from beak to claw and all menace disappeared. The eagle looked slightly uncomfortable in the new version, clutching his branches tightly with both feet, but we resubmitted it anyway, and it greased on through channels and won final approval.

Years later the eagle still pops out at me from unfamiliar places, and the design has even found its way onto the reverse side of the part-silver Eisenhower dollar. This coin was issued during John Connally's short stint as Secretary of the Treasury, and he was kind enough to pen me a short note in August 1971 stating, among other things: ". . . the design on the reverse is a rendering of the Apollo 11 insignia." As if I didn't know. The "designer" of the coin's eagle told the press that it was a "happy" eagle, but I don't know, it still looked uncomfortable to me. I just hoped it dropped that olive branch before landing. The eagle was reproduced also in a number of silver medallions, which the three of us commissioned,* to carry on the flight as commemorative pieces, and in three gold medallions for our wives.

The choice of an eagle as a motif for the landing led swiftly to naming the landing craft itself *Eagle*. It was appropriate, Neil and Buzz felt, and it sounded good on the radio.† The choice for the CSM was not as obvious, and I was frankly floundering for a name. One day I was chatting long distance with Julian Scheer, NASA's Assistant Administrator for Public Affairs in Washington, and he inquired whether we had yet found a name for our CSM. When I told him no, he said that "some of us up here have been kicking around *Columbia*." I gave a noncommittal grunt and the conversation went on to other things. *Columbia*, eh? It sounded a bit pompous to me, especially compared to its two named predecessors, *Gumdrop* and *Charlie Brown* (I liked *Gumdrop* but thought *Charlie Brown* was terrible). But I supposed we could

* To the tune of $2,731.74.
† Pilots worry about how their call signs will sound over the radio, which does not transmit either the very high or the very low frequencies in the human voice. This slight alteration sometimes renders a familiar sound unrecognizable. I remember well one fighter group call sign, "Flit Gun," which was always misunderstood by ground controllers when transmitted by the squeaky voice of our excitable group commander. "Roger, Six Gun," they would say, and he would tartly reply, "No, it's *Flit* Gun." "Roger, Six Gun." That would destroy him. "No, goddamn it, *Flit* Gun! *Flit! Flit!*" It was a pleasure to fly in his formation and share these military moments.

afford to be slightly pompous, and *Columbia* did have a lot of things going for it. One obvious tie-in was the close similarity to Jules Verne's mythical moon ship, which was shot out of an immense gun barrel, the *Columbiad*, near Tampa, Florida, in 1865 and was recovered in the Pacific Ocean after a lunar fly-by. Perhaps most significant, of course, was the close relationship between the word "Columbia" and our national origins.* The eagle was our symbol, but Columbia had almost become the name of our country, and it seemed, therefore, a suitable if somewhat jingoistic counterpart to *Eagle*. Finally, the lyrics "Columbia, the gem of the ocean" kept popping into mind—and they augured well for the recovery of a spacecraft which hopefully would float on the ocean. Since Neil and Buzz had no objections, and since I couldn't come up with anything better, *Columbia* it was, and I think not a bad choice. Thank you, Julian.

The plaque which was to be left on the moon, bolted to a leg of the LM, was another matter. "Here Men From Planet Earth/ First Set Foot Upon The Moon/July 1969 A.D./We Came In Peace For All Mankind," it said, and it was signed by President Nixon and the three of us. It also showed two circles, one the Western and the other the Eastern Hemisphere. Since it was tied to the leg of the LM, perhaps Neil and Buzz had participated in its design, but I don't think so, and I certainly had not. I don't recall seeing a replica of it until after the flight, and my signature on it came not from me directly, but from a signature machine in the astronaut office in Houston. I can spot that signature a mile away, with its sloppy overlapping loops.† In addition to the international flavor of the "for all mankind" inscription, a tiny silicon disk, an inch and a half in diameter, was being prepared. It contained miniaturized

---

* Should not the country have been named after Columbus, its discoverer? Instead, Columbia became the name of the ornament on top of the Capitol dome, a South American country, and a brand-new community in the Maryland suburbs near Washington, to name just a few.
† So, incidentally, can autograph hounds and stamp collectors. The serious collector, when writing in his request for astronaut signatures, frequently adds: "No machine signature, please. Return envelope unsigned if it cannot be signed by hand." Apparently a machine signature is worth less than nothing.

messages from seventy-three heads of state, as well as from Presidents Eisenhower, Kennedy, Johnson, and Nixon. This disk would be left on the moon, along with an American flag—the only flag so honored. For a while, NASA discussed leaving a flag from each UN member nation, and at MSC a Christmas-tree-type stand had been built which, when opened, displayed all the flags in a colorful cone. However, the screams from Congress in opposition to this scheme caused NASA quickly to back away from the idea. I agreed with the Congress, since the landing was being financed by the American taxpayers alone, and I felt they were entitled to one show of nationalism amid the international totems.

The weeks seemed to fly by in the spring of 1969. It seemed that we had a year's work still ahead of us in May, when suddenly we found ourselves the next crew to fly. Tom Stafford, John Young, and Gene Cernan departed the Cape from Launch Pad 39B* around noon on May 18. They arrived at the moon three days later, and became the fourth, fifth, and sixth humans to see it at close range. Stafford and Cernan were about to get an even better view, as their job was to descend to fifty thousand feet in their LM, *Snoopy*, and then return to John Young in the *Charlie Brown*. This was the whole point of the flight, to check out the LM in its native habitat, to swoop down over the Apollo 11 landing site and photograph it, and finally to demonstrate a relatively low-risk form of rendezvous. In the process the ground tracking network would have an opportunity to refine their model of the moon's peculiar gravitational field, studded as it was with mascons and other aberrations. Fortunately, all went according to plan and the rendezvous was a piece of cake. Neil, Buzz, and I breathed a sigh of relief, not only to have Tom and Gene return safely to the CSM, but also—more selfishly—to see the way cleared for 11 to attempt a landing

---

* Launch Complex 39 on Merritt Island was originally planned to encompass three different launch sites—A, B, and C. C was never built, and B was used for Apollo 10 only. All the other Saturn V flights departed from 39A, from which I conclude that B and C were really not required. Of course, when the Saturn V was first conceived, no one knew what damage the monster might inflict on the scorched concrete beneath it, or how popular space flight might become.

instead of merely repeating this dry run. Tom and Gene had descended, according to their radar, to a 47,000-foot perilune,* and had experienced no difficulty except for a brief tussle with the LM's control system, during which they had gyrated enough to cause Gene Cernan to exclaim for all the world to hear, "Son-of-a-bitch!" Cancel his Boy Scout membership and hire four more stenographers to handle the influx of mail. Oddly enough, the moon appeared different to Apollo 10 than it had to 8. Eight had seen black and white, with shades of gray separating the two. Ten, on the other hand, saw a lot of brown in the lunar surface, light tan at noon, and darker brown near sunrise and sunset. The photographs from either flight weren't much help in settling the argument, for by changing development processes and emulsions slightly, one could make the surface appear gray, or brown, or even green. It was a small point, but one which really piqued my curiosity. How could it have changed color in five months? How could six expert (?) observers be divided so evenly, three against three? Who was right?

We would find out soon enough, but in the meantime we had no time to contemplate the matter, for we were inundated with problems of our own—not call signs, or medallions, or protocol matters—but the thorny, hard, technical unknowns that still faced us. Paramount among these, in my mind, were the eighteen different rendezvous situations which could arise if the LM never made it to the lunar surface, or if it got there early or late, or departed crooked or straight. I was rapidly filling a book with notes, procedures, computer entries, and diagrams. I put everything together that I would need during the time that Neil and Buzz were gone, and called it a Solo Book; it ran 117 pages. I had practiced some of the eighteen cases in the simulator over and over again, but I didn't have time to practice the less likely ones, and I relied upon a supporting team of engineers who had read and reread them, and vouched for their practicality. In addition to rendezvous, there was a long list of other worries. Just as I had done on Gemini 10, I

* The lowest point in a lunar orbit, just as perigee is the lowest in the earth orbit.

assembled a black notebook and kept a section in it devoted to open items, i.e., those things that were still unsolved. As answers poured in, I crossed off the appropriate entry. By flight time, I had crossed off exactly one hundred items. Oddly enough, this was a smaller number than the 138 I had kept book on during Gemini 10, despite the fact that Apollo was a far more complex machine and despite the fact that I felt far busier this time. Perhaps on Apollo 11 I kept book only on the really necessary items, or perhaps I simply felt more pressured this time, because I felt the eyes of the whole world searing into my back.

My problems could be divided into two piles: those which Apollo 11 would be doing for the first time and for which there were no pat answers, and those which had been done before, but which, through sloth, inattention, or inability, I simply didn't understand well enough. One in the latter category was the mechanism that linked the CSM and LM together, the probe and drogue. When the CSM first docked with the LM, it weighed sixty-five thousand pounds and the LM thirty-three thousand pounds. The CSM gently placed its probe inside the LM's drogue, as an old lover might, and three tiny little prongs ("capture latches") around the rim of the probe engaged the LM's drogue and snapped into place. The two vehicles, whose combined weight was nearly 100,000 pounds, were then tenuously tied together by these little paper clips not four inches apart. As soon as the command module pilot (CMP, he will be called hereafter) detected this fact and further ascertained that the two vehicles were not swaying excessively, he flipped a switch. This switch caused nitrogen gas under pressure to escape from a small cylinder inside the probe, which pressure was used to retract the probe inside the CM, dragging the LM with it. The motion stopped when the CM's tunnel butted up against the LM's tunnel, at which point twelve mechanical latches were triggered and the LM and CM were held together in a vise grip, in the form of twelve contact points around the periphery of the tunnel wall. Now the CMP left his seat, stopped being a pilot for a while, and assumed the mantle of master mechanic. The probe, inside the tunnel, was in the way and had to be removed, as did the drogue, so that people could pass back and forth between the two machines; but before that could take place, there was an incredible check list to perform, full of mysterious notations such as "Preload Selector Lever—Rotate clockwise (away from orange stripe), Preload Handle-Torque counterclockwise to engage Extend Latch (red indicator not visible)," and on and on. If the thing refused to budge, I was supposed to get out the tool kit and dismantle it. Me, who couldn't repair the latch on my screen door. I hated that probe, and was half convinced it hated me and was going to prove it in lunar orbit by wedging itself intractably in the tunnel, forcing Neil and Buzz to make a dangerous space walk out through the LM front hatch and around to the CM side hatch. I practiced long hours in a simulated tunnel with a real probe and drogue, and complained that the space program needed more English majors to write check lists ("The capture latch release handle lock, for God's sake. Which one is that?"). Maybe we just needed better mechanics. I wasn't one.

The press always asked what part of the upcoming flight would be the most dangerous, and I always answered, that part which we had overlooked in our preparations; that was a truthful if somewhat evasive answer. There were actually eleven points along the Apollo 11 flight path that merited special attention, and I tried to distribute my training time in such a way that I was familiar enough with them to perform my routine chores smoothly and to understand the equipment design and its use well enough to cope with failures that might reasonably be expected to occur. Usually the three of us shared responsibilities, but a few were mine alone, and, of course, I could only monitor the LM operations. The eleven were:

1. *Launch* Obviously a hazardous time, with gigantic engines, explosive fuels, high temperatures and velocities, terrific wind blasts, and stringent guidance requirements combining to make a very tense eleven minutes from lift-off to earth orbit.

2. *TLI* Translunar injection, wherein the third-stage Saturn engine was reignited, causing us to depart the relatively stable situation of being in earth orbit and begin a trajectory that would hopefully just miss the moon three days later. If the engine stopped prematurely, we had some complicated scrambling to do to make it back to earth at all.

3. *T&D* Transposition and docking, the process of separating the CSM, turning it around 180 degrees, docking with the LM, and pulling it free from the carcass of the Saturn. I also include here clearing the probe and drogue from the interconnecting tunnel.

4. *LOI* Lunar orbit insertion, a two-burn procedure for slowing down enough to be captured by the moon's gravitational field, but not enough to crash into it. If the engine shut down prematurely during the first burn (the more important of the two), some really weird trajectories could result, and the LM's engine might have to be quickly pressed into service to return us to earth.

5. *DOI/PDI* Descent orbit insertion and powered descent initiation were the two LM burns which caused Neil and Buzz to depart my comfortable sixty-mile orbit and intersect the surface of the moon at the right spot. If not precisely performed, the LM would come down in the wrong place or—more likely—couldn't land at all, in which case some really zany rendezvous sequences could ensue.

6. *Landing* Could be very dangerous; we simply didn't know. Fuel was short, therefore timing was critical. Also, the properties of the lunar surface in that one spot might be poor. Worse yet, visibility and depth-perception problems could cause a crash instead of a landing. Thus has it been since the days of the Wright brothers.

7. *EVA* Walking on the moon might be physically taxing and overload the oxygen or cooling systems. There might be potholes, or even underground lava tubes which would cause the surface to collapse. Even more basic, any EVA puts man just

one thin, glued-together, rubber membrane away from near-instant death.

8. *Lift-off*   Only one engine, and it had better work properly—that is, provide enough thrust and provide it in exactly the right direction. If not, at best, the LM would limp up into an orbit from which I might be able to rescue it; at worst, Neil and Buzz would be permanent decorations among the rocks in the Sea of Tranquility.

9. *Rendezvous*   A piece of cake, *if* . . . a horror, *if* . . . which *if* would prevail? I would have my book with its eighteen variations on the theme tied to my neck, literally, as I waited in the CM to find out. Then, hopefully, docking and tunnel clearing again.

10. *TEI*   Transearth injection; we burn our one engine, which could get us home or leave us forever stranded in lunar orbit. No back-up this time, as the LM would be empty and gone at this point.

11. *Entry*   Diving into the earth's atmosphere at precisely the right angle was required for a successful splash, not to mention the flawless on-time performance of the parachute system and related claptrap. I would fly the entry phase, because, of course, I had to learn how to do it all in case I came back from the moon without Neil and Buzz.

If these were the major hurdles along the way, it didn't mean we could slight the other parts of the flight. It truly *was* the little things you ignored which could rise up and bite you. I thought that certainly some few would, because, with all that complicated gear up there for eight days, something had to break. What would it be, and would I be able to cope with it? That is what I spent my time worrying about, and my simulator hours practicing to overcome. I was also still struggling with the guidance and navigation system, as I was the navigator to and from the moon. My computer had been equipped with a slightly different program which M.I.T. called Colossus IIA; it was aptly named. I felt flea-sized in its presence, and punched the computer keys with a good deal of reverence, as I availed myself of the services of this giant. That is, until it and I started to disagree and it refused to give me the answers I sought. Then I lost my temper. Flash an "operator error" light at me, will you, you stupid goddamned computer, and I would sputter and stammer until the soothing voice of Tom Wilson* or one of the other instructors came over the earphones and unctuously explained how I had offended their precocious brat. The LM's computer program was called Luminary, and it was to flash crazy lights in Buzz's face practically all the way to the lunar surface, so I guess it was aptly named also. Anyway, Colossus IIA was my uncompromising companion, and I would never truly be alone in the command module. Now if I could only teach that son-of-a-bitch to remove the probe and drogue . . .

As May turned to June, I was spending more and more time at the Cape, and less and less in Houston and other places. I was still weekending in Houston, but had to leave early (my records show takeoffs as early as 4:55 A.M.) Monday morning to get down to the Cape for a full day's work. Therefore, it was nice when Pat and the kids came to the Cape for the first two weeks of June. We put up at a motel right on the beach, and Pat had an opportunity to unwind a bit from the turmoil of Houston, where Apollo 11 pre-flight activities were building into a frenzy of attention from press, friends, and interested strangers. Not that the Cape was quiet, but Pat was more or less incognito there, and there was the beach—that great blotter, whose serene presence always seems able to sop up the tensions of the Collins family. Mine needed a bit of sopping up, too, as our pace was accelerating. But what the hell, I figured I could put up with anything for just a few more weeks, and then that was it! My mind simply refused to function past Apollo 11. There might be more flights, but I would not be a part of them. Had not President Kennedy said to land on the moon and return, before the end of the decade? He hadn't said

---

* It was at about this time that I began calling Wilson "Hal," the name of the omniscient computer in the movie *2001*. Tom somehow didn't seem honored by this.

to do it over and over again, to explore it in the name of science. Not that I disapproved, quite the contrary, but I simply was putting too much of myself into Apollo 11 to consider doing it all over again at a later date; besides, the strain on my wife was not good and should end as soon as possible.

Consequently, one day when Deke and I shared a T-38 for the Houston-to-Cape run, and he said he wanted to "plug me in" to the sequence of later flight assignments, I politely declined. I told him if Apollo 11 aborted and fell into the ocean after takeoff, I would be back knocking on his door, but if we were able to fly it as planned, Apollo 11 would be my last space flight. I didn't have the foggiest notion what I would do after that, nor did I know what seat I might have later occupied. In all likelihood, however, I would have been back-up commander of Apollo 14 and would have walked the moon on 17, the last of the series. But in June 1969 I couldn't even count to seventeen, nor do I have any regrets today over the decision . . . Still, I watched Gene Cernan with more than average interest during that flight three and a half years later.

Once Pat and the kids returned to Houston, in mid-June, I really began to withdraw from the world and to spin a cocoon of invincibility around me. I *would* master Colossus IIA, I *would* learn all eighteen rendezvous cases, that probe *would* smoothly fold and be led like a lamb from the tunnel. There was talk of delaying the launch a month, to give us more training time, but I would have none of it. The problem really wasn't mine so much as Neil and Buzz's, as they were having a dickens of a time getting the LM simulator to hold together long enough to complete the series of elaborate rehearsals they had planned with Mission Control in Houston. Invariably, either party's computer would "bomb out" somewhere along the line and the whole exercise would be postponed; the loss of time was irretrievable, and Neil and Buzz were falling behind. With luck, they might still get all their training requirements completed by July 16, but just barely. Neil and Buzz both seemed ambivalent about it. No one likes to delay a launch, especially because he has been deemed unprepared; yet neither of them seemed to be speaking up in favor of a July launch. I couldn't tell whether they were being extra cautious or whether they truly would be unprepared to meet certain situations that might arise. I suspected the former, but it was not my place to make that judgment; I could only, when queried by Slayton, say hell yes, I would be ready by mid-July. Actually, slipping one month wouldn't be any big tragedy, but it did have grave implications, because if some hardware problem emerged at the last moment to prevent an August launch, we were in real trouble, as certain parts of the spacecraft exposed to corrosive fuels had to be returned to their factories for overhaul, and no one really knew, in terms of time or decreased reliability, what this disassembly and reassembly process entailed. Therefore, June 17 was a banner day, because I was delighted to hear (after a nine-hour formal meeting called a flight readiness review) General Sam Phillips announce that July 16 would be it. *Iacta alea est!* The die is cast! I was feeling good anyway, as just the day before I had passed a special physical exam called the T minus 30 physical (we even counted down in days). That night I moved into the Merritt Island crew quarters for the first time, with my bottle of gin and my bottle of vermouth, and a heavy load removed from either shoulder. One month to get ready, one month working at maximum capacity with minimum interference.

Which is why we moved into the monastic world of crew quarters. Generally we crew members enjoyed the bright lights and freedom of Cocoa Beach to the last possible moment. On Gemini 10, John and I had stayed on the beach until the last week, but this time it was different. We needed a month sealed off from the world, to live and relive the complex venture before us, and crew quarters was the only place to do that. Crew quarters abutted our offices in the huge assembly and test building on Merritt Island. With a special key, one gained access to a small living room and a windowless corridor, with small windowless bedrooms on either side of it, and then passed an exercise room and a sauna as the corridor made a 90-degree turn and emptied into a dining room, kitchen, and briefing room, complete with world map, closed-circuit-communications system, and storage rack for the hard hats needed to visit the launch pad. Only those invited by us could enter, plus the few who worked there, such as the super-efficient maids and Lew Hartzell, the cook. The maids wrote our names on our underwear, starched our polo shirts, and overwhelmed us with cleanliness. Lew had spent years as a tugboat cook, interspersed with forays into the social whirl of yachting, and when he was full of beer he recounted good stories of celebrities falling overboard and other excitement, but mostly he stayed in his kitchen and cooked. His handiwork could best be described as monolithic, with a solid tectonic base of meat and potatoes, upon which were piled salads, rolls, and, of course, a five-hundred-calorie dessert. It did no good to tell Lew that you were on a diet; he took no offense, he simply ignored this irrelevant information. The tugboat must plow ahead, even through heavy seas. Reaching the moon obviously necessitated heroic measures in the kitchen. More meat, more potatoes, more bread, more dessert! They would propel us to the moon.

Typically we got up between six and seven. For Gemini 10, which with my space walk was a more physical challenge, I had always run two miles in the morning, but this time I played it by ear: sometimes I ran and sometimes I didn't. Usually I had time only for a glance at the paper and a hurried breakfast before I was off to the simulator building a block away, or to some meeting in our offices outside the locked door. Generally we would have Lew send lunch to us in sealed containers, and we would stay near the simulators as we munched and answered our phone calls, for the secretaries always had a tidy pile of yellow government phone-message slips waiting for us as we exited the simulator at noon. "Oh really, Mrs. ——, you haven't received an invitation to the launch? Why, I can't understand that, anyone as dedicated to the space program as you have been!" Who the hell is in charge of this anyway, and why is this broad calling me? Maybe someone can do something for her, but I sure don't want to sic her on Julian Scheer up in Washington, as he has congressmen demanding his hide because he won't provide entire planeloads of constituents with passes to the launch and choice seats. Maybe she will go away. Next call—Bill Tindall from Houston, with several observant and really telling points about how we are going about our training and what specific problem areas we are apparently neglecting. The best coordinator of diverse technical opinions I have ever met, Tindall must be listened to; he knows what he is talking about, and we must do something about his criticism. Munch, munch. Lew's sandwiches are three inches thick. O.K., Fran, O.K., Hazel, no more calls, let me just sit here for a while and try to reconstruct what went wrong at this morning's simulation, while I work my way through Lew's submarine.

Occasionally I would pass Neil and Buzz, who were on separate but equal treadmills close by, but it was not until evening that our threesome would convene in crew quarters and have one precious drink before Lew pumped us full of victuals again. After dinner we would hit the books, sometimes independently and sometimes as a team, reviewing those phases of the flight plan which saw us acting in concert. Neil and Buzz required a much closer working relationship with each other than either had with me, but nonetheless I found them talking to me (especially Buzz) about events in which I was really not involved.

I especially recall one bad night following a particularly unfortunate simulation. Neil and Buzz had been descending in the LM when some catastrophe had overtaken them, and they had been ordered by Houston to abort. Neil, for some reason, either questioned the advice or was just slow to act on it, but in any event, the computer printout showed that the LM had descended below the altitude of the lunar surface before starting to climb again. In plain English, Neil had crashed the LM and destroyed the machine, himself, and Buzz. That night Buzz was incensed and kept me up far past my bedtime complaining about it. I could not discern whether he was concerned about his actual safety in flight, should Neil repeat this error, or whether he was simply embarrassed to have crashed in front of a roomful of experts in Mission Control in Houston. But no matter, Buzz was in fine voice, and as the Scotch bottle emptied and his complaints became louder and more specific, Neil suddenly appeared in his pajamas, tousle-haired and coldly indignant, and joined the fray. Politely I excused myself and gratefully crept off to bed, not wishing to intrude in an intercrew clash of technique or personality. Thank God, in the CM there were only me and Colossus IIA, and if that son-of-a-bitch mouthed off, I would turn off its power supply. Neil and Buzz continued their discussion far into the night, but the next morning at breakfast neither appeared changed, ruffled, nonplused, or pissed off, so I assume it was a frank and beneficial discussion, as they say in the State Department. It was the only such outburst in our training cycle.

Although Buzz never came out and said it in so many words, I think his basic beef was that Neil was going to be the first to set foot on the moon. Originally, some of the early check lists were written to show a co-pilot first exit, but Neil ignored these and exercised his commander's prerogative to crawl out first. This had been decided in April, and Buzz's attitude took a noticeable turn in the direction of gloom and introspection shortly thereafter. Once he tentatively approached me about the injustice of the situation, but I quickly turned him off. I had enough problems without getting into the middle of *that* one.

As the days grew short, my consciousness seemed to expand. I am told that we humans function at only a small fraction of our total capacity, and I claim no exception to this limitation, not even during early July of 1969. I believe that if normally I wheeze along at 20 percent of my capacity, then I was up to at least 25 percent, and it seemed to me that my ability to concentrate and absorb increased. Having gone through all this once before, in Gemini days, was a tremendous help, for a variety of reasons. In the first place, new and unexpected situations cause the unnecessary expenditure of a lot of nervous energy, and the crew quarters environment, instead of being strange and unfamiliar, was as comfortable as an old shoe. I even had the same bedroom as before Gemini, with the same photograph of an exquisite milk-skinned brunette in a dark-red one-piece bathing suit perched demurely on the stone steps of a crumbling building in what appeared to be an Italian hill village. She was my pin-up patron saint from before, and she reminded me that people did go from this room into the sky and return safely to earth. Having flown before was also valuable because there are some basics which, having been learned once, never need to be repeated, such as the best breathing technique to use under high-G loads. Even more important, my Gemini flight experience enabled me to visualize our planned in-flight activities page by page in the flight plan, and to determine where we were overloading ourselves and what to do about it.

On the other hand, there were profound differences between the flight of Gemini 10 and Apollo 11. Although Gemini had received worldwide attention, it had a provincial or local quality about it, and it was treated in a friendly, uncritical fashion—almost as if it were a sporting event. Sure, one could get killed, but even so—there was no overlay of international pressure on top of the element of personal risk. Apollo 11 was entirely different. We were our nation's envoys, we three, and it would be a national disgrace if we screwed it up. We would be watched by the world, including the unfriendly parts of it, and we must not fail. There was pressure to plan, to study, to concentrate, to explore each nook and cranny of my mind for some fatal flaw, something overlooked, something ill conceived, something I was supposed to do which I simply could not. To make it worse, only those of us inside the program knew the opportunities we had to fail, the unknowns in the equation; the rest of the world seemed to think it was already a *fait accompli*, but before *Columbia* became the Gem of the Ocean, there were a hundred close decisions and a thousand critical switch actuations facing us. A broken probe, a cracked engine nozzle, an electrical short, a crew lapse of attention, a jillion other things—and we would never make it. I don't know about Neil and Buzz, because we never discussed these things, but I really felt this pressure, this awesome sense of responsibility weighing me down, this completely negative sensation, this commandment which said, "Thou shalt not screw up." By flight time, I had tics in both eyelids, which went away as soon as we got airborne.

We also had a bit of comic relief, such as the time Chuck Berry (who billed himself as our "personal physician" although we rarely saw him) told the press that the President shouldn't have dinner with us in crew quarters before the flight because he might infect us with his germs. Had we been operating behind a germ-free barrier, this might have made a modicum of sense, but we were in daily contact with dozens of people, and one more (especially one whose physical condition is subject to constant checking) wouldn't have made a particle of difference. The President was really put on the spot by this pronouncement, because if anything happened to us during the flight, it would be his fault; even if nothing did, he would be charged with callous disregard of the professional advice of our "personal physician." Naturally he canceled. Other Berryisms included telling the press that one of us was probably going to get sick in the lunar receiving laboratory after the flight, and that we would have to be traced all our lives to make sure we weren't carriers of lunar organisms with extremely long incubation periods. Neither prediction came true.

Tom Paine was apparently germ-free, however, as he did come and have a quiet dinner with us in crew quarters. As NASA administrator, he must have felt much of the same pressure we did; yet neither side telegraphed the fact, and we had a relaxed and amiable chat. His main message—and a good one, I think—was that we were not to take any undue risks. If things didn't look right, we were to come on home, and he would see to it that we would get the next flight for another try at it. This removed the obvious risk of our letting our desire to be first on the moon cloud our judgment in analyzing hazards that might crop up along the way.

These diversions were rare, however, as July 16 approached.

How ready were we? In some areas, such as the rehearsal of our flight plan, I felt we were more than ready. In other, less important areas, we were woefully unprepared, but hopefully these were just frills. For example, most Americans relate Apollo to television. Our TV shows were vitally important links between them and us and greatly enhanced their understanding of what was taking place. Our TV camera was the eye of Apollo. It was also late getting delivered, and was a bloody nuisance of an afterthought that was not required for the safe completion of our flight; therefore we didn't want to fool with it. Reluctantly we agreed to turn it on a couple of times as specified in the flight plan, but we weren't happy about it, and we didn't practice with it, and we didn't rehearse any shows. We simply didn't have time to fool around with it; Neil and Buzz didn't even know how to turn it on or focus it, and my knowledge of it was pretty sketchy. In flight we found the best thing to do was use masking tape to stick the monitor unit on top of the camera so we could tell which way to point it, a jury rig that is ridiculous when compared to the mathematical precision and extensive rehearsals that accompanied our other preparations. But TV couldn't kill us, and a lot of other things could. As I recall, the last bit of advice we got about TV was something to the effect of: "Gee, I hope you guys will put on some great shows. You know, there will be a billion people watching, so don't screw it up, O.K.?"

My greatest source of reassurance came from the simulator. In the six months between January and July, I had accumulated four hundred hours in it, and it now felt like a comfortable home rather than a horror chamber. Sitting at their huge console outside, the instructors could insert a variety of make-believe malfunctions, and invariably I could catch them and fix them. Their console had hundreds of lights on it, and even one or two spares. In one of these blanks, they had gone to the trouble of printing "Collins looks good," and as soon as I staggered out after a training session, I would hurry over to see whether the "Collins looks good" light was burning cheerfully. If it wasn't, I'd turn it on myself. I didn't really appreciate the full significance of that light until years later when, in Washington, I discovered hordes of people whose careers were dedicated to lighting a "politician looks good" or "government leader looks good" light.

On Friday, July 11, we had our final medical, and on Monday, July 14, a final press conference. The medical exam was short and sweet, the press conference totally banal, both questions and answers. In between the two, on the weekend, I had the first fun I had had in a long time. I drove down to Patrick AFB, hopped into a T-38, and did acrobatics for an hour, once on Saturday and once on Sunday, once with Deke Slayton in the rear seat and once by myself. The purpose of these flights was not fun and games, but inner-ear conditioning. Since Borman had become sick during Apollo 8, and Schweickart during Apollo 9, we were mildly concerned about how we might prepare the fluid in our semicircular canals for the novel experience of weightlessness. The idea of the T-38 was not so much to fly weightless parabolas, but rather to perform a variety of violent acrobatic maneuvers, loops and rolls, to slosh that fluid around, in poor but hopefully adequate imitation of the sloshing motion induced by moving around the Apollo cabin in weightlessness. I don't have any difficulty making myself sick in a T-38, if I go about it the right way. An aileron roll produces the most abrupt change in position, but after long years, the inner ear, the brain, and the stomach apparently get used to this motion and say, "Ho hum, another aileron roll." Not so if the head is turned 90 degrees to either side before the roll and kept there. Now the inner ear senses, not a humdrum roll, but a very rapid heads-up or heads-down tumble; this is unexpected and, for me at least, sickening. So all I have to do is turn my head and do a couple of rolls, and I get a miserable feeling in the pit of my stomach. I always stopped short of throwing up, and interspersed these rolls with more conventional maneuvers and periods of gentle straight and level flight.

Physical conditioning, physiological conditioning, mental conditioning: nothing left to do. I was ready. I took Tuesday the

fifteenth off and browsed around crew quarters. I talked to Pat in
Houston, and read and reread her last-minute note, which included
the following poem, obviously not a last-minute thought:

> *To a Husband Who Must Seek the Stars*
>
> *In your eyes the first glad token*
> *As when first our love we proved,*
> *So your mind to mine has spoken*
> *Just as if your lips had moved.*
>
> *You are saying—yes, I know—*
> *That the lure of space beguiles.*
> *You are pleading—"Let me go,"*
> *Not unwilling, but with smiles.*
>
> *Can you love me, and still choose*
> *Whispers that I cannot hear?*
> *Late to love, how can I bear to lose*
> *Content for some inconstant sphere?*
>
> *Tell me how you see my role—*
> *To stay, to wait, yet yearn to go.*
> *Where is the comfort for my soul?*
> *You, my love, have helped me know:*
>
> *I'll be unafraid, undaunted.*
> *Yes, of course! I need not face*
> *Any peril; or be haunted*
> *By the hazards you embrace.*
>
> *I could have sought by wit or wile*
> *Your bright dream to dim. And yet*
> *If I'd swayed you with a smile*
> *My reward would be regret.*
>
> *So, for once, you shall not hear*
> *Of the tears, unbidden, welling;*
> *Or the nighttime stabs of fear.*
> *These, this time, are not for telling.*

Take my silence, though intended;
Fill it with the joy you feel.
Take my courage, now pretended—
You, my love, will make it real

I hope I will.

12

Houston, Apollo 11 . . . I've got the world in my window.
—Mike Collins, 28 hours and 7 minutes ground elapsed time

Gemini 10 was launched in the late afternoon, permitting a civilized schedule on launch day, but no such luck this time. Today, July 16, the launch window opens at 8:32 Houston time (the time we will keep on our watches), so Deke tumbles us out of bed shortly after four o'clock. A quick shave and shower, then into our exercise room, where Dee O'Hara is waiting to record a few last-minute physical measurements, such as weight and temperature. It's nice to have Dee around; she has been doing this since Mercury days, and she gives this most unusual morning a feeling of familiarity and comfort. Dee is a first-class nurse, a good friend, and a delightful person. She is always cheerful, but without the brittleness or hysteria so many cheerful Charlies bring to early-morning jobs. She does her work quickly, as usual, then fades away to be replaced by Lew and his final offering: steak and eggs, toast, and juice and coffee. An artist, Paul Calle, has been invited to sketch us during breakfast, as part of NASA's art program. As in Dee's case, Paul's intrusion is not resented, as he is obviously a professional, hopefully among other pros. We are hardly aware of him and his sketch pad, as the three of us munch our toast and chat amiably with Deke and Bill Anders. Launch breakfasts always have an air of studied casualness, and this one is no exception. We all know that Deke has programmed twenty-three minutes for it, or some such number, and that he will not permit a second's overtime, but we pretend otherwise, and anyone overhearing our conversation would think that we five were slightly bored at the prospect of another empty day.

I thank Lew for the feed and return briefly to my room. I brush my teeth very thoroughly, and recheck that my meager possessions are all packed. Someone will see that they get back to Houston. Some of them I have already earmarked (oh, great optimist!) for delivery to the lunar receiving laboratory, to await me in post-flight quarantine. I say goodbye to the brunette on the wall and wonder if I will ever see her again. Then I rejoin Neil and Buzz, and the three of us take the short trip upstairs to the suit room. On Gemini, we suited up in a house trailer near the launch pad, but this Apollo is fancy, and they have built an elaborate suit maintenance, storage, and donning facility near the crew quarters. Remembering the pain in my knee during Gemini 10, I waste no time getting suited up, and once inside the fish-bowl helmet, I begin breathing 100 percent oxygen and purging my body of bubble-producing nitrogen. As a result of the fatal fire, 100 percent oxygen is no longer used inside the CM on the launch pad; it has been replaced with a mixture of 60 percent oxygen and 40 percent nitrogen. (As the booster ascends, this mixture escapes through a vent hole and is replaced by pure oxygen, so that in orbit, when the atmosphere inside the spacecraft is stabilized at a pressure of five pounds per square inch, it is once again nearly pure oxygen.) However, inside the suit circuit, 100 percent oxygen is used at all times, and the sooner I get on it the better. Joe Schmitt is supervising our suit-donning, and Joe is good. He has been doing this since Al Shepard's flight: he has seen us all come and go, and change, but Joe remains Joe. He could easily slip something into a suit pocket and ask one of us to carry it to the moon for him, and he knows we would do it, but he would rather die than make such an out-of-line request. When we are all three ready, we plug ourselves into portable oxygen containers, and carrying them like heavy suitcases, we begin the long walk down to the transfer van, which in turn will take us to the launch pad. The corridors are crowded as we go, with some old friends and co-workers, but mostly with strangers. Apollo is so big we know only a tiny fraction of those who, like us, have years invested in this day: in this building alone there are hundreds of people who to some degree or another hold our lives in their hands, and yet we have never laid eyes on them. Now some of them are in the corridors to bid us a silent farewell. At least inside my helmet they are silent, as I hear only the squish of my awkward yellow rubber galoshes and the hiss of the oxygen coming into the suit. As we leave the building, there is a real mob scene, as this is the last "photo opportunity," as the PR men call it, and the courtyard outside our door is jammed with TV crews and their horrid lights, plus an army of still and motion-picture photographers. The security people have roped off a walkway for us, and we give jerky little waves to the photographers as we walk stiff-legged toward the van. Charlie Buckley, the head security man at the Cape, is there to greet us—another little pre-flight ritual. There are certain amenities to be observed, such as presenting Guenter Wendt, the czar of the launch pad, with a going-away present. Guenter has spent the past couple of weeks telling me what a great fisherman he is, and how he regularly plucks giant trout from the ocean. In return, I have located the smallest trout to be found in these parts, a minnow really, and have had it, uncured, nailed to a plaque and inscribed GUENTER'S TROPHY TROUT. I carry it now inside a brown paper shopping bag, which Charlie Buckley eyes suspiciously. I am a bit nervous about it myself. What if my awkward gloved hands drop it and the trout tumbles out in front of all those photographers? They are here to see us leave the earth, with dignity and perhaps a little pomp, but what if their cameras instead record an ungainly scramble after a tiny dead fish? What would Walter Cronkite say?

The van doors close behind us, and we proceed at precisely the rehearsed speed over the eight-mile route. We are surrounded by tourists in their cars, although it's still only 6:30 here at the Cape and 5:30 on my Houston wristwatch. They are nearly at a standstill, and in our carefully cleared lane we sail by them with dispatch. Well, that's one hassle we don't have to put up with today. Within a couple of minutes we turn off the main road, and now the people are behind us and my mood changes slightly. For the first time today, I have a chance to focus on the Saturn V and its cargo, and the reason we are up so early this morning. Frankly, up to this point I haven't really been sure we were going to launch today. I am not superstitious, but ever since our daughter Ann was born on my birthday, which happens to be Halloween, I have been conscious of dates, and for the past month or two I have been half convinced that we would slip a bit and launch the same day as Gemini 10—July 18. But now that notion seems crazy, for surely there is a sense of reality and finality about today that convinces me I will never again walk down that long corridor. No, today's the day.

It's a clear day, we can see that, and we are told that it's hot already with little breeze—a scorcher in the making. Last night the Saturn V looked very graceful, suspended by a cross fire of searchlights which made it sparkle like a delicate opal and silver necklace against the black sky. Today it is a machine again, solid and businesslike, and *big*. Over three times as tall as a Gemini-Titan, taller than a football field set on end, as tall as the largest redwood, it is truly a monster. It is parked next to a huge steel scaffold known as the launch umbilical tower, which is designed to hold the rocket and nurture it until the final second. The two partners make quite a contrast, the rocket sleek and poised and full of promise, the tower old, gnarled, ungainly, and going nowhere. We park at the

base of the tower and clamber out. The first elevator is waiting for
us with its doors already open. Something seems wrong, and
suddenly I realize what it is. The place is deserted! Every other
time I have been to the launch pad it has been a beehive of
activity, with workmen shouting at each other, equipment being
hoisted by crane, and all the other vital signs common to a big
construction site. Now it seems as if some dread epidemic has
killed all but those protected by pressure suits, except there are no
corpses and Joe Schmitt still looks healthy. Perhaps it is simply a
case of the air-raid siren having sounded and left the city deserted.
As the four of us ascend, I feel that more than the elevator door
has clanged shut behind me. I recall that there are one million
visitors here to watch the launch, but I feel closer to the moon
than to them. This elevator ride, this first vertical nudge, has
marked the beginning of Apollo 11, for we cannot touch the earth
any longer. I am treated to one more view, however, one last bit of
schizophrenia as I stand on a narrow walkway 320 feet up, ready to
board *Columbia*. On my left is an unimpeded view of the beach
below, unmarred by human totems; on my right the most colossal
pile of machinery ever assembled. If I cover my right eye, I see the
Florida of Ponce de Leon, and beyond it the sea which is mother to
us all. I am the original man. If I cover my left eye, I see civilization and technology and the United States of America and a
frightening array of wires and metal. I am but one adolescent in an
army which has received its marching orders. Neil has entered the
spacecraft, and I am next.

Apollo is a three-seater, and I am frequently asked which one
is mine. As in so many things in the space program, simple answers
don't work. I fly all three, whichever one seems to make sense at
any particular time. For launch the CMP (me) normally rides in
the center couch, but since Buzz had already been trained to ride
the center when I joined the crew, it made little sense for him to
have to learn another position. Therefore, I have learned the tasks
of the right-seat occupant for launch, Buzz stays in the center, and
Neil (as commanders always do) rides over on the left side, where

the abort handle is located. When Neil gets in, I give the trout to Guenter and his crew, and they frolic around a bit. Then I kick off my yellow galoshes, grab the bar inside the center hatch, and swing my legs as far as I can over to the right. After a couple of grunts and shoves, I finally manage to get my backside into the seat, with my head on a narrow rest and my legs up above me, my feet locked into titanium clamps. It's not very comfortable, especially in this suit, which is tight in the crotch, but I can put up with anything for two and a half hours—all we have left before launch. Joe is leaning over me busily, giving oxygen hoses, communications plugs, and restraining straps one last check; then he is gone. I barely have time to grab his hand before he leaves. Fred Haise is still with us. As a good back-up crewman, he was inside the CM when we got there, running some preliminary checks and certifying switch positions, and now he is down in the lower equipment bay, where we cannot reach, helping with last-minute preparations. Finally, Fred scrambles out and closes the hatch behind him. Now, hopefully, we will see no more people for eight days.

I am everlastingly thankful that I have flown once before, and that this period of waiting atop a rocket is nothing new. I am just as tense this time, but the tenseness comes mostly from an appreciation of the enormity of our undertaking rather than from the unfamiliarity of the situation. If the two effects, physical apprehension and the pressure of awesome responsibility, were added together, they might just be too much for me to handle without making some ghastly mistake. As it is, I am far from certain that we will be able to fly the mission as planned. I think we will escape with our skins, or at least I will escape with mine, but I wouldn't give better than even odds on a successful landing and return. There are just too many things that can go wrong. So far, at least, none has, and the monster beneath us is beaming its happiness to rooms full of experts. We fiddle with various switches, checking for circuit continuity, for leaks, and for proper operation of the controls for swiveling the service module engine. There is a tiny leak in the apparatus for loading liquid hydrogen into the Saturn's third stage, but the ground figures out a way to bypass the problem. As the minutes get short, there really isn't much for me to do. Fred Haise has run through a check list 417 steps long, checking every switch and control we have, and I have merely a half dozen minor chores to take care of: I must make sure that the hydrogen and oxygen supply to the three fuel cells are locked open, that the tape recorder is working, that the electrical system is well, and that the batteries are connected in such a way that they will be available to supplement the fuel cells, that we turn off unneeded communications circuits just prior to lift-off . . . all nickel-and-dime stuff. In between switch throws, I have plenty of time to think, if not daydream. Here I am, a white male, age thirty-eight, height 5 feet 11 inches, weight 165 pounds, salary $17,000 per annum, resident of a Texas suburb, with black spot on my roses, state of mind unsettled, about to be shot off to the moon. Yes, to the moon.

On Gemini 10 the launch had been a great event, for just to get into orbit meant a lot, but this time launch is only one link in a long and fragile daisy chain that encircles the moon. Our voyage has already begun because we are going to launch toward the east and thus take advantage of the earth's rotational velocity, and we are already moving toward the east at 900 miles per hour. There is a slight jolt as an access arm swings back away from the side of the spacecraft. This means that Guenter and his people have folded their tents and silently stolen away. I have to go by feel, because I can't see out at all. The two windows on my side have a cover over them, and I won't get a chance to see any sky until three minutes after lift-off, when we will already be sixty miles high. At that point we will jettison the launch escape tower, which is our means of rocketing away from an exploding booster, and with it will go the protective cover which is denying me a view. All these things I know, yet I don't feel the high excitement at the prospect of a rocket ride that I felt before Gemini 10. Neil and Buzz seem subdued too, as we go through our various checks.

There is plenty for us to do inside the CM, however, for it is one very dense, tightly designed, 12,500-pound package. The only part of this immense stack which will make the entire round trip, it is a cone eleven feet high and thirteen feet across the base. We are lying in three individual couches suspended on a joint frame, which is built to move independently of the rest of the structure, to cushion the impact of a possible emergency landing on a hard surface. In our bulky suits we touch each other at the elbows, and if we are not careful, our arms can interfere with each other as we reach for and grab various controls. Above our faces is the main instrument panel, packed with gauges and switches which must be accessible during the times we are burning one of our many engines (counting all the various little motors on the Saturn, the service module, and the command module, we have seventy-two engines—and me a single-engine pilot by inclination). There are other panels full of more switches and controls on the bulkheads to Neil's left and to my right. Then, below Buzz's feet, lies the lower equipment bay, home of the navigational equipment, the sextant, and the telescope, and the access way to the tunnel which will eventually lead to the LM, but which now points upward at empty sky. Beneath our couches there is a crawl space, where we will sleep in enclosed hammocks, and also an array of lockers containing food, clothing, and auxiliary equipment, such as the television camera. The right-hand side of the lower equipment bay is where we urinate (we defecate wherever we and our little plastic bags end up), and the left-hand side is where we store our food and prepare it, with either hot or cold water from a little spout. Most controls have been arranged for our convenience, but not all—some depend upon the paths of pipes external to our compartment, such as a valve for bypassing glycol coolant fluid, which is wedged into a recess in the wall below Neil's couch and which can only be turned with a special tool. Nearly every available cubic inch of space has been used, save for two great holes in the lower equipment bay which are reserved for the boxes of moon rocks to be brought back from the LM.

The walls of the spacecraft have been decorated with tiny squares of Velcro, giving it a pockmarked look. Velcro comes in two varieties, male and female, and when pressed against each other, they cling together in a fairly secure bond. The female half consists of a felt-like, loosely woven, fuzzy surface, and these patches are cemented onto each loose item, at various strategic spots. The walls of the spacecraft are adorned with hundreds of squares of the male variety, which is a coarse material from which protrude thousands of tiny stiff fabric hooks. These hooks intermesh with the wool of the female, and together they constitute a simple and practical solution to the problem of how to keep equipment from floating away when there is no gravity to hold things in place. If I want to use a light meter, and then put it "down" for a minute, I simply mesh the square of female Velcro on its side with the square of male Velcro adjacent to the window I am using, and it will stay there, unless my elbow knocks it loose. In addition to the Velcro, the instrument panels are adorned with tiny plastic rectangles, upon which various last-minute messages have been neatly lettered. BOIL > 50, says one, a typical reminder which says that if the adjacent radiator outlet temperature exceeds 50 degrees, the spacecraft's environmental control system will attempt to bring it back down by boiling some water. S-BAND AUX TO TAPE 90 SEC PRIOR TO DUMP says another, a memory crutch to save me some embarrassment in the operation of our tape recorder. These are personalized little notes that I have stuck on some panels. In addition, there are hundreds of other labels whose nomenclature and position are standard in all command modules. There are banks of circuit breakers, a cluster of forty-eight warning lights, two artificial horizons, and two keyboards for communicating with the computer. There are over three hundred of one type of switch alone. The paint is battleship gray, and despite the wear and tear of countless test hours, the interior appears brand spanking new.

At the moment, the most important control is over on Neil's side, just outboard of his left knee. It is the abort handle, and now it has power to it, so if Neil rotates it 30 degrees counterclockwise, three solid rockets above us will fire and yank the CM free of the service module and everything below it. It is only to be used in extremis, but I notice a horrifying thing. A large bulky pocket has been added to Neil's left suit leg, and it looks as though if he moves his leg slightly, it's going to snag on the abort handle. I quickly point this out to Neil, and he grabs the pocket and pulls it as far over toward the inside of his thigh as he can, but it still doesn't look too secure to either one of us. Jesus, I can see the headlines now: "MOONSHOT FALLS INTO OCEAN. Mistake by crew, program officials intimate. Last transmission from Armstrong prior to leaving launch pad reportedly was 'Oops.'"

Inevitably, as the big moment approaches, its arrival is announced by the traditional backward count toward zero. Anesthetists and launch directors share this penchant for scaring people, for increasing the drama surrounding an event which already carries sufficient trauma to command one's entire consciousness. Why don't they just hire a husky-voiced honey to whisper, "Sleep, my sweet" or "It's time to go, baby"? Be that as it may, my adrenalin pump is working fine* as the monster springs to life. At nine seconds before lift-off, the five huge first-stage engines leisurely ignite, their thrust level is systematically raised to full power, and the hold-down clamps are released at T-zero. We are off! And do we know it, not just because the world is yelling "Lift-off" in our ears, but because the seats of our pants tell us so! Trust your instruments, not your body, the modern pilot is always told, but this beast is best felt. Shake, rattle, and roll! Noise, yes, lots of it, but mostly motion, as we are thrown left and right against our straps in spasmodic little jerks. It is steering like crazy, like a nervous lady driving a wide car down a narrow alley, and I just hope it knows where it's going, because for the first ten seconds we are perilously close to that umbilical tower. I breathe easier as the ten-second mark passes and the rocket seems to relax a bit also, as both the

---

* Actually all three of us have profited from our previous space flights. The maximum heart rates we record during the Saturn's boost are: Armstrong 110 beats per minute, Collins 99, and Aldrin 88—all considerably below the values we recorded during equivalent periods of our Gemini flights.

noise and the motion subside noticeably. All my lights and dials are in good shape, and by stealing a glance to my left, I can tell that the other two thirds of the spacecraft is also behaving itself. All three of us are very quiet—none of us seems to feel any jubilation at having left the earth, only a heightened awareness of what lies ahead. This is true of all phases of space flight: any pilot knows from ready-room fable or bitter experience that the length of the runway behind him is the most useless measurement he can take; it's what's up ahead that matters. We know we cannot dwell on those good things that have already happened, but must keep our minds ever one step ahead, especially now, when we are beginning to pick up speed. There is no sensation of speed, I don't mean that, but from a hundred hours of study and simulation, I know what is happening in the real world outside that boost protective cover, even if I can't see it. We have started slowly, at zero velocity relative to the surface of the earth, or at nine hundred miles per hour if one counts the earth's rotational velocity. But as the monster spews out its exhaust gases, Newton's second law tells us we are reacting in the opposite direction. In the first two and a half minutes of flight, four and a half *million* pounds of propellant will have been expended, causing our velocity relative to the earth to jump from zero to nine thousand feet per second, which is how we measure speed. Not miles per hour, or knots, but feet per second, which makes it even more unreal.

The G load builds slowly past 4, but no higher; unlike the Titan, the Saturn is a gentleman and will not plaster us into our couches. The 4.5 are but a little smooch, letting us know that the first-stage tanks are about empty and ready to be jettisoned. Staging, it is called, and it's always a bit of a shock, as one set of engines shuts down and another five spring into action in their place. We are jerked forward against our straps, then lowered gently again as the second stage begins its journey. This is the stage which whisperers have told us to distrust, the stage of the brittle aluminum, but it seems to be holding together, and besides, it's smooth as glass, as quiet and serene as any rocket ride can be. We are high above the disturbing forces of the atmosphere now, and the second stage is taking us on up to one hundred miles, where the third stage will take over and drive us downrange until we reach the required orbital velocity of 25,500 feet per second. At three minutes and seventeen seconds after lift-off, precisely on schedule, the launch escape rocket fires (no longer being needed). As it pulls away from our nose, it carries with it the protective cover that has been preventing me from seeing out my windows. Now it's much brighter inside the cockpit, but there is nothing to see outside but black sky, as we are already above all weather, at two hundred miles downrange from the Cape and pointed up.

As each minute passes, Houston tells us we are GO (all is well), and we confirm that everything looks good to us. At nine minutes the second stage shuts down, and briefly we are weightless, awaiting the pleasure of the third-stage engine. Due to the heightened awareness that always comes at these important moments, my sense of time is distorted, and it seems to take forever for the third stage to light. Finally! Ignition, and we are on our way again, as the single engine pushes us gently back into our couches. This third stage has a character all its own, not nearly as smooth as the second stage, but crisp and rattly. It vibrates and buzzes slightly, not alarmingly so, but with just enough authority to make me delighted when it finally shuts down on schedule at eleven minutes and forty-two seconds. "Shutdown," Neil says quietly, and we are in orbit, suspended gently in our straps. The world outside my window is breathtaking; in the three short years since Gemini 10, I have forgotten how beautiful it is, as clouds and sea slide majestically and silently by. We are "upside down," in that our heads are pointed down toward the earth and our feet toward the black sky, and this is the position in which we will remain for the next two and a half hours in earth orbit, as we prepare ourselves and our machine for the next big step, the translunar injection burn which will propel us toward the moon. The reason for the heads-down attitude is to allow the sextant, in the belly of the CM, to point up at the stars, for one of the most important things I must do is take a couple of star sightings to make sure that our guidance and navigation equipment is working properly before we decide to take the plunge and leave our safe earth orbit.

Just as on Gemini 10, the first few minutes in orbit are busy ones, as a long check list must be followed to convert the spacecraft from a passive payload to an active orbiter. Between Bermuda and the Canary Islands, I work my way swiftly through a couple of pages of miscellaneous chores, opening and closing circuit breakers, throwing switches, and reading instructions for Neil and Buzz to do likewise. Then we all remove our helmets and gloves, and I fold down the bottom half of my couch and slip over it into the lower equipment bay. Here there are more switch panels, plus lockers full of equipment which I must unpack and distribute, and, of course, the all-important navigational instruments, the sextant and the telescope, which must be checked. I move slowly and cautiously, with no unnecessary head movements, for this is a phase of the flight I have been warned about. This is the first chance I will have to slosh and swirl that fluid in my inner ear, the first chance to make myself sick, and I desperately want to avoid *that*, not only on general principles, but specifically because I am the only one trained to perform the transposition and docking maneuver, which is essential to retrieving our LM from its position behind us, buried inside the top of the Saturn. Therefore, I move slowly, listening to my stomach as I go. So far so good, as I move over underneath Neil's couch and hand up to him a helmet stowage bag and a tool for turning a glycol valve. Then I check out our main oxygen pressure regulator, and unstow a couple of cameras for Buzz to use.

Buzz seems to have gotten up on the wrong side of bed this morning, or at least it seems to me he's more interested in slowing me down than in helping me get through my chores. He questions a glycol pressure reading: "O.K., now, is that normal for the discharge pressure to zap down low and to do that? Do you think, Mike?" I reassure him, and hold out a camera in front of him. "Buzz?" "Yes, just a second." I can't wait for him. "O.K., I'll just let go of it, Buzz; it will be hanging over here in the air." I've got to keep moving, as there are just so many minutes till TLI and so much work to be done. Let's see, the Canaries are behind us, we must be just about over our Tananarive Station on the island of Madagascar, which means Carnarvon, Australia, next; followed by the U.S.A. one last time; then around again to the middle of the Pacific Ocean, where the TLI burn will occur. In the meantime, I have to keep things moving. "O.K., Buzz, are you ready for the 16-millimeter?" This is the movie camera to record the transposition and docking maneuver. "Yes, how about a bracket?" "Neil will give you the bracket." "Now, let's see, you got an 18-millimeter lens on here, right?" "Yes." "So—do I push the thing all the way up? Is that right?" "Yes." "About with that white mark?" Lord, I don't have time to discuss camera brackets and lenses now, because it is time to take a couple of star sightings and to realign our inertial platform.

I ignore Buzz for the moment and swing around into position at my navigator's console in the middle of the lower equipment bay. I unstow and install two eyepieces, one for the sextant and one for the telescope, and I attach a portable handhold on either side of them. Handholds I need, and I let Neil and Buzz know it. "I'm having a hell of a time maintaining my body position down here; I keep floating up." No big problem, but annoying, as I jettison the protective covering over the optics and peer out through the telescope. What I see is disappointing, for only the brightest stars are visible through the telescope, and it is difficult to recognize them when they are not accompanied by the dimmer stars, which give each constellation its distinctive visual pattern. The situation is not helped by the fact that I am looking for Menkent and Nunki, two of the more nondescript Apollo navigation stars. Some stars are great: Antares, for instance, is not only very easy to find, just behind the head of the scorpion, but it has a distinctive reddish color as well. Menkent, on the other hand, is awfully tough to find unless the entire constellation Centaurus is clearly visible, and Nunki (in Sagittarius) is not much better. Unlike the Gemini, however, Apollo has a fancy computer tied to the optics, and now I call on it for help; it responds by swinging the sextant around until it points at where it thinks Menkent is. Aha! There it is, in plain view, and it's a simple task for me now to align cross hairs precisely on it and push a button at the instant of alignment. Now I repeat the process using Nunki, and the computer pats me on the back by flashing the information that my measurements differ from its stored star angle data by .01 degree. It displays this information as 00001. In M.I.T.-ese, a perfect reading of 00000 is called "five balls." I have scored "four balls one." Glenn Parker, one of the Cape simulator instructors, and I have bet a cup of coffee. On this, my first measurement, Glenn doesn't think I am going to do better than four balls two, but I think I'm going to get five balls. The bet is a standoff, and after relaying this data to the ground, I add, "And tell Glenn Parker down at the Cape that he lucked out. He doesn't owe me a cup of coffee." Houston has no idea what I'm talking about but dutifully agrees to pass the information on.

We are over Australia now, precisely one hour after lift-off, and with the star check behind me, I can breathe a little easier. Things are going extremely well, we are precisely on schedule, and now I have time to talk about camera brackets and other trivia. In fact, we are going to attempt to send the ground some television pictures as we approach the Baja California coast, and now I unstow the TV camera, its cable, and the small monitor set which allows us to see the same picture we are transmitting. It is still dark as we pass south of Hawaii, heading toward the second dawn of this day, and the first one for me from orbit in three years. As usual, the sun comes up with a rush, and as usual my uncouth mouth records the event. "Jesus Christ, look at that horizon! Goddamn, that's pretty, it's unreal." Neil agrees. "Isn't that something? Get a picture of that." "Ooh, sure I will. I've lost a Hasselblad. Has anyone seen a Hasselblad floating by? It couldn't have gone far, a big son-of-a-gun like that . . . I see a pen floating loose down here, too. Is anybody missing a ball-point pen?" After a lengthy search, I finally find the camera off in a corner, too late to record the brilliant flush of sunrise, but I am nonetheless pleased to have it in hand. The bulky 70-millimeter Hasselblad could turn into a dangerous projectile at the instant the rocket engine lights for the TLI burn. Just because something is "weightless" in space doesn't mean it has lost any of its mass. It still contains the same number of molecules and, if thrown, can do just as much damage when it hits something as it would do on earth.

As we swing across Mexico, we try to relay TV signals to Houston via the huge antenna at Goldstone, California, but we are at such a shallow angle that we only get through with a minute's worth. At least all the equipment seems to be working. We decide that the monitor unit dangling from its separate cable is unhandy; next time, we will tape it to the top of the TV camera.

Once the TV is stowed away securely, our attention—and that of our compatriots in Houston—is focused entirely on the upcoming TLI burn. Our orbital altitude and angle have been determined very precisely by now, and modified instructions have been radioed to the Saturn to account for the slight variation between our actual orbit and the pre-flight estimate. We need to know the exact moment to expect ignition (2:44:16) and the duration of the burn (5:47). Beyond that, we need to write down long lists of numbers which will tell us how to get home should some disaster strike *after* TLI and should we not be able to talk to the ground. At one time, it had been planned to equip the CM with a teleprinter, which would have been ideal for relaying columns of numbers, but a teleprinter was deemed an unnecessary frill, so now it is necessary for Bruce McCandless in Mission Control in Houston to read each digit, and for me to write them all down on my check list and read them all back to Bruce. By the time this chore is finished, we have whizzed past the United States and out over the southern Atlantic.

Less than an hour now to TLI, time for me to get strapped into the right-hand couch, with helmet and gloves back on, a little extra precaution in case the Saturn should blow up and damage the command module badly enough to cause us to lose cabin pressure. It doesn't make a hell of a lot of sense, really, because should the

CM be that badly damaged, certainly our service module engine (which we require to get down out of this orbit) would also be damaged beyond use, not to mention the damage to us. But anyway, the check list says helmet and gloves on, so that's the way it is. As we pass over western Australia, Houston relays to us—through Carnarvon—formal permission to go to the moon. The umbilical snipping ceremony carries about as much drama as asking for a second lump of sugar: "Apollo 11, this is Houston. You are GO for TLI." I answer, "Apollo 11. Thank you." There should be more to it. As we pass over Australia, we normally experience a long quiet spell over the Pacific, out of radio range, but not this time. Specially equipped jet transports circle below us in the darkness, serving as a communications link to Houston, and they will relay every last detail of the TLI burn to Houston's computers, which will immediately begin to chart our course, to keep track of our trajectory, and to let us know if corrections are necessary. NASA thinks of everything. I hope.

The Saturn is doing its thing now, preparing to pump hydrogen and oxygen to its engine, holding itself rigidly pointed in the direction the computers have selected. We have no control over the intricacies of its innards, we merely observe from some lights on our panel that the Saturn is counting down to ignition. When it finally lights, Neil says, "Whew!" I don't know about him, but I feel both relief and tension. We are on our way to the moon now, with one more hurdle behind us, but only if this thing continues to burn. If it shuts down prematurely, we will be in deep yogurt, on an odd-ball trajectory that will require some fancy computations on Houston's part and some swift and accurate work on ours, using our service module engine to get back to earth. "Pressure looks good," says Neil, referring to the Saturn's fuel (hydrogen) and oxidizer (oxygen) supply. Neil is in the left-hand couch with these gauges in front of him. Buzz is in the center with the computer. I am in the right seat with very little to do other than keep track of how long the engine has been burning. We have five windows in the CM, and they are numbered, from left to right. Neil has 1 at his left elbow and 2 directly in front of him. Three is a circular porthole in the hatch above Buzz's head. Four, directly in front of me, is small and wedge-shaped (matching 2), while 5 (matching 1) is at my right elbow, large and rectangular. Now I notice some strange lights out 5. "Flashes out window 5," I report. "I'm not sure . . . it could be lightning, or it could be something to do with the engine . . . continual flashes." Buzz comments on the Saturn's steering. "About 2 degrees off in pitch." Neil replies, "Yes, wouldn't worry too much about that." Houston joins the conversation. "Apollo 11, this is Houston. At one minute, trajectory and guidance look good . . . thrust is good."

I am amazed that I can see evidence of the engine operating, as it is mounted on the tail of the Saturn, 110 feet behind us. As far as I can recall, no one (except the Gemini crews who experienced firing the nose-mouthed Agena engine) has ever seen anything like the constant flashes and insistent fireflies I see out my window. I decide Neil might enjoy the show also, and, laughing, I tell him, "Don't look out window 1. If it looks like what I see out window 5, you don't want to look at it." "Why?" asks Buzz. "I don't see anything," says Neil. He doesn't? "These flashes out here?" "Oh, I see a little flashing out there, yes." Neil never admits surprise. I'll try it on Buzz. "You see that? Just watch window 5 for a second. See it?" "Yes, yes. Damn, everything's—just kind of sparks flying out there." I start to explain to him what I have seen. "Yes, that's—" when suddenly I feel a sudden lurch. "Oopsedo!" It's gone as quickly as it came, just as if the Saturn had abruptly shifted gears. We discuss it and agree that it is the result of a programmed shift in the ratio of fuel to oxidizer flowing into the engine. Marvelous machine! To think it's taking liquid hydrogen, stored at 423 degrees below zero, and liquid oxygen, at a mere 293 degrees below zero, and burning them seconds later at over 4,000 degrees. Houston makes a periodic report at about this time ("Everything's still looking good"), so we relax a bit more and start to enjoy the ride. It's pushing us back into our seats with almost the same force

we are accustomed to on earth (one G), although it feels like more
than that. It's still not smooth, "just a little tiny bit rattly," says
Buzz, but it's getting the job done, and our computer is spewing
out numbers which are very close to perfection. The shaking picks
up a little toward the end of the burn, and Buzz is worried that it
will dislodge the movie camera he installed in a bracket over my
head. "I hope that camera doesn't fall on your face." No problem
because "I checked it; it's locked in there pretty well. Won't hurt
this visor . . ." We are climbing out of darkness now, into the
dawn, and since we are pointed due east, the sun is coming directly
into our windows, especially 2 and 4. Neil has had the foresight to
install a piece of cardboard over his, to enable him to continue to
read the dials describing the Saturn's operation. He chortles, "Glad
I got my card up." I wish I had done the same. "I'm glad too. That
was a hell of a good idea. I can't see very much." It doesn't matter,
really, as we should be shutting down shortly. Five minutes and
forty-seven seconds, huh, and I have started the stopwatch on my
wrist at the instant of ignition. Just a few seconds to go . . .
there! "We have cut off," Neil tells us, and then congratulates the
ground. "Hey, Houston, Apollo 11. That Saturn gave us a magnificent ride." He's talking about all three stages of the Saturn
rocket, not just this last one, but he wasn't about to say how great
the first two were while the third still had a chance to bite us. Now
it's safe to be magnanimous because we are on our way!

We started the burn at one hundred miles altitude, and had
reached only 180 at cutoff, but we are climbing like a dingbat. In
nine hours, when we are scheduled to make our first mid-course
correction, we will be fifty-seven thousand miles out. At the instant
of shutdown, Buzz recorded our velocity as 35,579 feet per second,
more than enough to escape from the earth's gravitational field. As
we proceed outbound, this number will get smaller and smaller
until the tug of the moon's gravity exceeds that of the earth's, and
then we will start speeding up again. It's hard to believe we are on
our way to the moon, at twelve hundred miles altitude now, less

than three hours after lift-off, and I'll bet the launch-day crowd down at the Cape is still bumper to bumper, straggling back to the motels and bars.

Inside the command module it is now time to play musical chairs, switching seats for the transposition and docking maneuver. I will fly it from the left couch, with Neil in the center and Buzz over on the right. This is my first chance to get my hands on the controls and I am looking forward to it, even if it is nothing like flying a zippy fighter or the small Gemini. The command and service modules together weigh around sixty-five thousand pounds when the service module is full of fuel, as it is now. My first task is to separate the command and service modules from the Saturn and proceed away from it a safe distance; then to turn around and face it. The more slowly I do so, the less fuel it will cost. However, I don't want to turn around *too* slowly because I don't want that monster behind me to be out of sight any longer than is necessary. In the simulator I have worked out a compromise which involves separating at a relative velocity of only one half a mile per hour, and then after fifteen seconds starting the 180-degree turn at a rate of 2 degrees per second. Slow, deliberate work, which should bring the Saturn into view in a minute or so, while I coast out some seventy-five feet. I will fly it manually, except that I have asked the computer to help me during the turnaround, to keep the proper rate of turn going. When the time comes, I push a button to free us from the Saturn, and with my left hand I push a small control handle forward. Thrusters mounted around the periphery of the service module fire, and we feel their barely perceptible thrust as we move away. When my panel tells me I have the proper speed, I relax my hand and the thruster firing ceases. Now we coast for a few seconds, and then I start a pitch-up maneuver, by rotating my right wrist upward. When the motion seems firmly established, I take my right hand off the stick and push a button on the computer keyboard. PRO, it says, proceed with what it has been told to do. For some reason it balks this time, and I have to hit PRO a couple of more times to make it continue to turn around. In the process it is firing thrusters unnecessarily as the motion stops and starts, and I am perplexed and disgusted by it.

By the time we finally get turned around, we have drifted at least a hundred feet away from the Saturn, and it is going to cost more fuel to get back. In addition, the gadget which keeps track of my speed relative to the Saturn is reading a preposterous number, so I really don't know how fast I am moving, although my eyeballs tell me I am drifting away from the Saturn ever so slowly. So I take a guess at it, and hose out a bit of fuel with my left hand, holding the craft steady with my right. As we close, I can see the LM nestled in its container atop the Saturn like a mechanical tarantula crouched in its hole. Its one black eye peers malevolently at me; it is the drogue, into which I must insert our probe. Not being able to see the probe out my window (it is down and to the right, out of my field of view) is no problem. I am peering out through an optical sight which displays cross hairs in the sky in front of me. At the appropriate place on the LM is mounted a three-dimensional cross. When my cross hairs are precisely superimposed on it, the probe and drogue will be in perfect alignment, and it is then simply a question of holding that position and driving the probe into the drogue with just the right emphasis. It is similar to aerial refueling of aircraft, except that with no airflow or turbulence up here, the process is smoother. As I get close, the visibility is beautiful; the sun is back over my shoulder, and now the LM fills my window with unlikely-looking gold foil, flat gray surfaces, tubular legs, and iridescent windowpanes. Above all, that stand-off cross beckons me, and I tense up as it comes closer. Both hands are working now, my left deciding whether we move up or down, left or right, in or out; my right holds us steady and pointed in the right direction, making corrections in our pitch, roll, or yaw angles in response to what my eyes tell my hands. Buzz and Neil are helpless spectators, since only I am in a position to align hairs and cross. It must be a strange sensation for them to be so close to this great lumbering scrap pile in the sky, and to know we are about to collide with it, but when, where, and how? We are so close now that the exhaust gases from our rocket control motors are impinging on the LM and causing its thin skin to ripple rhythmically, like wind blowing over a Kansas wheat field. It all looks good. We must be only inches away now, and I have time for just one last correction. Hold it. Slightly off in roll, no matter, no time. Thrust toward it. Ah! That felt good, that gentle kiss. Computer mode to free. Fire one gas bottle to retract the probe. Bang! The latches slam into position down there in the tunnel and we are docked. "That wasn't the smoothest docking I've ever done," I say, fishing for compliments. Neil bites. "Well, it felt good from here." Actually, it was fine except for the extra fuel I must have used. My gas gauge isn't that precise, but I estimate that we have used eighty pounds of propellants rather than the fifty for which I had planned. Shouldn't matter, really, still . . . I know well from Gemini 10 how easy it is to gulp down extra fuel, and I want to save every last drop in case of a freakish rendezvous situation around the moon.

I slide out of my seat now and ease on down into the lower equipment bay, preparing to inspect the tunnel. First I must remove the tunnel hatch, check the probe and drogue hardware which holds LM to CM, and connect an electrical plug which will supply power from the CM to the LM. The hatch comes off easily, and as I stick my head into the tunnel, the pungent odor of some burned substance surrounds me. It smells as if it could be charred electrical wire insulation, but all the exposed wires appear in brand-new condition, not discolored as they would be if overheated. I simply can't find any source for the smell, which appears to be getting better, so I assume it is due to some past condition that no longer exists. Perhaps something overheated in the dense lower atmosphere during launch, or perhaps some of the rocket fumes from the launch escape tower motors were trapped in the tunnel. I ignore the smell as best I can and proceed with my list, checking each docking latch by hand, jiggling it to make sure it has really seated properly. I count twelve good ones, but could I have skipped over one and checked another one twice? I check them all again, and they are all engaged: the probe and drogue system has successfully jumped its first hurdle. Now we throw the appropriate switch on the main panel, the LM is gently sprung loose from the Saturn, and we proudly back off with our prize securely fastened to our nose. The Saturn is nearly empty now, and nearly dead. Its trajectory will be changed just one more time under ground control, to allow it to miss the moon and enter a solar orbit.

Our job now is to get a safe distance away from it, and we do this by firing our big service module engine for the first time. This baby is the one that will slow us down enough, when we arrive at the moon, to be captured by its gravity instead of whizzing on by. Once captured, we will stay captured, until the engine agrees to give us the extra speed needed to come home. All our equipment is important, but this engine is vital, obviously and dramatically so, and we are eager to have a look at it. This burn is only three seconds long, but that is enough to give us a fairly comprehensive look at the engine, its combustion chamber, its propellant supply system, its nozzle-swiveling capability, and so on. Since it produces twenty thousand pounds of thrust and our combined vehicles weigh one hundred thousand pounds, when it lights it gives us only a slight boot in the tail, one fifth of a G, but it's a reassuring feeling nonetheless. In the three brief seconds all looks good to us, and the ground confirms that their telemetered information reveals no flaws. Now we really are on our way to the moon, precisely on course at twenty thousand miles out. God, I'm hungry. No wonder, it's after 2 P.M., Cape time.

Before lunch, however, a few chores remain. The first is a pleasant one, getting out of our pressure suits. We help each other unzip and remove the bulky suits, thrashing around like three great white whales inside a small tank, banging into the couches and instrument panels as we struggle with the bulky garments. One by one we get them off and neatly folded and stuffed into storage bags under the center couch. Now the place seems much larger, for our bodies are smaller, and infinitely more comfortable. We dress in white two-piece nylon jump suits.

My next task involves realigning our inertial platform for the second time, and again, with help from the computer in pointing the sextant, it goes swiftly and well. Five balls! How about that, sports fans? But it doesn't pay to be a smartass, and Magellan gets a nasty start when he moves on to his next task—measuring the angles between five selected stars and the earth's horizon. A couple of stars I can see fine, like Altair, but with them I have difficulty finding that spot on the horizon which is directly below them, the substellar point as it is called. In other cases, such as Enif, the star is not bright enough to be readily seen. Finally, I wade through it all, but the results are not very accurate and I am discouraged. This exercise is for practice, really, as we would not have to rely on such measurements unless we lost radio contact with the ground, but nonetheless it is a shock to find I am still a lousy navigator, not much improved since Gemini days. At least I have time to eat now, but I find I am not nearly as hungry as I thought I would be. I don't feel sick, but somehow all these zero-G acrobatics have upset the delicate relationship between my gastrointestinal system and my brain, which usually results in a clamor for more food than pocketbook, reason, or even Lew could provide. My left knee is also hurting, just as it had on Gemini 10, but not as badly this time; I must have done a better job of denitrogenating, and having had the problem once before helps immeasurably this time, as I fully expect it to go away within a couple of hours. The entire flight so far has been much less traumatic than Gemini 10 was at this stage, even though there are massive differences in the significance of the two flights, from a historical and a safety point of view.

Out from behind the shadow of the earth, we are into the constant sunlight. In a way, there is constant darkness as well, for it depends on which way one looks. Toward the sun, nothing can be seen but its blinding disk, whereas down-sun there is simply a black void. The stars are there, but they cannot be seen because, with sunlight flooding the spacecraft, the pupil of the eye involuntarily contracts, and the light from the stars is too dim to compete with the reflected sunlight, as both enter the eye through the tiny aperture formed by the contracted pupil. No, to see the stars, the pupil

Suddenly, east of Scorpio, it appeared like a great ripe grape in the sky, our long-awaited moon

Close up, we sped over hill and dale, delighted by its sunny slopes

but sobered by its grim and forbidding craters

The Sea of Tranquility looked familiar enough, with the landing site in the center, at the edge of darkness, but there were mysteries aplenty

Why the straight line?

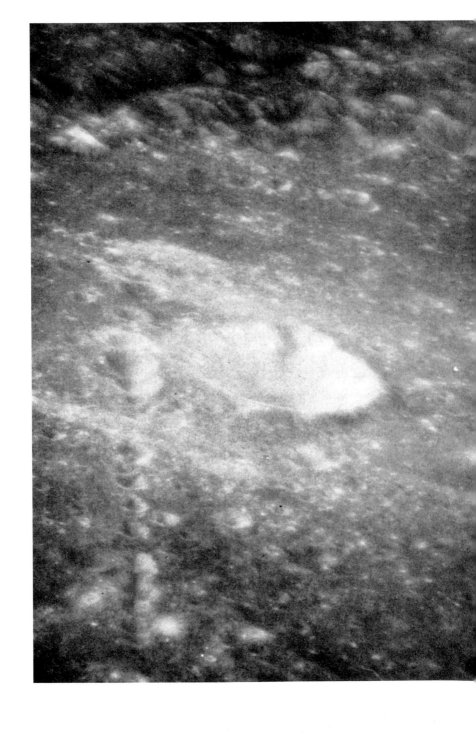

What formed these strange craters?

Do you see what I see?

The best sight of all was seeing it get
smaller and smaller as we headed for home

must be allowed to relax, to open wide enough to let the starlight form a visible image on the retina, and that can be done only by blocking out the sunlight. In practical terms, that means putting metal plates over all five windows, and then pointing the telescope at exactly the right angle, an angle which is not only away from the sun but which also does not permit any sunlight to bounce off the LM or CM structure into the telescope's field of view. Under these conditions the eye slowly "dark adapts" itself, and the brighter stars gradually emerge from the void. After a few minutes the familiar patterns of the constellations become recognizable (assuming you are fortunate enough to have familiar constellations in the part of the sky you have been forced to use to escape the sunlight), now the navigator can continue with his work.

What do we call this strange region between earth and moon? Cislunar space is the most common term, but that doesn't say much. Is it day or night? Since we humans generally define night as that time when our planet is between our eyes and the sun, I suppose this must be considered constant daytime, but it sure looks like night out of several of my windows. My body ignores all these considerations and stubbornly clings to the twenty-four-hour cycle it has known for thirty-eight years. This circadian rhythm will make me drowsy at the same time my family in Houston becomes drowsy, in the late evening, and alert again as the sun rises above the Houston horizon. My wristwatch knows best, or at least better than my eyes do.

Our next task also concerns the sun, but this time the problem is heat. Our service module is vulnerable if we hold any one fixed position; the side of it pointing at the sun will become too hot, and the side in the shade will become too cold. Too hot means propellant tank pressures rising dangerously high; too cold means radiators freezing. To prevent either, we must position ourselves broadside to the sun, and then rotate slowly, like a chicken on a motorized barbecue spit. Easier said than done, though, as it requires a very precise sequence of computer-assisted thruster firings to achieve a pure roll motion. If not done properly, pitch and yaw

motions will ensue, just like a top wobbling crazily at the end of its
spin, and then we must stop and begin all over again. We roll very
slowly, at three tenths of one degree per second (or one complete
turn each twenty minutes); once this motion is established, we are
free to relax and watch earth and moon alternate in a stately
promenade past our windows. The moon doesn't appear to be
getting much bigger, but the earth is shrinking noticeably. As bedtime approaches, home scarcely fills one small window. What it
lacks in size, however, it makes up in brilliance; as more and more
blackness become visible all around it, the intensity of its reflected
sunlight seems to increase, at least by comparison. We humans are
accustomed to watching the moon and thinking of it (when it's
full at least) as being very bright. In scientific terms, however, it is
a dullard: its albedo is .07, which means that it reflects only 7
percent of the sunlight that strikes its surface. The earth, on the
other hand, has an albedo four times that of the moon, caused
primarily by the reflected brilliance of clouds and water, and this is
what I see now. White clouds, blue water, with but a faint trace of
green jungles, and only a slightly more noticeable smear of rust as
North Africa slowly rotates into view. The window is cold and has
a slight film of condensation on it, which throws the earth a little
out of focus and seems to diffuse its light somewhat.

I sense no motion on our part, other than the rotation required to distribute the sun's heat evenly, and the earth, like the
minute hand on a watch, is not turning fast enough for the eye
really to track its motion. But the effect is clear: we are hanging
weightless in the void, while the earth turns slowly as it recedes
above us or below us or to the side of us—I know not which. It is a
totally different sensation than being in the race track of an earth
orbit. I am conscious of distance this time, not speed, and distance
away from home. It is a sobering, almost melancholy, sight, this
shrinking globe, and for the first time in my life I think I know
what "outward bound" means.

I have promised myself, and conveyed to Neil and Buzz, that
we must conserve our energy during the first three days of the flight

so as to be in peak condition to perform our intricate and demanding chores in lunar orbit. We must not become excited by the view or remain overwrought by today's adventures. We must pretend that the flight has not begun yet, that it will not begin until we get to the moon and begin our preparations for landing. All this is easy to say, but hard to do. Fortunately, it really helps that all three of us have been in space before, and I sense a feeling of quiet awareness inside the CM, rather than the jubilation and apprehension which I am sure would have infected three rookies. As bedtime approaches (fourteen hours after lift-off, 10:30 P.M. in Houston), we tidy up our new home, positioning each switch with care, fastening covers over the windows to blot out the circling sun, and finally we stretch out in the darkness. Neil and Buzz are in light mesh sleeping bags under the left- and right-hand couches; I am up above in the left-hand seat with a lap belt to keep me from floating off and a miniature headset taped to my ear, in case the ground should call. My knee is already feeling better, and I am cautiously optimistic about our day's work. By and large, things have gone extremely well, and we seem to be on our way to the moon with two healthy vehicles. There are many, many links in the daisy chain which may yet break, but at least three of the biggest and most fragile ones (launch, TLI, transposition and docking) are behind us. We are on our way, and it is necessary that we sleep.

The next thing I hear is Houston's insistent call, summoning the faithful to the temple for another day. "Apollo 11, Apollo 11, this is Houston. Over." "Good morning, Houston, Apollo 11." "Roger, Apollo 11. Good morning. When you're ready to copy, 11, I've got a couple of small flight-plan updates and your consumable update, and the morning news, I guess. Over." Jesus, no coffee, no amenities, simply pull out pen and paper and copy what they say. The news is pretty thin: the Russian space probe, Luna 15, has preceded us to the moon by a couple of days. We are not concerned by its presence, but apparently the people on the ground are, and the long-distance wires between Washington and Moscow have nearly gotten burned up making sure that our two trajectories will not intersect somewhere around the moon. The chances of their so doing are about equivalent to my high-school football team beating the Miami Dolphins, but no matter, it is a theoretical possibility and something the diplomats can massage, as a dog might worry a bone. The rest of the news is more interesting: Agnew thinks we should put a man on Mars by the end of the century. Right on, Spiro! The Mexican immigration officials are refusing passage to American hippies unless they first bathe and cut their hair. Right on, amigos! President Nixon has declared Monday, July 21, a federal holiday in celebration of what, hopefully, we will do on Sunday the twentieth. Right on, Mr. President, but don't you think you are just a trifle premature? Le Figaro waxes poetic: "The greatest adventure in the history of humanity has started." At least they have the sense to say "started." The House of Lords has been assured that a miniature submarine will not "damage or assault" the Loch Ness monster. How nice for him, and all this before breakfast. Shut up, Houston, will you—for just a little while?

I scurry around making coffee for us. The coffee is dehydrated, in little plastic bags that also contain sugar and cream (or whatever we have specified) and can be found in a stowage compartment down in the left side of the lower equipment bay. I hunt through this locker until I find the correct three bags, and then I attach them, one by one, onto the hot-water spigot, fill them, and knead them until all coffee and condiments have been dissolved. A check valve prevents the hot fluid from escaping. I pass out two, and then start to suck on my own, through a tube at the end opposite the check valve. It is lousy coffee, but at least it's lukewarm and familiar, and reminds me vaguely of earth mornings.

Today should be a quiet day, with only routine housekeeping chores and one mid-course trajectory correction to be accomplished. The housekeeping is mostly my responsibility, since Neil and Buzz have had other preoccupations during their training and have not concerned themselves with the nit-picking details of keeping the CM healthy. There is a sizable list of things to be done: fuel cells purged, batteries charged, waste water dumped, carbon dioxide canisters changed, food located and prepared, drinking water supply chlorinated, etc. Each item on the list is minor indeed, but together they add up to a sizable investment in terms of our labor. We are a hundred thousand miles out now, and the earth appears ridiculously small—not much larger than the stopwatch on my wrist. We are nearly halfway to the moon, and our speed has been decreasing ever since TLI, so that it is a mere fifty-four hundred feet per second, about one seventh of what it was at this time yesterday.

I remember last December, during the flight of Apollo 8, my five-year-old son had one, and only one, specific question: who was driving? Was it his friend Mr. Borman? One night when it was quiet in Mission Control I relayed this concern of his to the spacecraft, and Bill Anders promptly replied that no, not Borman, but Isaac Newton was driving. A truer or more concise description of flying between earth and moon is not possible. The sun is pulling us, the earth is pulling us, the moon is pulling us, just as Newton predicted they would. Our path bends from its initial direction and velocity after TLI in response to these three magnets. Up until now the earth's influence has been dominant, but by late tomorrow the moon will take over and our speed will begin to increase again. In the meantime, we have to correct our course slightly, as we have slowly been drifting off since TLI. For the three brief seconds of service module engine firing, Mike Collins will be driving instead of Sir Isaac Newton. Three seconds' worth! I wonder at the precision of this journey, which people keep comparing with Columbus's. I recall that as his crew grew more and more restless, with no land in sight, and as the pressure to turn back increased, Columbus is supposed to have doctored the daily log to show that the *Niña* really hadn't traveled all that far, and therefore it was reasonable still to be out of sight of land. Imagine my trying to doctor our flight plan, in case the moon proved to be more than a three-day journey. What would I say to the Houston computers?

My appetite is returning now, and I look forward to lunch-time. A couple of items on our menu, such as cream of chicken soup and salmon salad, are really delicious by anyone's standards, and even the less palatable things, like peanut cubes, are certainly edible. Besides, for a week, who cares? I remember being stranded on a remote West Virginia farm as a teenager for three or four days with nothing to eat except corn on the cob, three times a day, washed down by well water. I still like corn on the cob, but ah, this salmon salad! I'll rate it four spoons any day.

After lunch has restored my spirits and the mid-course correction has restored our trajectory to perfection, we have plenty of time to prowl around this strangely shaped compartment known as the command module. On earth it has always rested on its heat-shield base, tapering upward toward the tunnel at the apex of its triangular shape. Now there is no up or down, and without gravity the cabin assumes an entirely different character. This can't be the same command module I spent so many hours in at Downey and the Cape. It seems much more spacious now, and its parts some-how seem to be stuck together at different angles. As I slide over the center couch into the lower equipment bay, my legs unexpectedly curl around into the tunnel ("up," that would be on earth), so that instead of finding my face against the navigator's panel, I discover that I'm looking the other way, back toward the side hatch and its circular window. It takes some getting used to. The tunnel on earth is simply waste space overhead, but now it turns into a pleasant little nook where one can sit (?), crouch (?), well anyway, where one can stay, out of the way of the other two. I find that corners and tunnels have a lot to recommend them, because to stay in place, one must wedge himself between surfaces or else use something like a lap belt to keep from floating off and banging into equipment or compatriots.

Neil and Buzz are spending some time reviewing their future lunar module activities, and I burn up a little excess energy by running in place. I find a spot in the lower equipment bay which suits my purpose, allowing me to brace my arms overhead against one bulkhead, holding my body steady while my feet pound against another flat surface (the "floor"). Since I still have biomedical sensors attached to my chest, I decide to find out how heavy a load I can put on my heart by simply kicking my legs. "Hey, you got any medics down there watching? I'm trying to do some running in place . . . and I'm wondering . . . whether it brings my heart rate up." Neil joins me now, and the two of us jog along like fools while Buzz gets out the TV camera and points it at us. Houston reports, "Mike, we see about a ninety-six heartbeat now." "O.K., thank you . . . that's about all that is reasonable, without getting hot and sweaty." Enough foolishness. With the certain prospect of no shower for the next six days, there is no point in working myself into a lather, although I do feel better. My lower back seems to stiffen up and ache slightly in weightlessness, and now I'm more comfortable.

We switch the TV camera into position to point out the window at the earth 130,000 miles away. Neil describes in great detail everything from the polar ice cap to bands of clouds near the equator. I am holding the camera as steadily as I can, framing the world in the window. I wonder what it must look like to the millions watching their TV sets. I'll bet the North Pole is at the top of their screens. Now is my big chance to give millions of people vertigo! Slowly, I turn the camera 180 degrees in my hands, announcing as I do, "O.K., world, hang on to your hat. I'm going to turn you upside down." "Roger," says Charlie Duke in Mission Control, and I chortle, "You don't get to do that every day."

By the time we get the TV equipment packed up, it's time for bed, and all three of us are relaxed and ready for a long snooze. This time it is my turn under the left couch, zipped loosely inside a floating hammock, and I am comfortable indeed, much more so than last night or during any of my three Gemini nights. It is a strange sensation to float in the total darkness, suspended by a cobweb's light touch, with no pressure points anywhere on my body. Instinctively, I feel I am lying on my back, not my stomach, but I am doing neither—all normal yardsticks have disappeared, and I am no more lying than I am standing or falling. All that can be said with any certainty is that I am stretched out, with my body in a straight line from head to toe. The reason I think of myself as lying on my back is that the couch and main instrument panel are in front of me, not behind me, and if the spacecraft were in the white room at Downey, it would be positioned in such a way that gravity would force my backside up against the bulkhead behind me. Hence I am "lying" on my back. It reminds me of the damn-fool psychologist who ordered the inside of the very early command modules painted brown below an arbitrary line on the wall and blue above it. This was to give us a reassuring earth-sky analogy. But what happens if we roll the spacecraft upside down, we asked him, or turn our bodies inside it, won't we then feel out-of-sorts by your own reasoning? The interior of old *Columbia* is painted battleship gray all over, not that it makes any difference in the total darkness.

The next thing I know, Buzz is talking on the radio and I realize it is "morning"—or at least that eight hours or so have passed. Houston has decided we don't need a mid-course correction today, since our trajectory is still near perfect. I really feel good as I begin our third "day" in space. O.K., if everything between earth and moon is perpetual day, what *do* I call a twenty-four-hour increment of time? And how do I describe the past nine hours, if not "a good *night's* sleep"? With the cancellation of the mid-course, we have plenty of time to putter around, and about the only big job I have is to remove the probe and drogue from the tunnel, in order to clear a path for Neil and Buzz into the LM, so that they can check it over. This whole procedure is televised and takes an hour and a half. The ground seems to enjoy the TV a lot, judging from the comments coming from Houston, and I guess it must be eerie for the layman to see us floating in all directions past the endless panels of switches. I finally realize why Neil and Buzz have been looking strange to me. It's their eyes! With no gravity pulling down on the loose fatty tissue beneath their eyes, they look squinty and decidedly Oriental. It makes Buzz look like a swollen-eyed allergic Oriental, and Neil like a very wily, sly one.

After the TV show, we amuse ourselves by playing some music we have brought with us on a miniature tape recorder, and we eat and fuss about with our housekeeping chores. It is July 18, the third anniversary of the flight of Gemini 10, and a very quiet day.

Day 4 has a decidedly different feel to it. Instead of nine hours' sleep, I get seven—and fitful ones at that. Despite our concentrated attempt to conserve our energy on the way to the moon, the pressure is overtaking us (or me at least), and I feel that all of us are aware that the honeymoon is over and we are about to lay our little pink bodies on the line. Our first shock comes as we stop our spinning motion and swing ourselves around so as to bring the moon into view. We have not been able to see the moon for nearly a day now, and the change in its appearance is dramatic, spectacular, and electrifying. The moon I have known all my life, that two-dimensional, small yellow disk in the sky, has gone away somewhere, to be replaced by the most awesome sphere I have ever seen. To begin with, it is *huge*, completely filling our window. Second, it is three-dimensional. The belly of it bulges out toward us in such a pronounced fashion that I almost feel I can reach out and touch it, while its surface obviously recedes toward the edges. It is between us and the sun, creating the most splendid lighting conditions imaginable. The sun casts a halo around it, shining on its rear surface, and the sunlight which comes cascading around its rim serves mainly to make the moon itself seem mysterious and subtle by comparison, emphasizing the size and texture of its dimly lit and pockmarked surface.

To add to the dramatic effect, we find we can see the stars again. We are in the shadow of the moon now, in darkness for the first time in three days, and the elusive stars have reappeared as if called especially for this occasion. The 360-degree disk of the moon, brilliantly illuminated around its rim by the hidden rays of the sun, divides itself into two distinct central regions. One is nearly black, while the other basks in a whitish light reflected from the surface of the earth. Earthshine, as it's called, is sunlight which has traveled from the sun to the earth and bounced off it back to the moon. Earthshine on the moon is considerably brighter than moonshine on the earth. The vague reddish-yellow of the sun's corona, the blanched white of earthshine, and the pure black of the star-studded surrounding sky all combine to cast a bluish glow over the moon. This cool, magnificent sphere hangs there ominously, a formidable presence without sound or motion, issuing us no invitation to invade its domain. Neil sums it up: "It's a view worth the price of the trip." And somewhat scary too, although no one says that.

Before we become too engrossed in the mystery of the moon, however, Houston restores some of our perspective with an earful of terrestrial chatter. We must check the flow of fluid through our secondary coolant system by bringing a second radiator onto the line. This is a test I have long opposed ("If the goddamned primary system is working O.K., why screw around with testing the secondary?"), but I have been overruled in a series of meetings. After I have successfully put the system through its paces, I guess I grump a little bit and tell Houston, "Well at least we don't have to have any more meetings on the subject." "The flight director says 'Ouch,' " replies the CAPCOM. Have I offended Cliff Charlesworth? "No, no 'ouch' intended. I enjoyed every one of those meetings," I lie. Houston rewards us with the day's news, mostly baseball and other trivia. They do mention that Pravda has referred to Neil as the "Czar of the Ship," which title I heartily endorse for the remainder of the flight, and there is one amusing (at least to me) story about my son Michael. "What do you think," Michael is asked, "about your father going down in history?" "Fine," says Michael; and after considerable pause, "What *is* history, anyway?"

Houston is on to bigger things now, pumping us up with all the last-minute information we require before disappearing out of sight around the left-hand side of the moon. We have to know how to get into lunar orbit, and if trouble develops, how to get out of it—all without help from the ground. Our line-of-sight radio dictates that we can talk only to those people we can see; behind the moon, we will see no one. For the last fourteen hours we have been in the lunar sphere of influence, and our velocity has gradually picked up from a low of three thousand feet per second to its present seventy-six hundred feet per second. To be captured by the moon's gravity, we must slow down, by 2,917 feet per second to be exact, and we do this by burning our service module engine for six minutes and two seconds. Known as LOI₁, or the first lunar orbit insertion, this burn will put us in an elliptical orbit around the moon. Four hours later, we will attempt LOI₂, which should achieve a sixty-mile circular orbit.

As we ease on around the left side of the moon, I marvel again at the precision of our path. We have missed hitting the moon by a paltry three hundred nautical miles, at a distance of nearly a quarter of a million miles from earth, and don't forget that the moon is a moving target and that we are racing through the sky just ahead of its leading edge. When we launched the other day, the moon was nowhere near where it is now; it was some 40 degrees of arc, or nearly 200,000 miles behind where it is now, and yet those big computers in the basement in Houston didn't even whimper but belched out super-accurate predictions. I hope. As we pass behind the moon, finally, we have just over eight minutes to go before the burn. We are super-careful now, checking and rechecking each step several times. It is very much like the de-orbit burn of Gemini 10, when John Young and I must have checked our directions thirty times. If only one digit got slipped in our computer, the worst possible digit, we could be turned around backward and be about to blast ourselves into an orbit around the sun, instead of the moon, thereby becoming a planet the next generation might discover as the last one has discovered Pluto. No thanks.

When the moment finally arrives, the big engine instantly springs into action and reassuringly plasters us back in our seats. The acceleration is only a fraction of one G, but it feels good nonetheless. For six minutes we sit there peering intent as hawks at our instrument panel, scanning the important dials and gauges, making sure that the proper thing is being done to us. When the engine shuts down, we discuss the matter with our computer and I read out the results: "Minus one, minus one, plus one. Jesus! I take back any bad things I ever said about M.I.T." What I mean is that the accuracy of the overall system is phenomenal: out of a total of nearly three thousand feet per second, we have velocity errors in our body axis coordinate system of only one tenth of one foot per second in each of the three directions. That is one accurate burn, and even Neil acknowledges the fact. "That was a beautiful burn," he says, and I echo, "Goddamn, I guess! . . . I don't know if we're at sixty miles or not, but at least we haven't hit that mother." Buzz queries the computer as to our orbit. "Look at that, 169.6 by 60.9," and I reply, "Beautiful, beautiful, beautiful, beautiful!" "You want to write that down or something?" Buzz wants to know. Why not? "Write it down just for the hell of it; 170 by 60, like gangbusters." Buzz is precise. "We only missed by a couple of tenths of a mile." I am elated. "Hello, moon, how's the old back side?" We have arrived.

Once LOI$_2$ has passed, and established us in a near-circular orbit, averaging sixty miles altitude above the surface, we have a chance to examine the old back side, the front side, and parts in between. We are especially keen to study the landing site, the actual spot, just as we have been studying the photos of it for the last few months. Neil sums it up for Houston. "It looks very much like the pictures, but like the difference between watching a real football game and watching it on TV. There's no substitute for actually being here." Neil and Buzz also call out the familiar features along tomorrow's landing approach path: Mount Marilyn (named after Jim Lovell's wife), Boot Hill, Duke Island (named after Charlie Duke), Diamondback and Sidewinder (two sinuous rills etched in the Sea of Tranquility which look exactly like rattle-snakes), and so on, right up to the landing site itself. The Sea of Tranquility is just past dawn, and the sun's rays are intersecting its surface at a mere 1 degree angle. Under these lighting conditions, craters cast extremely long shadows, and to me the entire region looks distinctly forbidding, with no evidence that any part of its surface is smooth enough to park a baby buggy, never mind a lunar module. I can't resist commenting on it ("rough as a cob"), but all three of us know (we do?) that it will look a lot smoother tomorrow as the sun angle climbs toward the 10 degrees it will reach by the time of landing.

If anything, the rear side of the moon looks even rougher than the front. It doesn't have any "flat" *maria*, or seas, as the front does, but is a continuous region of "highlands," an uninterrupted jumble of tortured hills, cratered and recratered by 5 billion years of meteorite bombardment. There is no atmosphere surrounding the moon to produce clouds or smog or otherwise obscure the surface, so the details are uniformly clear. CAVU, as pilots describe a perfect day on earth: clear and visibility unlimited. The only thing that changes is the lighting, as our spacecraft passes from sunshine into earthshine, that eerie region of reflected sunlight, and then into total darkness. The feeling is more like circling in earth orbit than hanging suspended in cislunar space, as the past three days have been spent, but there are marked differences as well. First, we are traveling only one fifth as fast as we would in earth orbit, because the moon has much less mass than the earth and therefore produces a weaker gravitational pull, which, in turn, means that we require a slower orbital velocity to counterbalance this gravitational force with our own centrifugal force. However, since the moon is much smaller (2,160-mile diameter vs. the earth's 7,927), we get around it almost as fast, taking two hours for one orbit instead of ninety minutes. Also, because we are in a lower orbit (you can't orbit the earth at sixty miles because of its atmosphere), we get a noticeable sensation of speed. It's not quite as exhilarating a feeling as orbiting the earth, but it's close. In addition, it has an exotic, bizarre quality due entirely to the nature of the surface below. The earth from orbit is a delight—alive, inviting, enchanting—offering visual variety and an emotional feeling of belonging "down there." Not so with this withered, sun-seared peach pit out my window. There is no comfort to it; it is too stark and barren; its invitation is monotonous and meant for geologists only. Look at this crater, look at that one, are they the result of impacts, or volcanism, or a mixture of both?

As three amateur geologists, it doesn't take us long to get caught up with the mystery of the place and the fascination of discovering new craters on the back side. "What a spectacular view!" exclaims Neil. I agree. "Fantastic. Look back there behind us, sure looks like a gigantic crater; look at the mountains going around it. My gosh, they're monsters." Neil points out another one, even larger, and I'm even more impressed by it. "God, it's huge! It is enormous! It's so big I can't even get it in the window. That's the biggest one you have ever seen in your life. Neil, God, look at this central mountain peak. Isn't that a huge one? . . . You could spend a lifetime just geologizing that one crater alone, you know that?" Neil doesn't sound taken with *that* idea. "You could," he grunts. I hasten to add, "That's not how I'd like to spend my lifetime, but—picture that. Beautiful!" Buzz pipes up. "Yes, there's a big mother over here, too!" "Come on now, Buzz, don't refer to them as big mothers; give them some scientific name." Buzz ignores me and goes on. "It sure looks like a lot of them have slumped down." "A slumping big mother? Well, you see those every once in a while." Buzz decides to descend to my level. "Most of them are slumping. The bigger they are, the more they slump—that's a truism, isn't it? That is, the older they get." This conversation is taking a turn for the worse, and it will only be another few minutes until the earth pops up over the moon's rugged rim, so I drop the subject and start talking about camera and gimbal angles, as we all wait for our old friend to reappear. Still . . . the possibilities of weightlessness are there for the ingenious to exploit. No need to carry bras into space, that's for sure. Imagine a spacecraft of the future, with a crew of a thousand ladies, off for Alpha Centauri, with two thousand breasts bobbing beautifully and quivering delightfully in response to their every weightless movement . . . and I am the commander of the craft, and it is Saturday morning and time for inspection, naturally . . .

I am wrenched back to reality by the sudden appearance of the earth, a truly dramatic moment or two that we all scramble to record with our cameras. It pokes its little blue bonnet up over the craggy rim and then, not having been shot at, surges up over the horizon with a rush of unexpected color and motion. It is a welcome sight for several reasons: it is intrinsically beautiful, it contrasts sharply with the smallpox below, and it is home and voice for us. This is not at all like sunrise on earth, whose brilliance commands one's attention; it is easily missed, and therefore all the more precious to us, as we have anticipated its appearance and prepared for it. As it bursts into view, Houston starts talking, and we are back to business as usual.

One of the few mysteries from Apollos 8 and 10 involves the color of the surface of the moon. Eight had said simply black-gray-white, while 10 had said black-brown-tan-white, and we are to arbitrate the issue. We have discovered a bit of truth on either side of the argument—it seems to depend on the sun angle. At dawn and dusk, we have to vote with the Apollo 8 crew. It is dark gray, with some white, but no other colors—a darkened monochromatic plaster of Paris. On the other hand, near noon the surface assumes a cheery rose color, darkening toward brown on its way to black night. We vote with Apollo 10 in the late morning and early afternoon. We report all this to the ground, and then turn to navigational matters.

We need to know as much about the surface as possible, including how far it is below us, and one way of improving this measurement is by pointing the sextant at one piece of real estate and measuring our angle to it as we whiz by. I have picked a crater in the Foaming Sea (*Mare Spumans*) and have named it KAMP, in honor of my children and wife (Kate, Ann, Michael, and Patricia). KAMP will be helpful to Neil and Buzz tomorrow, in that *Mare Spumans* must be crossed before *Mare Tranquillitatis*, so that any improvement in the knowledge of their altitude over

Mare Spumans will be most useful as they descend toward the landing site. I make five marks on KAMP and incorporate them into my computer. It swallows down this information and uses it to improve its catalog of the moon's vital statistics. It's not long before these mathematical computations destroy the wonder of it all, or at least balance the frivolity of the big mothers, and I am forced to admit, "Amazing how quickly you adapt, why it doesn't seem weird at all to me to look out there and see the moon going by, you know."

This day is about over, our fourth day out from the earth, and I only want to rest now. Everything we have done so far amounts to zilch if we make a serious mistake, and we all know it, although Neil and Buzz seem less prone to admit our vulnerability than I do. I can only muse, "Well, I thought today went pretty well. If tomorrow and the next day are like today, we'll be safe." My wristwatch tells me it's already a few minutes past midnight in Houston. That makes it July 20, lunar landing day. If we were bullfighters, we would call it the moment of truth, but all I want is a moment of no surprises.

13

Houston, Tranquility Base here, the *Eagle* has landed.

—Neil Armstrong, 3:18 p.m., Houston time, July 20, 1969

"Apollo 11, Apollo 11, good morning from the Black Team."
Could they be talking to me? It takes me twenty seconds to fumble
for the microphone button and answer groggily, "Good morning,
Houston . . . You guys wake up early." "Yes . . . looks like you
were really sawing them away." They can tell when we are in a
deep sleep by monitoring our heart rates, which dip down to forty
or so when we are really zonked. "You're right," I tell them, and
then inquire about my machine. "How are all the CSM systems
looking?" "Looks like the command module is in good shape.
Black Team has been watching it real closely for you." ". . . appreciate that, because I sure haven't." I guess I have only been
asleep five hours or so; I had a tough time getting to sleep, and now
I'm having trouble waking up. Neil, Buzz, and I all putter about

fixing breakfast and getting various items of equipment ready for transfer into the LM.

Houston adds to the confusion by keeping up a steady chatter on the radio, reading us the day's news. "Among the large headlines concerning Apollo this morning, there's one asking that you watch out for a lovely girl with a big rabbit. An ancient legend says a beautiful Chinese girl called Chang-O has been living there for four thousand years. It seems she was banished to the moon because she stole the pill of immortality from her husband. You might also look for her companion, a large Chinese rabbit, who is easy to spot since he is always standing on his hind feet in the shade of a cinnamon tree." Jesus Christ, am I imagining all this? Here I am, half asleep, trying to fix a tube full of coffee, about to watch two good friends depart for the crater fields of the moon, there to join a Chinese rabbit under a cinnamon tree! There are just too many things going this morning, and I have to force myself to stick with the activities of the flight plan. "A three-ring circus. I got a fuel-cell purge in progress and am trying to set up camera and brackets, watch an auto maneuver . . ." I grump groggily.

Now it is time for Neil and Buzz to get dressed, and they begin by pulling their lunar underwear out of storage bins. These garments are *liquid*-cooled, with hundreds of thin, flexible plastic tubes sewn into a fishnet fabric. The back pack they will wear on the lunar surface will pump water through these tubes, cooling their bodies much more efficiently than could be done simply by blowing cool oxygen over them. I don't need water-cooled underwear because I don't have any back pack, and because hopefully I won't be working that hard, but I do require a pressure suit, so all three of us struggle into them, helping each other with inaccessible zippers and generally checking the condition of each other's equipment. What would we do if, for example, Neil's zipper broke, or his helmet somehow refused to lock on to his neck ring? He couldn't venture out onto the lunar surface that way, that's for sure, nor could he allow Buzz to, because he would perish as soon as the LM door was opened to the vacuum of space. He couldn't stay in the CM and let Buzz land by himself, because the LM requires simultaneous manipulation by two people. I couldn't take his place because I was not trained to fly the LM. Perhaps he and I could switch suits, but I doubt that he could fit into mine, and mine can't accommodate a back pack. From such fabric are nightmares woven. Fortunately, everything seems to fit together, and I stuff Neil and Buzz into the LM along with an armload of equipment.

Now I have to do the tunnel bit again, closing hatches, installing drogue and probe, and disconnecting the electrical umbilical running into the LM. I am supposed to rig the TV camera to shoot out one of my windows to show the departure of the LM, but I decide I am too busy preparing for undocking to fool with it. I inform Houston, "There will be no television of the undocking. I have all available windows either full of heads or cameras, and I'm busy with other things." Generally one discusses these things with Houston and follows their advice, but this time I'm *telling* them, not asking them, and they must sense this, because they immediately reply, "We concur."

I am on the radio constantly now, running through an elaborate series of joint checks with *Eagle*. In one of them, I use my control system to hold both vehicles steady while they calibrate some of their guidance equipment. I check my progress with Buzz. "I have five minutes and fifteen seconds since we started. Attitude is holding very well." "Roger, Mike. Just hold it a little bit longer." "No sweat, I can hold it all day. Take your sweet time. How's the czar over there? He's so quiet." Neil chimes in, "Just hanging on—and punching." Punching those computer buttons, I guess he means. "All I can say is, beware the revolution," and then, getting no answer, I formally bid them goodbye. "You cats take it easy on the lunar surface; if I hear you huffing and puffing, I'm going to start bitching at you." "O.K., Mike," Buzz answers cheerily, and I throw the switch which releases them. With my nose against the glass of window 2 and the movie camera churning away over in window 4, I watch them go. When they are safely clear of me, I inform Neil, and he begins a slow pirouette in place, allowing me a look at his outlandish machine and its four extended legs. "The *Eagle* has wings!" Buzz exults.

It doesn't look like any eagle I have ever seen. It is the weirdest-looking contraption ever to invade the sky, floating there with its legs awkwardly jutting out above a body which has neither symmetry nor grace. Everything seems to be stuck on at the wrong angle, which I suppose is what happens when you turn aeronautical engineers loose designing a vehicle which always flies in a vacuum and hence requires no streamlining. I make sure all four landing gear are down and locked, report that fact, and then lie a little, "I think you've got a fine-looking flying machine there, *Eagle*, despite the fact you're upside down." "Somebody's upside down," Neil retorts. "O.K., *Eagle*. One minute . . . you guys take care." Neil answers, "See you later." I hope so. When the one minute is up, I fire my thrusters precisely as planned and we begin to separate, checking distances and velocities as we go. This burn is a very small one, just to give *Eagle* some breathing room. From now on it's up to them, and they will make two separate burns in reaching the lunar surface. The first one, called descent orbit insertion (DOI, for short), will take place behind the moon and will serve to drop *Eagle*'s perilune to fifty thousand feet at a point 16 degrees east of the landing site. Then, when they reach this spot over the eastern edge of the Sea of Tranquility, *Eagle*'s descent engine will be fired up for the second and last time (power descent initiation, or PDI), and *Eagle* will lazily arc over into a twelve-minute, computer-controlled descent to some point at which Neil will take over for a manual landing.

After DOI, the LM swoops down farther and farther below me, picking up speed as it goes, so that at the time of PDI it will be about 120 miles in front of me. After PDI, the situation changes very swiftly. As *Eagle* slows down, I will start gaining on it and will whiz by overhead, so that I will be about two hundred miles ahead of it at the instant of touchdown. I will try to keep them in sight as long as possible, because if they have to abort, it would be nice for me to know where they have gone, helping to determine which one of my eighteen rendezvous cases applies.

As we swing around the right edge of the moon after DOI, I regain contact with Houston before *Eagle* does, as I am considerably higher now. I have to remember that I am *Columbia* for the next twenty-four hours, and I should stop calling myself Apollo 11. "Houston, *Columbia*. Reading you loud and clear." "Roger, Mike. How did it go?" They want to know about DOI, naturally enough. "Listen, babe, everything's going just swimmingly. Beautiful." I am wide awake now and have lost this morning's feeling of being rushed. Things feel good, with the CSM shipshape and the LM also apparently in good condition. "Great," responds the ground. "We're standing by for *Eagle*." "O.K., he's coming along."

My navigational equipment is really working well, which adds to my confidence. On my last two platform alignments, I got five balls and four balls one, which is always nice, and I have been using my sextant to track a lunar landmark and the LM. All this work has been going extremely well, almost effortlessly, and the marks made on the LM in particular give me confidence about tomorrow's rendezvous. I don't really need to be able to make accurate marks on the LM today, but the fact I can do so (and old Colossus IIA confirms the fact) bodes well for tomorrow. As *Eagle* approaches PDI, I am still hanging in there, peering out through the sextant at a minuscule dot. The LM is nearly invisible and looks like any one of a thousand tiny craters, except that it is moving. Finally, as it passes the hundred-mile mark, I lose it. I rub my eyes in relief at the end of this practice session. I have been sighting with my right eye, my left being covered by a small black plastic patch held on by an elastic string. I find that the muscles in my left eyelid get weary squinting shut, so I have used an eyepatch since Gemini days.

The best thing I can do now is to keep quiet, as Houston and *Eagle* have lots to discuss as power descent initiation arrives and the final descent begins. At the beginning of the burn, Neil and Buzz are pointed heads up, feet forward, and see nothing but black sky. This awkward position is necessary to get the best communications with the earth and to confirm the accuracy of their trajectory; after that, they will roll over and start checking the parade of familiar landmarks past their window. Then it's up to Neil to find a spot smooth enough to put it down. At five minutes into the burn, when I am nearly directly overhead, *Eagle* voices its first concern. "Program Alarm," barks Neil, "it's a 1202." A 1202? What the hell is that? I don't have all the alarm numbers memorized for my own computer, much less for the LM's, so I don't have the foggiest notion how bad it is in terms of continuing the descent. I jerk my own check list out of my suit pocket and start thumbing through it, but before I can find 1202, Houston says, "Roger, we're GO on that alarm." No problem, in other words. My check list says 1202 is an "executive overflow," meaning simply that the computer has been called upon to do too many things at once and is forced to postpone some of them. I guess it means the same thing for the LM, as M.I.T. has designed both computer programs. A little farther along, at just three thousand feet above the surface, the computer flashes 1201, another overflow condition, and again the ground is superquick to respond with reassurances. Good work on someone's part.

Now it begins to sound like a ground-controlled approach into fog, as Buzz calls out altitude and velocity to Neil, who has his eyes glued on the scene out the window. "Six hundred feet, down at nineteen [feet per second]." "Four hundred feet, down at nine." "Three hundred feet . . . watch your shadow out there." "Two hundred feet, 4½ down." "One hundred feet, 3½ down, nine forward. Five percent [fuel remaining]." "Forty feet, down 2½, kicking up some dust." That sounds good, just a bit of dust at forty feet. "Thirty seconds," says Houston. That's how much fuel they have left. Better get it on the ground, Neil. "Contact light!" sings out Buzz, and then a bunch of gibberish concerning shutting down their engine. They have arrived! "We copy you down, *Eagle*," says Houston, half question and half answer. Neil makes it official. "Houston, Tranquility Base here, the *Eagle* has landed." Whew! I re-establish radio contact with Houston and they inform me of the landing. "Yes, I heard the whole thing . . . Fantastic!" Neil explains why he nearly ran out of gas. "The auto targeting was taking us right into a football-field size . . . crater . . . with a large number of big boulders . . . It required . . . flying manually over the rock field to find a reasonably good area." Christ, I don't care if he landed on top of a gigantic anthill, just as long as they are down in one piece.

My command module chores now include an extra task: finding the LM on the surface. If I can see it through my sextant, center my cross hairs on it, and mark the instant of superposition, then my computer will know something it doesn't know now: where the LM actually is, instead of where it is supposed to be. This is a valuable—not vital, but valuable—piece of information for "Hal" to have, especially as a starting reference point for the sequence of rendezvous maneuvers which will come tomorrow (or sooner?). Of course, the ground can take its measurements as well, but it really has no way of judging where the LM came down, except by comparing Neil and Buzz's description of their surrounding terrain (lurain?) with the rather crude maps which Houston has. But I am far past the landing site now, about to swing around behind the left edge of the moon, so it will be awhile before I get my first crack at looking for the LM. It takes me two hours to circle the moon once.

Meanwhile, the command module is purring along in grand shape. I have turned the lights up bright, and the cockpit reflects a cheeriness which I want very much to share. My concerns are exterior ones, having to do with the vicissitudes of my two friends on the moon and their uncertain return path to me, but inside, all is well, as this familiar machine and I circle and watch and wait. I have removed the center couch and stored it underneath the left one, and this gives the place an entirely different aspect. It opens up a central aisle between the main instrument panel and the lower equipment bay, a pathway which allows me to zip from upper hatch window to lower sextant and return. The main reason for removing the couch is to provide adequate access for Neil and Buzz to enter the command module through the side hatch, in the event that the probe and drogue mechanism cannot be cleared from the tunnel. If such is the case, we would have to open the hatch to the vacuum of space, and Neil and Buzz would have to make an extravehicular transfer from the LM, dragging their rock boxes behind them. All three of us would be in bulky pressurized suits, requiring a tremendous amount of space and a wide path into the lower equipment bay. In addition to providing more room, these preparations give me the feeling of being proprietor of a small resort hotel, about to receive the onrush of skiers coming in out of the cold. Everything is prepared for them; it is a happy place, and I couldn't make them more welcome unless I had a fireplace. I know from pre-flight press questions that I will be described as a lonely man ("Not since Adam has any man experienced such loneliness"), and I guess that the TV commentators must be reveling in my solitude and deriving all sorts of phony philosophy from it, but I hope not. Far from feeling lonely or abandoned, I feel very much a part of what is taking place on the lunar surface. I know that I would be a liar or a fool if I said that I have the best of the three Apollo 11 seats, but I can say with truth and equanimity that I am perfectly satisfied with the one I have. This venture has been structured for three men, and I consider my third to be as necessary as either of the other two.

I don't mean to deny a feeling of solitude. It is there, reinforced by the fact that radio contact with the earth abruptly cuts off at the instant I disappear behind the moon. I am alone now, truly alone, and absolutely isolated from any known life. I am it. If a count were taken, the score would be three billion plus two over on the other side of the moon, and one plus God only knows what on this side. I feel this powerfully—not as fear or loneliness—but as awareness, anticipation, satisfaction, confidence, almost exultation. I like the feeling. Outside my window I can see stars—and that is all. Where I know the moon to be, there is simply a black void; the moon's presence is defined solely by the absence of stars.

To compare the sensation with something terrestrial, perhaps be-ing alone in a skiff in the middle of the Pacific Ocean on a pitch-black night would most nearly approximate my situation. In a skiff, one would see bright stars above and black sea below; I see the same stars, minus the twinkling, of course, and absolutely nothing below. In each case, time and distance are extremely important factors. In terms of distance, I am much more remote, but in terms of time, lunar orbit is much closer to civilized conversation than is the mid-Pacific. Although I may be nearly a quarter of a million miles away, I am cut off from human voices for only forty-eight minutes out of each two hours, while the man in the skiff—grazing the very surface of the planet—is not so privileged, or burdened. Of the two quantities, time and distance, time tends to be a much more personal one, so that I feel simultaneously closer to, and farther away from, Houston than I would if I were on some remote spot on earth which would deny me conversation with other humans for months on end.

My windows suddenly flash full of sunlight, as *Columbia* swings around into the dawn. The moon reappears quickly, dark gray and craggy, its surface lightening and smoothing gradually as the sun angle increases. My clock tells me that the earth is about to pop into view, and I prepare for it by positioning my parabolic antenna so that it points at the proper angle. Sure enough, here comes the earth on schedule, rising swiftly above the horizon, and shortly thereafter I can tell from one of my many gauges that the antenna has locked onto its signal and conversation should be possible. The three big antennas on earth are located at Honey-suckle Creek in eastern Australia; near Madrid, Spain; and at Goldstone Lake in the Mojave Desert, not too far from Las Vegas. As the earth turns, Houston shifts its control from one to the other, the hand-off being based on which one is pointing most directly at the moon. Since they are nearly evenly spaced around the globe, one antenna is always in excellent position. I don't know which one I may be talking through, although I suppose I could tell either by looking at my flight plan or by carefully examining that blue and white pea out there, but the fact is I really don't care. I call them all Houston, which simplifies it. "Houston, *Columbia.* How's it going?" "Roger . . . we estimate he landed about four miles downrange . . . we'll have a map location momentarily. Over." Now they send me up a bunch of numbers which I punch into my computer. Colossus IIA knows how to use this ground estimate to point my sextant, but as I whiz by overhead, I can't see a darn thing but craters. Big craters, little craters, rounded ones, sharp ones, but no LM anywhere among them. The sextant is a powerful optical instrument, magnifying everything it sees twenty-eight times, but the price it pays for this magnification is a very narrow field of view, only 1.8 degrees wide, so that it is almost like looking down a gun barrel. The LM might be close by, and I swing the sextant back and forth in a frantic search for it, but in the very limited time I have, it is possible to study only a square mile or so of lunar surface, and this time it is the wrong mile.

I am sixty nautical miles above Tranquility Base, traveling at about thirty-seven hundred miles per hour. If Neil or Buzz were able to see me, they would find that I would come up over their eastern horizon, pass almost directly overhead, and disappear below their western horizon. The entire pass takes thirteen minutes, but most of that interval is not available to me, because I need to be looking down through my sextant at a steep angle. If I consider 45 degrees as the minimum acceptable angle, then I have only two minutes and twelve seconds of useful time, as I swing from 45 degrees on one side of the LM to 45 degrees on the other. A busy two minutes indeed. During the thirteen minutes, I can talk directly to the LM. At other times, when I am on the front side of the moon (but out of sight of the LM), I can talk to them via earth, if Houston's switches are set properly to relay our conversation. At 186,000 miles per second (the speed of light), it takes about one and one quarter seconds for the radio waves to reach the earth and an equal time for them to be relayed back to the moon. This delay can have some interesting side effects, as, for example, in the seconds following the LM's touchdown, when Houston told

Neil, "Be advised there's lots of smiling faces in this room and all over the world. Over." Neil responds, "Well, there are two of them up here." It takes two and a half seconds for me to hear this; as soon as I do, I say, "And don't forget one in the command module"; but in the meantime Houston has heard Neil and has answered him, "Roger. That was a beautiful job, you guys." I was more than a little embarrassed to hear their message coming in as I was mouthing mine. It sounded like I was asking them not to forget to compliment me for doing a beautiful job in the command module, instead of merely adding my smiling face to the list.

Although I can't see the LM, I *can* listen, as Neil and Buzz describe what no men have seen before—the view from the surface of another planet. I can't help interrupting. "Sounds like it looks a lot better than it did yesterday at that very low sun angle. It looked rough as a cob then." "It really was rough, Mike," Neil replies. "Over the targeted landing area, it was extremely rough, cratered, and large numbers of rocks that were . . . larger than five or ten feet in size." "When in doubt, land long," I say, using the pilot's cliché about never landing short of the runway. "So we did," he replies simply.

Things must be going extremely well, for Neil and Buzz want to forgo a scheduled four-hour nap in favor of proceeding immediately out onto the lunar surface. I thought they might, as this has been a topic of debate for some months. It seems ridiculous to expect them to unwind at this stage of the game and suddenly fall asleep; on the other hand, if they do go EVA now and struggle back into the LM dog-tired a few hours later, and *then* are confronted with an emergency requiring immediate lift-off and rendezvous, they would be so shot that they would probably make a lot of mistakes, and rendezvous is not a very forgiving phase of flight. But anyway, Houston agrees, and so do I, for whatever that is worth.

Houston also has its eye on me, and lets me know it. "*Columbia*, Houston. We noticed you are maneuvering very close to gimbal lock. I suggest you move back a way. Over." Gimbal lock, one of the command module pilot's most familiar enemies, is a condition caused by maneuvering to a direction that prevents the three gyroscopes in the inertial platform from moving freely. To prevent damage to the gyroscopes, the system "freezes," causing the platform to be rendered useless, until the command module pilot can run through an elaborate procedure to restore the platform and hence regain the vital knowledge of "up-down" and "left-right" which the platform provides. Gimbal lock is a pet peeve among astronauts, especially those of us who flew the Gemini spacecraft, which avoided the problem by providing four gimbals instead of Apollo's three. We felt that we were being unnecessarily restricted in the maneuvers we could perform because NASA's Apollo right hand had ignored its Gemini left. Now I can't resist letting my annoyance show. "How about sending me a fourth gimbal for Christmas?" They don't know what I am muttering about. "You were unreadable. Say again, please," but what the hell, this is no time for sour grapes, so I let it drop. "Disregard." Houston is chattering away now about waste water dumps, battery charges, and other things they want to make sure I won't forget as I depart "over the hill" into my own forty-eight-minute universe. In case I have not heard, they also let me know about the plan to go EVA before sleeping. "Sounds good to me," I say. "Tell them to eat some lunch before they do." The Jewish mother is in orbit.

Houston's final transmission is different and is a bit disturbing. "*Columbia*, Houston. We show your EVAP OUT temperature running low. Request you go to manual temperature control and bring it up. You can check the procedures in ECS MAL 17. Over." Translated into English, they are telling me that something is wrong with the system for regulating the temperature of the coolant fluid which is necessary to keep all my fancy equipment operating at the proper temperature. If the system gets too cold, it is possible for the radiators sunk in the skin of the service module to freeze, and then I will have serious problems. Further, they are telling me to dig into my library and refer to Environmental Control System Malfunction Procedure 17. As soon as all the noise in my headset subsides, I do just that. It feels like one of the simulations at the Cape, in which I have been given a pretended problem to solve, only this time it's for real: I really am behind the moon by myself, and there will be no debriefing over coffee to critique what I am about to do. For some reason I cannot explain, I do not wade through Malfunction 17. It is an involved procedure, and it seems like radical surgery to me. Instead, I simply check all my switch positions and flip the one which controls the offending temperature from automatic to manual, then back to automatic again. Somehow I feel very confident about this machine, and I am not too concerned about this problem. Let's give the machine a chance to cure itself. As the quiet minutes tick by, I keep one eye on the temperature gauge as I go about my other housekeeping duties, such as dumping excess water produced by the fuel cells. By golly, everything seems to be returning to normal, and when I finally swing around into view of the earth, I am able to report, ". . . whatever the problem was, it seems to have gone away without any changing of J52 sensors or anything like that. My glycol evaporator outlet TEMP is up above fifty now, and it's quite comfortable in the cockpit; so we'll talk more about that one later."

Meanwhile, I must get prepared for my second Easter egg hunt, squinting through the sextant in search of the LM below. Again I have no luck, just more indistinguishable craters and no glint of sunlight off metallic skin. "Do you have any topographical cue that might help me out here?" Houston sends up some more vague descriptions of craters, but they're no help. Also, I can't even hear the LM this time, which is strange. The pre-flight agreement was that all LM transmissions would be automatically relayed to me; apparently they aren't doing it. "Houston, *Columbia*. Could you enable the S-band relay at least one way from *Eagle* to *Columbia* so I can hear what's going on?" "Roger, there's not much going on at the present time, *Columbia*. I'll see what I can do about the relay." Goddamn. I want to hear what's going on. "O.K. I haven't heard a word from those guys, and I thought I'd be hearing them through your S-band relay." I don't care so much right now, because they are still about two hours away from depressurizing the LM, but when they get out on the surface, I want to be able to hear them. What will Neil say, for instance? He hasn't confided any magic first words to me, but I'll bet he has some. Neil doesn't waste words, but that doesn't mean he can't use them; he nearly always rises to an occasion, and if ever man had anything to say, this is the time. I want to hear him!

Houston confirms that my coolant problem seems to have solved itself, and I pass out of sight with a feeling of confidence in *Columbia* and a feeling of growing anticipation of what is going on down there in *Eagle*. The back side is really peaceful this time, with a quality of guaranteed silence that is uniquely satisfying. When I get around on the front side again, I find Neil and Buzz engaged in equipment check-out, and still over an hour from stepping on the surface. Rats! I will probably be on the back side when they get out. Another try at seeing them through my sextant (no luck) and another peaceful back-side pass; then around once more and try to get the radio working. "Reading you loud and clear. How's it going?" "Roger. The EVA is progressing beautifully. I believe they are setting up the flag now." The American flag! "Great!" "I guess you are about the only person around that doesn't have TV coverage on the scene." "That's all right, I don't mind a bit. How is the quality of the TV?" "Oh, it's beautiful, Mike. It really is." "Oh gee, that's great! Is the lighting halfway decent?" "Yes, indeed. They've got the flag up now and you can see the Stars and Stripes on the lunar surface." "Beautiful, just beautiful." Just let things keep going that way, and no surprises, please. Neil and Buzz sound good, with no huffing and puffing to indicate they are overexerting themselves.

But one surprise at least is in store, and a very impressive one at that. Houston comes on the air, not the slightest bit ruffled, and announces that the President of the United States would like to talk to Neil and Buzz. "That would be an honor," says Neil, with characteristic dignity. "Go ahead, Mr. President. This is Houston. Out," says Bruce McCandless, the CAPCOM, as if he instructed Presidents every day. The only clue to how he must feel comes in his use of the word "out," which we are all taught in telecommunications protocol but which we practically never use. "Out" has a formality and finality that renders its use most unusual. Perhaps it should be reserved for Presidents.

The President's voice smoothly fills the air waves with the unaccustomed cadence of the speechmaker, trained to convey inspiration, or at least emotion, instead of our usual diet of numbers and reminders. "Neil and Buzz, I am talking to you by telephone from the Oval Office at the White House, and this certainly has to be the most historic telephone call ever made . . . Because of what you have done, the heavens have become a part of man's world. As you talk to us from the Sea of Tranquility, it inspires us to redouble our efforts to bring peace and tranquillity to Earth . . ." My God, I never thought of all this bringing peace and tranquillity to anyone. As far as I am concerned, this voyage is fraught with hazards for the three of us—and especially two of us—and that is about as far as I have gotten in my thinking. Peace and tranquillity indeed; I wish I had time to digest that, and decide in my own mind whether it's true or not; in the meantime, I am proprietor of this orbiting men's room and there are other demands on my time.

Neil, however, pauses long enough to give as well as he receives. "Thank you, Mr. President. It's a great honor and privilege for us to be here, representing not only the United States but men of peace of all nations, and with interest and a curiosity and a vision for the future. It's an honor for us to be able to participate here today." The President responds, "And thank you very much, and I look forward—all of us look forward—to seeing you on the *Hornet* on Thursday." "I look forward to that very much, sir," Buzz pipes up, before Houston abruptly cuts off the White House and returns to business as usual, with a long string of numbers for me to copy for future use. My God, the juxtaposition of the incongruous: roll, pitch, and yaw; prayers, peace, and tranquillity. What will it be like if we really carry this off and return to earth in one piece, with our boxes full of rocks and our heads full of new perspectives for the planet? I have a little time to ponder this as I zing off out of sight of the White House and the earth in my proud and solitary vigil.

The next time around, I am more concerned than I have been before. The *Hornet*, indeed; these guys may never see *Columbia*, much less the *Hornet*. "How goes it, anyway?" "Roger, *Columbia* . . . the crew of Tranquility Base is back inside . . . everything went beautifully. Over." "Hallelujah!" Well, that's a big one behind us: no more worrying about crashing through into hidden lava tubes, or becoming exhausted, or the front door sticking open, or the little old ladies using weak glue, or any of that! Whew! Now all we have to do is grab some shuteye, and get the top half of *Eagle* up here where it belongs; then we can haul ass! Meanwhile, it's two in the morning in Houston, and it's been a long day (tougher than yesterday, but perhaps not as tough as tomorrow). It's time for me to douse the lights and get some sleep. Sleep? Alone by myself? You'd better believe it. These are familiar surroundings, not the bewildering jungle of switches I once regarded with awe, but old friends now, just part of *Columbia*. As I scurry about, blocking off the windows with metal plates and dousing the lights, I have almost the same feeling I used to have years ago when, as an altar boy, I snuffed the candles one by one at the end of a long service. Come to think of it, with the center couch removed, *Columbia*'s floor plan is not unlike that of the National Cathedral, where I used to serve. Certainly it is cruciform, with the tunnel up above where the bell tower would be, and the navigation instruments at the altar. The main instrument panels span the north and south transepts, while the nave is where the center couch used to be. If not a miniature cathedral, then at least it is a happy home, and I have no hesitation about leaving its care to God and Houston as I fade away into a comfortable snooze.

"*Columbia, Columbia*, good morning from Houston." "Hi, Ron." Ron Evans has had the night shift so far this flight, and I haven't talked to him very much. Now he's got the early-morning duty, and he's trying to make sure I'm awake. "Hey, Mike, how's it going this morning?" "How goes it? . . . I don't know yet, how's it going with you?" I'm still groggy. "Real fine here, *Columbia* . . . we're going to keep you a little busy here . . ." he apologizes. Don't I know it! A space day always seems to start with a bang, with no time even to take a pee before the switch throwing begins. Today is rendezvous day, and that means a multitude of things to keep me busy, with approximately 850 separate computer key strokes to be made, 850 chances for me to screw it up. Of course, if all goes well with *Eagle*, then it doesn't matter too much, as I merely retain my role as sturdy base-camp operator and let them find me in my constant circle. But if . . . if . . . if any one of a thousand things goes wrong with *Eagle*, then I become the hunter instead of the hunted. Furthermore, the roles can become reversed with no warning at any point along that 850-step path, so I must get up on my tiptoes and stay there, all day long. Neil and Buzz will lift off in slightly over three hours; Ron hasn't awakened them yet. He wants me to get a head start on them by tracking a lunar landmark one last time, to update my computer prior to lift-off. By the time they get up, I have whizzed on by them and I am halfway through breakfast, fully awake. Things must be stirring in Houston too, and I imagine there is quite a crowd of people hunched over every available console in Mission Control. Jim Lovell, Neil's back-up, comes on the air with some unaccustomed formality. "*Eagle* and *Columbia*, this is the back-up crew. Our congratulations for yesterday's performance, and our prayers are with you for the rendezvous. Over." "Thank you, Jim." "Thank you, Jim." Neil and Buzz answer quickly, and I add, "Glad to have a big room full of people looking over our shoulder."

When the instant of lift-off does arrive, I am like a nervous bride. I have been flying for seventeen years, by myself and with others; I have skimmed the Greenland ice cap in December and the Mexican border in August; I have circled the earth forty-four times aboard Gemini 10. But I have never sweated out any flight like I am sweating out the LM now. My secret terror for the last six months has been leaving them on the moon and returning to earth alone; now I am within minutes of finding out the truth of the matter. If they fail to rise from the surface, or crash back into it, I am not going to commit suicide; I am coming home, forthwith, but I will be a marked man for life and I know it. Almost better not to have the option I enjoy. Hold it! Buzz is counting down: "9 — 8 — 7 — 6 — 5 — . . . abort stage . . . engine arm ascent . . . proceed . . . beautiful . . . thirty-six feet per second up . . ." Off they go: their single engine seems to be doing its thing, the thing earthlings have been insisting it could do for half a dozen years, but it's scary nonetheless. One little hiccup and they are dead men. I hold my breath for the seven minutes it takes them to get into orbit. Their apolune is forty-seven miles and their perilune is ten miles. So far so good. Their lower orbit ensures a satisfactory catch-up rate, and they will be joining me in slightly less than three hours, if all goes well.

In the meantime, my hands are full with the arcane, almost black-magical manipulations called for by my solo book. The book is attached by an alligator clip to my helmet tie-down strap, and I am religiously following it line by line, checking off each item—no matter how picayune—as I go. I have locked on to them with my electronic ranging device, which is part of my VHF radio. It reports they are 250 miles behind me, and then promptly breaks lock. I fiddle with it; it reacquires the LM briefly and then breaks lock again. Each time it does this, I must inform Colossus IIA to ignore the data coming from the VHF ranging ("Verb 88, enter") or to start paying attention to it again ("Verb 87, enter"). In between these computer key punches, I keep my eye glued to the sextant, which Colossus IIA is pointing at where it thinks the LM ought to be. Sure enough, there is a tiny blinking light in the darkness, and I am able to align my sextant precisely with it and hit the mark button several times. With both VHF ranging and sextant angular data coming in, I am in good shape. Colossus IIA knows precisely where the LM is now, and if for some reason the LM's thrusters act up, I can perform the mirror image of the LM maneuvers and catch it, instead of being caught by it. The continual interruption of data from the VHF ranging is annoying but not serious. We are both on the back side of the moon now, and it is time for the LM to make its first overtaking maneuver, raising itself into a circular orbit some fifteen miles below mine. I am prepared to make the burn if they cannot, and I nervously count them down. "Forty-five seconds to ignition." "O.K.," says Buzz, and then, "We're burning . . . burn complete, Mike." I punch their numbers into my computer, and it does some calculating and reports our orbits: my apolune is 63.2 nautical miles, my perilune, 56.8. Theirs is now 49.5 by 46.1. In theory, I should be 60 by 60, and they should be 45 by 45, but these numbers are well within acceptable bounds, and everything looks good so far.

There is a strange noise in my headset now, an eerie woo-woo sound. Had I not been warned about it, it would have scared hell out of me. Stafford's Apollo 10 crew had first heard it, during their practice rendezvous around the moon. Alone on the back side, they were more than a little surprised to hear a noise that John Young in the command module and Stafford in the LM each denied making. They gingerly mentioned it in their debriefing sessions, but fortunately the radio technicians (rather than the UFO fans) had a ready explanation for it: it was interference between the LM's and command module's VHF radios. We had heard it yesterday when we turned our VHF radios on after separating our two vehicles, and Neil said that it "sounds like wind whipping around the trees." It stopped as soon as the LM got on the ground, and started up again just a short time ago. A strange noise in a strange place.

Buzz and I are working on a new problem now, measuring any difference in the plane of our orbits. Reassuringly, we both agree that our orbital paths are tilted at precisely the same angle, that we are close enough to perfection that we require no sideways burn at this time to align ourselves more accurately. The LM does make a very small in-plane correction, to reduce its altitude variations as it overtakes me. Shortly thereafter, I pass over the landing site for the first time since they have departed. What a relief! *"Eagle, Colum-*

bia passing over the landing site. It sure is great to look down there and not see you!" Not that I ever did see them on the surface, but to pass over and know they are not stranded down there is worth the price of the entire Apollo program to me.

The LM is fifteen miles below me now, and some fifty miles behind. It is overtaking me at the comfortable rate of 120 feet per second. They are studying me with their radar and I am studying them with my sextant. At precisely the right moment, when I am up above them, 27 degrees above the horizon, they make their move, thrusting toward me. "We're burning," Neil lets me know, and I congratulate him. "That-a-boy!" We are on a collision course now, or at least we are supposed to be; our trajectories are designed to cross 130 degrees of orbital travel later (in other words, slightly over one third of the way around in our next orbit). I have just passed "over the hill," and the next time the earth pops up into view, I should be parked next to the LM. As we emerge into sunlight on the back side, the LM changes from a blinking light in my sextant to a visible bug, gliding golden and black across the crater fields below. "I see you don't have any landing gear." Of course, only the top half, called the ascent stage, of *Eagle* is returning; the descent stage sits at Tranquility Base for all time, its last (and best) function having been to serve as launch pad. "That's good," chortles Neil. "You're not confused which end to dock with, are you?" Then he adds, "Looks like you are making a high side pass on us, Michael," using fighter-pilot terminology. Buzz sees me too. "O.K., I can see the shape of your vehicle now, Mike." So close, yet so far away: all that remains is for them to brake to a halt using the correct schedule of range vs. range rate. My solo book tells me that at 2,724 feet out, they should be closing at 19.7 feet per second; at 1,370 feet, 9.8 feet per second, etc. While they are doing this, they must make certain they stay exactly on their prescribed approach path, slipping neither left nor right nor up nor down. John Young and I both know that fuel-guzzling whifferdills result if one is not extremely careful, and this is what concerns me now. The sextant is useless this close in, so I close up shop in the lower equipment bay, transfer to the left couch, and wheel *Columbia* around to face the LM.

Goddamn, it looks good! I can look out through my docking reticle and see they are steady as a rock as they drive down the center line of that final approach path. I give them some numbers. "I have 0.7 mile and I got you at thirty-one feet per second." Buzz replies, "Yes—yes, we're in good shape, Mike; we're braking." Jesus, we really *are* going to carry this thing off! For the first time since I was assigned to this incredible flight six months ago, for the first time I feel that it *is* going to happen. Granted, we are a long way from home, but from here on, it should be all downhill. Bigger and bigger the LM gets in my window, until finally it nearly fills it completely. I haven't touched the controls. Neil is flying in formation with me, and doing it beautifully, with no relative motion between us. I guess he is about fifty feet away, which means the rendezvous is over. "I got the earth coming up . . . it's fantastic!" I shout at Neil and Buzz, and grab for my camera, to get all three actors (earth, moon, and *Eagle*) in the same picture. Too bad *Columbia* will show up only as a window frame, if at all. Within a few seconds Houston joins the conversation, with a tentative little call. "*Eagle* and *Columbia*, Houston standing by." They want to know what the hell is going on, but they don't want to interrupt us if we are in a crucial spot in our final maneuvering. Good heads! However, they needn't worry, and Neil lets them know it. "Roger, we're stationkeeping."

Neil has turned *Eagle* around now, so that its black spot (the drogue used in docking) is directly facing me. Control passes from *Eagle* to *Columbia* at this point, as per our training sessions. It is easier to fly the docking maneuver from the command module; although it can be done from the LM, it is awkward for Neil, because he would have to crane his neck to see out an overhead window, whereas I can look straight ahead in more conventional fashion. So I sight through my reticle and align my probe with

Eagle's drogue, just as I did five days ago when I pulled the LM loose from the Saturn. There are a few differences, mainly that the puny little LM ascent stage, nearly empty, weighs less than six thousand pounds now instead of the nearly thirty-three thousand both stages weigh when they are full of fuel. However, I am not the slightest bit worried as I draw closer and closer. The alignment looks very good indeed at the instant of contact, which I feel as a barely perceptible little nudge.

As soon as we are engaged by the three little capture latches, I flip a switch which fires one of my nitrogen bottles to start the retraction cycle, to pull the two vehicles together. When I do, I get the surprise of my life! Instead of a docile little LM, suddenly I find myself attached to a wildly veering critter that seems to be trying to escape. Specifically, the LM is yawing around to my right, and we are misaligned by about 15 degrees now. I work with my right hand to swing *Columbia* around, but there is nothing I can do to stop the automatic retraction cycle, which takes some six or eight seconds. All I can hope for is no damage to the equipment, so that if this retraction fails, I can release the LM and try again. Things are moving swiftly now, as I wrestle with my right-hand controller. We are veering back toward center line now, and get there, and *bang*, the docking latches slam shut, and miraculously all is well again. Whew! I explain it to Neil and Buzz. "That was a funny one. You know, I didn't feel a shock, and I thought things were pretty steady. I went to RETRACT there, and that's when all hell broke loose." Oops, one never swears on the radio—perhaps that's why I swear so much otherwise. There's no point in worrying about it now. It's time to hustle on down into the tunnel and remove hatch, probe, and drogue, so Neil and Buzz can get through the tunnel. Thank God, all the claptrap works beautifully in this its final workout. The probe and drogue will stay with the LM and be abandoned with it, for we have no further need for them and don't want them cluttering up the command module. The first one through is Buzz, with a big smile on his face. I grab his head, a hand on each temple, and am about to give him a smooch on the forehead, as a parent might greet an errant child; but then, embarrassed, I think better of it and grab his hand, and then Neil's.

We cavort about a little bit, all smiles and giggles over our success, and then it's back to work as usual, as Neil and Buzz prepare the LM for its final journey and I help them transfer equipment into *Columbia*. We also have to go through an elaborate vacuum-cleaning procedure to make sure that everything returning from the LM is free of loose dust or dirt. The microbe people have insisted on it, to keep any lunar bugs in the LM, and we go along with it as best we can, feeling slightly ridiculous. We also are pumping oxygen from *Columbia* into the LM and thence overboard, so any bugs would have to swim upstream to get into *Columbia*. Finally comes the punch line, the reason for making the trip. Buzz announces, "Get ready for those million-dollar boxes. Got a lot of weight. Now watch it." I have seen the two lunar rock boxes before, at the Cape: they are shiny little metal caskets about two feet long, built with a fancy sealing system that preserves the rocks in their original environment—the lunar vacuum—without exposing them to our atmosphere and any chemical modifications it might cause.

After we get the rock boxes zippered inside white fiberglass fabric containers, I have a chance to quiz Neil and Buzz about those parts of their experience this back-side absentee missed. "How about that lift-off from the moon; what did it feel like?" "There was a little blast, then we started moving . . . the floor came up to meet you . . . maybe half a G or two thirds of a G." "And the landing was no problem, because as I understand it, the dust did not engulf you but sprayed out parallel to the surface, is that so?" "Yes." "And the dust can be light tan or dark battleship gray? What do you think it is . . . basalt dust?" No commitment there. "Well, do the rocks all look the same?" No, there are differences, they say; some have "little sparkly stuff" in them, and they had time enough to take samples carefully from the most interesting specimens they could find. "Great, great . . . man, that's beautiful . . . that'll keep those geologists jumping for years." My curiosity about things geologic is easily quelled; besides, it is time to get on with other things, such as dumping the LM and heading for home.

When the time comes to jettison *Eagle*, I flip the necessary switches, there is a small bang, and away she goes, backing off with stately grace. With her goes the probe and drogue, thank God, and I simply can't express my pleasure at not ever having to fool with them again! In fact, the whole goddamned LM has been nothing but a worry for me, and I'm glad to see the end of it. Neil and Buzz, on the other hand, seem genuinely sad: old *Eagle* has served them well and deserves a formal or at least a dignified burial. Instead, it is to be left in orbit, while Houston watches its systems slowly die. Then its carcass will be an orbiting derelict for days, or weeks, or months—until finally its orbit deteriorates and it crashes forlornly into the lunar surface. Just to make sure *Eagle* is out of our way, I back off from it, firing our small thrusters to change our speed by two feet per second, so we can forget *Eagle* now and get on with preparations for the next major hurdle in our young lives.

TEI. Transearth injection. NASA jargon has an uncanny knack of disguising the meaning of even the most obvious things. This is the get-us-home burn, the save-our-ass burn, the we-don't-want-to-be-a-permanent-moon-satellite burn, and they call it TEI. It involves firing our big SPS engine for two and a half minutes over on the back side of the moon, adding a bit over three thousand feet per second to our velocity, just the right amount to break the bond of lunar gravity and to put us on a trajectory which should slice through the earth's atmosphere two and a half days from now. We are keyed up and ready to make the burn one revolution early, but Houston isn't too keen on that idea, so we stick with our original schedule. It's just as well. This way we will have plenty of time to get set, and it gives the ground a cheery and chatty interlude. Houston asks me, "How does it feel up there to have some company?" "Damn good, I'll tell you!" "I'll bet. I bet you'd almost be talking to yourself up there after ten revs or so." "No, no. It's a happy home here . . . as a matter of fact, it'd be

nice to have a couple of hundred million Americans up here . . .
let them see what they're getting for their money." I suppose a
small commercial is permitted.

Next, Houston pours out a potpourri of news and congratula-
tory messages (I wish they'd wait until *after* TEI). ". . . Prime
Minister Harold Wilson . . . the King of Belgium . . . Premier
Alexei Kosygin . . . Mrs. Robert Goddard . . ." Now that's one
to think about: did Robert Goddard anticipate this moment? Is
that what kept this quiet scientist going all those years out on the
New Mexico desert, firing one liquid-fueled rocket after another,
back in the twenties when everyone probably thought he was nuts?
We also hear quotes from our three wives. "Fantastically marvelous" doesn't sound like Pat, but that's what they say she said. We
also hear about football and baseball, and Thor Heyerdahl and his
papyrus boat, and even President Nixon, who apparently is headed
out toward the *Hornet* to watch our return. Then, abruptly,
Houston abandons the English language and plunges back into the
technical phrases and numbers upon which we are dependent.
"TEI 30, SPS/G&N: 36691; minus 061, plus 067 135 23 4149;
Noun 81, plus 32020, plus 06713, minus 02773 181 054 013; Noun
44, Ha is N/A, plus 00230 32833 228, Delta Vc 32625 24 1510
355 . . . your set stars are Deneb and Vega, 242 172 012 . . .
the horizon will be on 11-degree mark at Tig minus two minutes . . ." I read all this back, digit by digit. Some of these
numbers are merely important, but others are supremely important; there must be no slip of mind or mouth or hand as they are
recorded.

As we ease around the western edge of the moon into our own
quiet zone, hopefully for the last time, we run through our last-
minute checks prior to igniting our engine. Our one and only
engine. We go about our work very carefully indeed, with meticulous attention to each last detail. More than anything else, the
direction we are pointing concerns us, silly as that may sound.
With all our sophistication, and all the confirmed numbers in our
computer, we still want to see it for ourselves, and that is difficult

to do, as we won't break out into sunlight until a few minutes before ignition. "I see a horizon. It looks like we are going forward," I say with a nervous laugh. "Shades of Gemini," answers Neil, referring to the care we used to take with Gemini retrofire burns to get out of orbit. "It is most important that we be going forward," I insist, carried away by laughter now. I don't know what's so goddamn funny, but Buzz also thinks something is, as he goes through a little pantomime, reciting rocket fundamentals, ". . . Let's see—the motor points this way and the gases escape that way, therefore imparting a thrust that-a-way." "Beautiful-looking horizon," says Neil. He is in the center couch, with the computer; Buzz is over on the right with fuel cells and other electrical claptrap; I am in the left couch and will "fly" it, in the sense that I can take over manually if the automatic steering fails, or shut the engine down, or handle various other problems that may arise. Buzz is reading the check list and I am throwing the switches as he calls them out. "Just about midnight in Houston town," says Neil, a strange detached comment at this stage of the game. "Yes," I answer. I don't know what the hell time it is in Houston, I just want his attention back in the cockpit. "O.K., coming up on two minutes," sings out Buzz, and I confirm that the horizon is in perfect position in my window. "Beautiful." Buzz counts the last few seconds: "5 — 4 — 3 — 2 —" and when she lights I reply, "Burn! a good one . . . nice! . . . pressures are good . . . busy in steering, but it's holding right in there."

Ten seconds later, Buzz wants more detail. "How is it, Mike?" I don't blame him; over on his side, it's hard to tell what's going on. "It's pretty busy in roll, but it's holding in its deadband . . . it's possible we have a roll thruster problem, but if we have, it's taking it out. No point in worrying about it . . . coming up on one minute, chamber pressure's holding right on a hundred . . . gimbals look good, total attitude looks good. Rates are damped out . . . still a little busy . . . how's that nitrogen pressure? O.K.?" "Yes," says Neil. "Good . . . two minutes . . . hits the end of that roll deadband, it really comes crisply back. O.K., the chamber pressure's falling off a little bit, now it's going back up. Chamber pressure's oscillating just a tad . . . brace yourself! . . . standing by for engine off." It seems to me it's not shutting down on schedule. My pre-flight agreement, worked out in tedious meetings with the engine and trajectory experts, is that I will allow it two extra seconds and then shut it down manually if another gauge confirms that we have burned forty feet per second too much. The latter gauge is changing numbers so fast I can't really make that determination properly, but when the two seconds are up, I snap two switches shut and we are weightless again. The computer says the burn was perfect, so I guess it shut itself down just as my fingers touched the switches, or else I was just lucky to grab them at precisely the right instant. Either way, it doesn't matter; the main thing is, we're on our way home! "Beautiful burn! SPS, I love you, you are a jewel. Whoosh!"

Now Buzz leads us through the litany of the check list again, turning off all those switches we turned on for the burn, and then we whip out our cameras and start taking pictures of the lunar surface, just like tourists leaving Venice who suddenly discover they have three rolls of film left. It's nearly time for Houston to join us, and Neil asks, "Anybody got any choice greetings they want to make to Houston?" Not me; all I want to do is talk about the TEI burn. ". . . The best burn I've ever seen in my life, I'll tell you! I guess you guys have seen two good ones today," referring to the LM ascent engine which lifted them from Tranquility Base. "Oh, a couple," murmurs Neil, and Buzz adds more emphatically, "Yes, we sure as hell have . . . hey, I hope somebody's getting a picture of the earth coming up." With the earth comes Charlie Duke's unmistakable Carolina accent. He sounds excited. "Hello, Apollo 11, Houston. How did it go?" "Time to open the LRL doors, Charlie," I reply. "Roger. We got you coming home. It's well stocked." I hope it is, I hope somewhere in that huge quarantine building in Houston they have stocked a little bit of vermouth and a lot of gin. The last time I saw it, all it had was white mice. Those same mice are waiting for us now or, more accurately, for our moon rocks; if the mice get sick from contact with the rocks, we may stay in that LRL a hell of a lot longer than twenty-one days, and we'll need more than gin.

Meanwhile, we have two and a half days to sweat out. It took us three days to get here, but our return trajectory is swifter; even so, I expect the next two days to be long ones indeed. Right now we are still tourists, plastered up against the windows as we climb steeply up from the lunar surface. We approached the moon from the west in its penumbra, that eerie shadow zone that made it appear a ghostly globe, with illuminated rim but barely discernible surface. It is just as impressive now, but in a totally different way. We are departing from its eastern side, and it glares brilliantly in the sunlight. We can see it all now, from pole to pole and edge to edge, and we can clearly differentiate between the *maria* and the highlands. Both are cratered, but the seas do seem calm by comparison with the tortured uplands. The *maria* are darker too and seem more neutral gray than the golden hills which surround them. It seems like a cheery place, not the scary one I first saw two days ago, but cheerier yet is the notion that we are leaving it. I have absolutely no desire to come back.

My next concern, of course, is the accuracy of our return-to-earth trajectory. Our own ability to navigate home independent of Houston is very poor when we are close to the moon, so for the time being we are dependent on earth tracking of our position. "How does that tracking look, or is it too early to tell?" "Stand by, Mike . . . looking real good." That's nice. The next call from Houston indicates that Deke Slayton has grabbed the mike away from Charlie Duke. "Congratulations on an outstanding job. You guys have really put on a great show up there. I think it's about time you powered down and got a little rest, however. You've had a mighty long day . . . I look forward to seeing you when you get back here. Don't fraternize with any of those bugs en route except for the *Hornet*." Get some rest? Who cares about rest, although I suppose we must be tired. Neil reports he got only three hours' sleep in the LM last night, and Buzz got four; I suppose I must have gotten five hours in my more comfortable machine. However, we still have a few things to take care of: moon photography, optional, of course; but then the mandatory chores. The platform must be realigned, the spacecraft put into its broadside roll to distribute the sun's heat evenly, lithium hydroxide canisters must be changed, the oxygen-tank heaters need to be turned on for a while, the water supply must be chlorinated . . . when I get to the latter task, I realize just how tired I am. I patiently explain to Houston how many chlorine ampules I have left in my storage box and how there are not enough unused ones left to keep using them at the present rate, etc., only to have them point out to me that I have overlooked a second supply cabinet full of them. Dunce! I must have been going flat out for seventeen hours now, but that shouldn't cause me to forget what supplies we have on board. The moon is nearly five thousand miles away as we pack it in for the day. "Good night, Charlie. Thank you," says Neil, and Buzz echoes, "Good night, Charlie. Thank you." "Adios," I add. "Adios," says Charlie. "Thanks for a great show, you guys." As long as everybody is congratulating everybody else, I might as well get in the last word. "Thanks again for a great job down there." Lights out.

What a beautiful night's sleep. We awake of our own accord, and when I open one eye, I find Buzz puttering around and Neil apparently still zonked. We strike up a casual conversation with the ground, report that we each got eight or more hours' sleep, and verify that old *Columbia* is humming along in good shape. Houston reports the instant at which we leave the lunar sphere of influence. This means simply that despite the fact we are only thirty-four thousand nautical miles from the moon, and still 174,000 away from the earth, the earth's pull has become dominant, and the mathematical equations now recognize that fact. "Mark," they say, "you're leaving the lunar sphere of influence, over." "Roger," I reply. "Is Phil Shaffer down there?" He's the one who, on Apollo 8, somehow gave the press the idea that the spacecraft physically jumped at this point, and then had a hell of a time trying to unconvince them. No, Shaffer's not on duty, but someone else is ("We've got a highly qualified team on in his stead"). "Roger, I wanted to hear him explain it again to the press conference . . . tell him the spacecraft gave a little jump as it went through . . ." "Thanks a lot," says Houston sarcastically. "Dave Reed is sort of burying his head in his arms right now."

Now it's news time again, and Houston reads up a long blurb having to do with the international implications of our flight and a lot of less weighty matters. The main thing I get from it is that President Nixon will be watching our entry into the earth's atmosphere from the bridge of the *Hornet*. I can feel my stomach muscles tightening. Just one big one left, and with the President of the United States there to watch. I'd better not screw it up. Neil, unruffled, calls for the Dow Jones averages, and the ground describes a minor adjustment to our trajectory. Our flight plan shows four adjustments, or mid-course corrections, on the way to the moon and three on the way back to earth. In fact, we required only one to get there, and should require only one to get home. This next one will be small, just eleven seconds of firing using our small thrusters. It is scheduled to change our velocity by 4.8 feet per second, out of a total of 4,075 feet per second. Now that's precision, and I appreciate it. I want to stay absolutely in the center of our entry corridor, which means slicing back into the atmosphere at an angle of 6.5 degrees below the horizon. Too shallow, and we skip back out; too steep, and we burn up. Each sounds as grim as the other, and I don't want to edge .01 degree in either direction; 6.5 please!

The ground is fussing at Buzz now about one of the biomedical sensors taped to his chest. Apparently it is not making a good contact, and Buzz dutifully goes through their suggested procedure of removing and reapplying it. I don't think I would bother: the bloody things are a nuisance and an unneeded encumbrance. Who needs to monitor our heart rates constantly, anyway? Obviously they are no busier in Houston than we are up here, and their next call confirms that fact. "For $64,000, we're still trying to work out the location of your landing site, Tranquility Base. We think it is located on LAM–2 chart at Juliet 0.5 and 7.8 . . . we are wondering if Neil or Buzz had observed any additional landmarks . . . which would confirm or disprove this." No wonder I couldn't find the LM; nobody seems to know where the bugger came down!

While they are discussing West Crater and Cat's Paw and other mysteries, I am making like a French chef in the lower equipment bay. Lunch today includes cream of chicken soup, one of my favorites, and I hum happily as I attach a dehydrated bag to the hot-water tap and fill it up with five or six ounces. Then I knead the packet carefully, until all lumps have disappeared, and slice the end of it open with my surgical scissors, unfolding a small tube through which I suck up the ambrosia. I share my good fortune with Houston. "My compliments to the chef. The food is outstanding. This cream of chicken soup I give at least three spoons." "O.K., cream of chicken, three spoons" comes back a bewildered voice, apparently not familiar with the conventions for ranking restaurant food and service. Perhaps if I told him the service was surly, he'd understand. What I would like right now is a little drink, say a nip of cognac. Suddenly my mind flashes back to something Deke said before launch, when we had been discussing the theoretical advisability, on future flights, of adapting the mariners' rum-ration philosophy. Deke had indicated it wasn't such a far-fetched idea after all; he had said . . . what had he said? . . . I have forgotten. Still, if I can't even find the spare chlorine ampules, is it possible I have overlooked some cognac? After all, Jules Verne provided Chambertin for his crew, could Deke do less? Perhaps he has caused a small bottle of cognac to be included somewhere among our abundant provisions. I rummage through various out-of-the-way storage boxes, but all I can find is more food, spare underwear, tissue paper, flashlights, film magazines, tools, and medical kits (I *know* the medics wouldn't shake loose any cognac). No luck. Ah, well, I should have known better. I'm not going to let it destroy my good humor, so I move over to a window and watch the grand scene outside. "Nice to sit here and watch the earth getting larger and larger and the moon smaller and smaller." "Roger," says Houston. They say Roger to everything.

We may not have any cognac on board, but we do have some other unusual things: two great big flags, for example; American flags that are supposed to fly over the House of Representatives and the Senate. We also have a "sporting license," strange terminology if I have ever heard it, for this little yellow card is an authorization to attempt to set flight records which will be recognized by the Fédération Aéronautique Internationale. Presumably my twenty-eight hours alone in lunar orbit will be one such record, but apparently without this card on board, the results would be unofficial. The card doesn't tell me what kind of sporting I am limited to, but . . . in the command module? All it says is that "This license must be produced to take part in any sporting event governed by FAI regulations." We also have a stamp kit, including a first-day cover commemorating the issuance of a new ten-cent stamp, showing an astronaut at the foot of the LM ladder, about to sample the lunar surface. With the envelope is an ink pad and a cancellation stamp, which says   . Never mind that it is JUL

22; this is the first chance we have had to get to it. We try the cancellation out first, inking it and printing it in our flight plan three times until we get the hang of it, and then we apply it gingerly to the one and only envelope, which we understand the Postmaster General will put on tour. The die from which the stamp was printed is also on board *Columbia*.

In addition we carry a small hand-held tape recorder, besides the very complex one built into the spacecraft. The small one is used to record verbal notes during those times when we don't want to take the time to plug ourselves into the big one. In addition to recording, it can play tapes, and someone has recorded music and other sound effects for us. I told Neil and Buzz before the flight I didn't care what kind of music they requested, it suited me fine, and the result is bland, popular selections with "moon" in them wherever possible. My favorite, which I've never heard before, is "Everyone's Gone to the Moon," or at least that's the line the vocalist keeps repeating. It's very restful. There is also some strange electronic-sounding music, a favorite of Neil's called "Music out of the Moon." Neil claims it was recorded about twenty years ago, but I never heard of it either. Finally, there is a jangling cacophony of bells, whistles, shrieks, and unidentifiable sounds at the end of the tape. We amuse ourselves now by pushing our radio trans-mitter button and holding the screeching tape recorder next to a microphone. This gets an immediate reaction. "Apollo 11, Hous-ton. You sure you don't have anybody else in there with you?" I feign innocence. "Houston, Apollo 11, say again, please?" "We had some strange noises coming down on the downlink, and it sounded like you had some friends up there." I know that they have just changed shifts in Mission Control, with the Green Team going off duty and the White Team straggling in from wherever. "Where do the White Team go during their off hours, anyway?"

Now we must stop fooling around and unpack the TV camera. It's prime evening time in Houston, and we are scheduled for a show. We have not rehearsed, nor have we acquired any affection for the TV camera and what it represents. Neil can't show the scientists the rocks because they are locked in their vacuum containers, but he does the next best thing—show them the boxes the rocks are stored in. Buzz continues with a demonstra-tion of food preparation, culminating with his slopping ham spread on a piece of bread. Then he spins a small can in midair, demon-strating the principle of gyroscopic action.

When my turn comes, I fill a spoon with water to demon-strate weightlessness in a way kids might understand. "I am going to show you how to drink water out of a spoon, but I'm afraid I filled the spoon too full, and if I'm not careful I'm going to spill water right over the sides. Can you see the water slopping around on the top of the spoon, kids?" Houston is my straight man.

"That's affirmative, 11." "O.K. Well, as I said, I was going to show you, but I'm afraid I filled it too full and it's going to spill over the sides. I'll tell you what. I'll just turn this one over and get rid of the water and start all over again. O.K.?" "O.K.," says Houston. I slowly turn the spoon "upside down," but, of course, the water stays right in it. "As you can see, up here we don't know where 'over' is. One 'up' is as good as another. That really is water, though. I'll show you." Carefully I put the upside-down spoon in my mouth and swallow the water. God, what a ham! Then I go on to demonstrate the operation of our water gun, trying to squirt a half ounce into my mouth from a distance. "It's sort of messy. I haven't been at this very long. It's sort of the same system that the Spaniards used to drink out of wineskins at bullfights, only I think this is even more fun." The hell I do. "Well, be seeing you, kids." I am reminded of the New York radio man who used to read the Sunday funnies to the kids, and who inadvertently kept his microphone keyed for one last accidental sentence. "Well, that ought to hold the little bastards for another week." But I restrain myself, and am rewarded by a "Thank you from all us kids in the world."

We wind up the program by showing the earth out our window. The last time we tried this, Charlie Duke couldn't tell the difference between earth and moon, on his black-and-white TV set in Mission Control, but for us there is a dramatic difference: the moon is a dull homogeneous disk, while the earth is a delightful blue-and-white half circle, growing larger and glistening brighter and more full of promise as each hour goes by.

As this quiet day approaches an end, we pack up the TV camera and query Houston about the weather and how things are going on the home front. The weather is lousy in Houston, which is normal, but seems to be holding its own in our recovery area in the mid-Pacific, which is what counts. The gals are all off at some party, we are told. Good. I expect they have had a rougher couple of days than we have had, and I hope Pat is unwinding now. We are all in good shape and are trying to stay that way. We are even exercising periodically, trying to keep our hearts from getting too lazy in this benign, zero-G environment. It's Buzz's turn now, and Charlie Duke lets us know how alert Houston is. "Buzz, you brought the surgeon right out of his chair. We see you exercising . . . we've got your heartbeat way up."

They don't have anything better to do, according to me. "Say, the old White Team's really got a busy one tonight, huh?" Charlie concurs. "Oh boy, we're really booming along here with all this activity. Can barely believe it." "What are you doing, sitting around with your feet up on the console, drinking coffee?" He laughs. "You must have your X-ray eyes up. You sure can see a long way." Mission Control is a big room. Traditionally, during busy flight phases it is jam-packed with workers at their consoles "in the trench," and onlookers in the glassed-in viewing room up at the rear. Not now. I estimate "two people in the viewing room and that's more than is in the trench." Charlie takes a count. "We've got eight in the viewing room, and let's see, about six in the trench right now. And this is the highlight of the day. Buzz's exercise for the surgeon." That's how I like space flight to be, slow and easy and no excitement, please. Old *Columbia* is ticking along like a fine Swiss watch.

The next day is our eighth in space; since we will be landing early the ninth morning, we have just over one day left as we awaken. My digital clock says it is 168:03—168 hours and three minutes after lift-off from Cape Kennedy. Re-entry is scheduled for about 195 hours. We idle the morning away by chatting with Houston about the weather and the chinch bugs which are eating my yard, and we receive such gems of news as the fact that some family has named its newborn daughter Module in honor (?) of us. We joke about filling Apollo 12 full of spaghetti to sustain Al Bean on his upcoming lunar voyage, and we note that it seems to be getting slightly chilly inside *Columbia*. I even get to poke a bit of fun at Cliff Charlesworth, the flight director for whom I worked on Apollo 8. He's head of the White Team still, but Bruce Mc-Candless has taken my job as CAPCOM. "How's old White,

Bruce? Did he ever let you go get a cup of coffee when we were over on the back side?" Bruce is a diplomat. "Oh, things have been going pretty smoothly down here. He's really not that hard to get along with." "He must be mellowing . . . he always used to make me sit at the console through the back-side passes, just for training." Charlesworth would never grab the microphone himself, that would be violating protocol, but he's not going to let his mouthpiece remain silent. "Well, the word we have here is, that was because whenever you came back, you had to be retrained," says McCandless, with a smirk in his voice. "Touché," I grunt, defeated.

The conversation changes to technical matters, slightly ridiculous ones at that. "Do you have Change Lima for your entry operations check list dated 23 July? Over." Bruce has to be kidding! This *is* the twenty-third of July, and we took off the sixteenth; and how the hell could they change our check list anyway, after we took off? Neil perks up his ears. "I'm not sure we hung around long enough to pick that one up . . . How can you make changes after lift-off?" I pipe up, "You sure you don't mean June?" "Negative." So Bruce reads up an involved procedure in order to, as he puts it, "reduce the oxygen pressure in your manifold and eliminate the oxygen bleed flow through the potable and waste water tanks during descent. Over." "O.K.," I groan. Jesus, they could fly this command module one hundred times, and on the one hundred and first, some engineer somewhere would come up with a procedure that he was convinced should replace all its predecessors. When I have finished writing it all down, Houston rewards us with an interesting tidbit. "You are now 95,970 miles out from earth. Over." "Right in our own back yard," says Neil, and Buzz asks for details. "Trying to come downhill a little bit now. What's our velocity?" "Your velocity is 5,991 feet per second," says Bruce, and then, after a pause, "And you are indeed coming downhill."

Lo and behold, it's TV time again, but for once we have anticipated it and I have given at least an hour's thought to what I might say. So have Neil and Buzz. This will be our last TV show, and although we didn't want the bloody thing on board in the first place, this time we are going to try and make the goddamn tube work *for* us; we are using this last opportunity to MAKE OUR STATEMENT! We don't rehearse, but five minutes of talk confirms the fact that each of us has a different bone to pick and that we needn't worry about overlap in our messages. Neil starts it off, like a good MC. "Good evening. This is the commander of Apollo 11. A hundred years ago, Jules Verne wrote a book about a voyage to the moon. His spaceship, *Columbiad*, took off from Florida and landed in the Pacific Ocean after completing a trip to the moon. It seems appropriate to share with you some of the reflections of the crew as the modern-day *Columbia* completes its rendezvous with the planet earth and the same Pacific Ocean tomorrow. First, Mike Collins."

"Roger," I say, getting off to a bad start. "This trip of ours to the moon may have looked, to you, simple or easy. I'd like to assure you *that* has not been the case. The Saturn V rocket which put us in orbit is an incredibly complicated piece of machinery, every piece of which worked flawlessly. This computer up above my head has a thirty-eight-thousand-word vocabulary, each word of which has been carefully chosen to be of the utmost value to us, the crew." I am wedged into the lower equipment bay to hold myself steady, and I have written a few notes on a small card (like a teleprompter) and stuck it on one of the couch struts just to the right of the camera lens into which I am peering. I continue, in what I hope is smooth, TV commentator fashion. "The switch which I have in my hand now has over three hundred counterparts in the command module alone, this one single switch design. In addition to that, there are myriads of circuit breakers, levers, rods, and other associated controls. The SPS engine, our large rocket engine on the aft end of our service module, must have performed flawlessly or we would have been stranded in lunar orbit. The parachutes up above my head must work perfectly tomorrow or we will plummet into the ocean. We have always had confidence that all this equipment will work, and work properly, and we continue to have confidence that it will do so for the remainder of the flight. All this is possible only through the blood, sweat, and tears of a number of people." Jesus, I am worked up now and there is a big lump in my throat. My brain must have gotten soft as a grape in the past three years I have spent lying on my back in white rooms. I just hope I can get through the rest of it. "First, the American workmen who put these pieces of machinery together in the factory. Second, the painstaking work done by the various test teams during the assembly and retest after assembly. And finally, the people at the Manned Spacecraft Center, both in management, in mission planning, in flight control, and last but not least, in crew training. This operation is somewhat like the periscope of a submarine. All you see is the three of us, but beneath the surface are thousands and thousands of others, and to all those, I would like to say, Thank you very much." Shit, I left out the simulator people and the Cape and a few others, but it's too late now. Maybe they will consider themselves included under the umbrella of some of those other categories.

Now it's Buzz's turn. "Good evening. I'd like to discuss with you a few of the more symbolic aspects of the flight of our mission. Apollo 11. As we've been discussing the events that have taken place in the past two or three days here on board our spacecraft, we've come to the conclusion that this has been far more than three men on a voyage to the moon; more, still, than the efforts of a government and industry team; more, even, than the efforts of one nation. We feel that this stands as a symbol of the insatiable curiosity of all mankind to explore the unknown. Neil's statement the other day upon first setting foot on the surface of the moon, 'This is a small step for a man, but a great leap for mankind,' I believe, sums up these feelings very nicely. We accepted the challenge of going to the moon; the acceptance of this challenge was inevitable. The relative ease with which we carried out our mission, I believe, is a tribute to the timeliness of that acceptance. Today, I feel we're really fully capable of accepting expanded roles in the exploration of space. In retrospect, we have all been particularly pleased with the call signs that we very laboriously chose for our spacecraft, Columbia and Eagle. We've been particularly pleased with the emblem of our flight, depicting the U.S. eagle bringing the universal symbol of peace from the earth, from the planet earth to the moon, that symbol being the olive branch. It was our overall crew choice to deposit a replica of this symbol on the moon. Personally, in reflecting on the events of the past several days, a verse from Psalms comes to mind to me. 'When I consider the heavens, the work of Thy fingers, the moon and the stars, which Thou hast ordained; What is man that Thou art mindful of him?' "

Neil winds it up. "The responsibility for this flight lies first with history and with the giants of science who have preceded this effort; next with the American people, who have, through their will, indicated their desire; next with four administrations and their Congresses, for implementing that will; and then, with the agency and industry teams that built our spacecraft, the Saturn, the *Columbia*, the *Eagle*, and the little EMU, the space suit and back pack that was our small spacecraft out on the lunar surface. We would like to give special thanks to all those Americans who built the spacecraft; who did the construction, design, the tests, and put their—their hearts and all their abilities into those craft. To those people, tonight, we give a special thank you, and to all the other people that are listening and watching tonight, God bless you. Good night from Apollo 11."

A good note upon which to end this day, except that Houston tells me that one of my biomedical sensors has worked loose, and that the weather in our recovery area is full of thunderstorms and they are going to move our splash point 215 miles to the east in a search for clearer skies and calmer seas. I pay no attention to the sensor ("Well, I promise to let you know if I stop breathing"), but the bit about the weather is bad news, because I have not had time in my training to practice this type of entry. As long as the computer keeps working, well O.K., another 215 miles doesn't make much difference, but if I have to take over and fly it manually, that is different. To get that extra range will require a great soaring arc after our initial penetration into the atmosphere, and the difference between soaring an extra 215 miles and skipping out of the atmosphere altogether is slim indeed. I may just land short, whether or not it disappoints or embarrasses President Nixon on the *Hornet*. That will be something to sort out tomorrow, but best say adios to the White Team now and get some sleep. "Thank you very much, Bruce. It's been a pleasure working with you." "Have a nice trip down," Bruce responds, not mentioning the lousy 215 extra miles. As I drift off to sleep, my mind is uneasily full of speeches and entry simulations. The speeches were O.K., I believe, mine more superficial than the others but more heartfelt, I thought; anyway, not bad for three engineering test pilots. I wonder what my favorite crew of philosopher-priest-poet might have said, or their back-up crew of psychiatrist-philologist-philistine? Or has a philistine somehow slipped onto the real crew?

I also wonder at the lack of communication, or at least the strange form of communication, among the three of us. We seem to speak only of technical minutiae, yet I could have predicted that Neil would emphasize the history of science in his remarks and that Buzz would focus on the symbolism of the flight. I have come to know them by osmosis or some other mysterious transfer process, rather than by direct communication. Especially Neil, who never transmits anything but the surface layer, and that very sparingly. I like him, but I don't know what to make of him, or how to get to know him better. He doesn't seem willing to meet anyone halfway. In frustration, I am tempted to turn to astrology; despite the fact I think it is a hoax, more than once I have found myself thinking of Neil as a "typical" Leo—proud and distant ruler of jungle and spacecraft. Buzz, on the other hand, is more approachable; in fact, for reasons I cannot fully explain, it is *me* that seems to be trying to keep *him* at arm's length. I have the feeling that he would probe me for weaknesses, and that makes me uncomfortable. I wonder whether all Apollo crews are satisfied to remain amiable strangers? It is certainly true that astronauts begin as competitors, rather than compatriots, but we three no longer have to worry about being picked over others for the *big flight*; it has happened, and we should be able to lower a few barriers now. The next moon-landing crew (Pete Conrad, Dick Gordon, Alan Bean) seems to me to have relaxed with each other; at least they project a spirit of cheery camaraderie that suggests a much closer personal relationship than that which exists between Armstrong and Collins and Aldrin. A closer relationship, while certainly not necessary for the safe or happy completion of a space flight, would seem more "normal" to me. Even as a self-acknowledged loner, I feel a bit freakish about our tendency as a crew to transfer only essential information, rather than thoughts or feelings. But, let's not knock it: everything is going extraordinarily well, and we've got to keep it that way. It's just that if we really carry this thing off, we are going to find good use for all the mutual support we can muster in the hectic months to come. As the members of the first lunar landing crew, we will be besieged, and we three alone will be able (if anyone is able) to control what happens to us. We have already noticed in these past six months a perceptible shift in attitude by those around us, including the other astronauts. Not hostility exactly, but a touch of envy, and an unwillingness to listen to any of our problems or frustrations. "Are you kidding? You guys got it made!" seems to be the standard reply, and perhaps that is true. I certainly don't blame them for feeling that way, right or wrong, but it does create a special relationship among the three of us. I don't have any idea what Neil and Buzz intend to do after the flight (or me for that matter), but whatever it is, we should support each other, and I'm not sure we have yet built the basis for that support.

When morning of our ninth day rolls around, we are ready for entry in more ways than one. We can almost "feel" the increasing pull of gravity as we race downhill toward that forty-mile-wide entry corridor, bracing ourselves for the deceleration we welcome. Then, too, we smell, and we are more than ready to vacate this grubby command module without a backward glance, despite all it has done for us. All three of us might be considered fastidious men.

I am probably the sloppiest, and I consider myself neat. Neil *is* neat. Buzz is not only neat, but almost a dandy. When he is decked out in full civilian regalia, he is a sight to behold. On more than one occasion I have seen him and his newly pressed iridescent suit festooned with more totems than one would believe possible. Once I counted ten. To begin with the lapel, there was the small "X" pin designating membership in the Society of Experimental Test Pilots. Second, a white handkerchief protruded from the breast pocket, onto which was pinned the small gold astronaut emblem, a rocket trailing three tails escaping through an orbital ring. Third, a tie clasp fashioned from a pair of Air Force pilot wings. Fourth, dangling on a chain suspended from the wings, what appeared to be a Phi Beta Kappa key. Fifth and sixth, a miniature Gemini spacecraft on one cuff and an Apollo on the other. Seventh, a gold wedding ring inscribed "Joan and Buzz." Eighth, a West Point class ring. Ninth, a Masonic emblem embedded in the stone of the ring, and finally, his mother's wedding ring on his other hand. It would surprise me not one whit to see him add earrings to express some new commitment, using one of the few portions of his anatomy that remains unadorned.

Meanwhile, this fastidious man and his two equally picky companions must slop about in crowded and ever more smelly surroundings. The right side of the lower equipment bay, wherein are located old launch day urine bags, discarded washcloths, and worse, is now a place to be avoided. The drinking water is laced with hydrogen bubbles (a consequence of fuel-cell technology which demonstrates that H_2 and O join imperfectly to form H_2O). These bubbles produce gross flatulence in the lower bowel, resulting in a not-so-subtle and pervasive aroma which reminds me of a mixture of wet dog and marsh gas. It seems degrading for *Columbia* to reach this smelly-old-man stage; I prefer to think of it as a ripe mango ready to fall from the tree—but in any event, it's time to get it on the ground, to end the indignity of having bowel movements in public, and the sooner the better. Things which were fun a couple of days ago, like shaving in weightlessness, now seem to be a nuisance. There is no sink in which to wash the hair, or even enough water to rinse the face. Instead, one must wipe the lathered face dry with tissues, and then itch and scratch for a couple of hours to get rid of the last few whiskers. Even brushing the teeth seems a bother, with no place to spit out the toothpaste, which is supposed to be swallowed. All of these things are small potatoes, but they do produce—in me at least—an overlay of irritation and impatience.

However, there are so many interacting considerations in space flight that one cannot hurry, for even the simplest tasks can become complicated in ways that a reasonable planner would never foresee. For example, a platform alignment must not follow a urine dump. What? Well, the tiny globules of urine shine like stars in the sunlight and disguise the real stars in the sextant, and so one must wait ten minutes or so until they slowly disperse. Although I am sure Neil and Buzz have a similar list of frustrations, they don't confide in me, or me in them; we have enough technical trivia to fill our quota of words as the days go by, and no one seems inclined to share anything more than that. It has been that way through our pre-flight training, and I expect the same pattern will continue after the flight.

Outside *Columbia*, all seems to be going well. The crescent earth glistens larger and more invitingly as each hour goes by, and Houston reports fairly benign conditions in the landing area, namely scattered clouds at two thousand feet, eighteen-knot winds, and waves three to six feet high. In deference to the waves, all three of us take an anti-motion-sickness pill, for we know that the keel-less command module wallows disastrously on the water, and there is no point in throwing up if we can avoid it. Houston, realizing that I am poorly trained to fly the extra-range procedure required to skip over the stormy area to where the *Hornet* is now parked, reads me the back-up procedures as a refresher. I repeat them: "O.K., it sounds straightforward enough. Understand constant G back-up procedure, lift vector up until MAX G and then lift vector down; then modulate bank angle until G dot equals zero. Maintain G dot equals zero until subcircular, then roll 45 degrees and hold until drogue time. Over." "O.K., that's mighty fine, Mike." Mighty fine, my ass; if I have to fly it that way, I guarantee I won't come down in sight of the boat. On the other hand, the computer has been working flawlessly for the past eight days, and I trust old Colossus IIA to drop us onto the carrier deck. In similar fashion, we trust *Columbia* to the extent that we have our pressure suits packed away under the couches, so that we will enter the atmosphere in our shirt sleeves. As Buzz recites the lengthy check list and marks off each item, I remark comfortably, "This entry time line is my kind of time line—nice and slow." We have time to check everything at least three times, and we do, for as our journey draws to a close, the consequence of a screw-up looms as large as life, literally.

Jim Lovell must still be haunting Mission Control, trying to keep us full of good cheer, for he comes on the radio now. "This is Jim, Mike. Back-up crew is still standing by. I just want to remind you that the most difficult part of your mission is going to be after recovery." All I can think to say is, "Keep the mice healthy." What I really would like to say is, "Stop the world, I want to get on," but somehow it doesn't sound right. Houston would reply, "Say again, Apollo 11?" and I would be embarrassed. The only other thing I feel compelled to say is that "It's a pleasure to be able to waste gas." I have been hoarding maneuvering fuel for the entire flight, but there is no point in *that* any more, so now I make my maneuvers crisply, with wanton disregard for the extra fuel I may be hosing out. For the first time in two space flights, I find myself in the delicious condition of being fat on fuel—fat as a forty-pound robin. Houston responds to my good mood with a final "Have a good trip and—remember to come in BEF." Wise asses! BEF means blunt end forward; in other words, heat shield forward, which is the only way *to* come in without burning up. Just as on Gemini, we will be looking behind us as we put our blunt heat shield forward, allowing it to absorb the tremendous frictional heat of the atmosphere by a process of controlled erosion, called ablation. Although our velocity is still well below that of a Gemini orbit, we are really starting to pick up speed now. We are scheduled to hit the entry corridor at minus 6.48 degrees (just .02 degree shallower than if we were on a perfect trajectory) at a speed of 36,194 feet per second (as opposed to Gemini's mere twenty-five thousand feet per second). Our computer is set to steer us to 169 degrees west longitude and 13 degrees north latitude, empty sea about eighty miles southwest of the Hawaiian Islands.

In the meantime, we have just one major chore left to perform, namely, jettisoning our faithful storehouse, the service module. "It's been a champ," I say as it departs. Truly, it has served us well. At launch we weighed six *million* pounds, now what's left of *Columbia* weighs in at a mere eleven *thousand*. The first and second stages of the Saturn, which fell into the sea, account for most of the weight. Then there is the empty third stage of the Saturn, in orbit around the sun, and the LM descent stage at Tranquility Base, *and* the ascent stage of *Eagle* abandoned in lunar orbit, and the service module, which is about to burn up, entering the atmosphere without a heat shield, *and* finally there is us—just eleven thousand pounds of us. I have assumed the position now, BEF, and about all we can do is wait and worry that we have overlooked something. With the service module gone, *Columbia* flies like a fighter, reacting vigorously as my right hand keeps it pointed properly. One thruster—the left yaw thruster—doesn't seem to be working properly, but I can make do without it. All three of us are very quiet, as we lie in our couches listening to the whir of the machinery for one final hour. Because of electrical inverters, hydraulic pumps, and other equipment, *Columbia* has been a noisy if happy home for us, but now it seems quieter somehow—more like a chapel than a machine shop.

Barely perceptible at first, the deceleration is heralded by a light which comes on at .05 G, and by the beginnings of a spectacular visual display out the window. We are in the center of a sheath of protoplasm, trailing a comet's tail of ionized particles and ablative material as we plummet obliquely through the upper atmosphere. The ultimate blackness of space is gone (for me, forever), to be replaced by a wispy tunnel of colors: subtle lavenders, light blue-greens, little touches of violet, surrounding a central core of orange-yellow, and surrounded, in turn, by the black void. It is very much like a Gemini re-entry, except that I don't see any tiny pieces of heat shield whizzing by or any other pinpoints of light, only the diffused colors. The radio is silent now and will be for four minutes or so, since the ionized sheath prevents the passage of radio signals. I have one eye on my navigational instruments and the other out the window. Buzz is taking pictures of our tail, and Neil is keeping me informed of the numbers that the computer is belching out in a steady stream. We are right on target, and I breathe an extra sigh of relief when our velocity drops below satellite speed. In other words, we don't have enough energy left to skip back out of the atmosphere; we have been recaptured by the earth's gravity, and we are guaranteed to come down somewhere on its surface. The G forces are squashing us now, and it is slightly uncomfortable to have such a heavy hand on the chest, but 6.5 Gs is not bad, even after eight days of no G at all, and it doesn't last long. The view out the window is breathtaking. The intensity of illumination has increased dramatically, flooding the cockpit with white light of a startling purity. Our fiery trail has expanded to the extent that its edges can no longer be seen. Instead, we seem to be in the center of a gigantic electric light bulb, a million watts' worth at least, flooding the entire Pacific basin with light. Our fireball must be spectacular as seen from the pre-dawn murk below, but the rainbow hues are ours alone, too subtle to penetrate the thick lower atmosphere.

When it is time for our two drogue parachutes to deploy, they do so, and Buzz announces the fact. I have been busy with my instruments and have not noticed the two sixteen footers go, but now I can see them flailing about out the window, stabilizing us enough to allow safe inflation of the three mains. A small jerk, and there they are! God, they are a sight to behold, huge orange-and-white blobs, each eighty feet in diameter, bundled together in a reassuring triad. We can survive a water landing with only two good ones, but three looks oh so much better!

I have bet Neil a beer that when we hit the water *Columbia* will remain upright and will not topple over into what we call the Stable II position. Actually, the line up should be Buzz and I vs. Neil, because extremely quick action will be required by Buzz and me to prevent the eighteen-knot winds from dragging the parachutes sideways and pulling us over in the process. At the instant of touchdown, Buzz must push in a circuit breaker below his right elbow, and then I must throw a switch to jettison the parachutes. Naturally, we don't want to fool with this electrical circuit until we are actually on the water. As we descend, Buzz and I discuss it, and his finger is delicately poised on the appropriate button as the instant of contact approaches. Splat! Like a ton of bricks we hit, and Buzz's hand is jerked away from the circuit-breaker panel. By the time he regains his bearings and finds the right circuit breaker again, it is too late. I flip my switch, but I can feel us going over as I do so. Blast! Lost again, and now we are trapped in here for an extra ten minutes or so while we pump up small air bags on our sunken nose, bags which, when full, will change our center of gravity enough to heave us back upright. In the meantime, we are in a topsy-turvy little world. Not only has gravity returned with its unaccustomed heaviness but it is pulling us in the wrong direction. We are hanging by our straps, with our couches behind and above us, and the main instrument panel is down below instead of up over our heads. Thank God, we don't have pressure suits on, for their bulk would make it next to impossible to move around, and their super insulation would give us heat stroke, as our air-conditioning system is no longer effective. We finally right ourselves, after an unpleasant few minutes, and prepare to get out and join the swimmers, who are encircling *Columbia* with a steadying "flotation collar," which is lashed to a life raft. We each take another motion-sickness pill, not because we feel sick, but because of the dreadful implications of getting sick later inside the biologi-

cal isolation garment, either strangling on our own vomit or breaking the germ barrier. When we are all set, we open the side hatch
briefly, and Lieutenant Clancy Hatleberg, who has been trained as
our decontamination assistant, throws three BIGs in to us. We
close the hatch again and set about putting them on.

I don mine in the lower equipment bay; it is my first attempt
to stand upright against gravity. I feel slightly swollen in the feet
and lower legs, and just a tiny bit lightheaded, but even inside this
heaving compartment, buffeted by eighteen-knot winds, I feel
good. Surprisingly, I feel better now than I did right after Gemini
10. I don't know whether it is because of the restful last two days,
or the fact that I have not had a pressure suit on, or that the
command module is a lot more commodious than the Gemini;
more likely, it is a combination of these and other factors. Whatever the reason, eight days of Apollo weightlessness seem to have
left me in better condition than did those three frantic Gemini
days. There is plenty of yakking on the radio now. After looking
around at Neil and Buzz clambering into their BIGs, I report,
"This is Apollo 11. Tell everybody, take your sweet time. We're
doing just fine in here. It's not as stable as the *Hornet*, but all
right . . ." No point in the swimmers or helicopter pilots getting
worked up into a frenzy; we didn't travel half a million miles to
have an accident on the water.

When we are zipped up inside our BIGs, we reopen the hatch,
inflate our water wings, and jump into the raft alongside. After a
tussle with the hatch, Clancy and I finally get it locked, and then
we all set about spraying each other with disinfectant and wiping
each other down with cloths saturated with an iodine solution and
with sodium hypochlorite. No lunar bugs can survive such a bath,
we like to believe, although what prevents them from escaping into
the sea I don't really know. Gentle waves break occasionally over
the side of the raft and keep our BIGs wet while we are scrubbing
each other. The sea water feels cool, and I really regret I cannot
scoop up a handful of it and pour it over my face. Even through my
misty visor, the water looks wonderfully inviting, purplish blue and

sparkling clean. It's good to be back, and what better place to end our journey than on the ocean. The colors are right—the blue and white of the whitecaps matching the blue and white of water and clouds that define the earth at a great distance.

The helicopter is overhead now and impatiently scoops us up one by one, in a little wire mesh basket on the end of a dangling cable. The pilot must be a good fisherman, for when he feels a tug on his line, up he yanks—ready or not. On board the chopper we have a chance to try our legs; I practice walking in the cramped space, and even do a few deep knee bends. Inside the BIG I can communicate with no one, but through my streaked visor I can see Buzz doing the same and Neil watching quizzically. Inside the BIG it is also deucedly hot, with no ventilation system, and I am thankful we have but a short ride to the boat. By the time we land on the *Hornet's* flight deck, I am really burning up, and if the end were not in sight, I would rip this faceplate off for a breath of cool air, germs or no germs. We are lowered, helicopter and all, below deck by a gigantic elevator. Then the helicopter door slides open and I stumble out, to the accompaniment of a brass band. Inside the goddamn BIG, I'm not only roasting now but almost blinded by a fogged visor. Fortunately, someone has painted lines on the hangar deck and I just follow them, waving at a vaguely perceived crowd of sailors off my starboard bow. Sure enough, the lines lead to a low door and I pop inside it—and find myself in the MQF.

The mobile quarantine facility is simply a glorified trailer without wheels, modified with filters, water tanks, etc., to provide a biological barrier between those inside and the 3 billion outside. Five of us are inside, the other two being Bill Carpentier and John Hirasaki. Bill is a flight surgeon and John a mechanical engineer; between them they will tend our happy home for the next three days. They are both good choices—quiet, flexible, unobtrusive. John will be busy as hell with housekeeping, cooking, and other MQF chores, but his main responsibility will be toward *Columbia* and its cargo, which will be hauled from the ocean and connected to the MQF by a plastic tunnel. John has been trained to render harmless all *Columbia*'s propulsion and other systems, and to remove the rock boxes and film, sterilize their containers, and pass them to the outside world through an airlock. Bill, in addition to giving us daily physical exams, is the bartender, not that we can't pour our own. More important, Bill is a delightful person with an offbeat sense of humor ("A flight surgeon is someone to hold your hand until the doctor gets there.") Bill is a Canadian and John is of Japanese extraction; they form an unlikely but most welcome combination to ease us back onto the earth.

Right now a shower tops my list of things to do, and we take turns (not too long, for how much hot water is a house trailer likely to have?), and then I shave. The three of us are really spiffy now, brushed and combed, clean in blue flying suits adorned with NASA and Apollo 11 patches, plus a button saying HORNET PLUS THREE, the motto of the carrier's crew for this particular cruise. It feels great to be clean for a change. We are looking for something to do, and it's not long coming. We are summoned to the end of the trailer, and parting the curtains, we see that the hangar deck has been arranged for some sort of ceremony—the first of many, I would guess. Lest there be any doubt, the band plays "Ruffles and Flourishes," and in marches none other than President Nixon, looking very fit and relaxed as he stands by a microphone just outside our window. In a jovial mood he jokes a bit with us, noting that Einstein's theory of relativity decrees that we have aged a bit less than our fellow earthlings during the days we have been speeding through space, and that he has invited our wives—and us—to dinner. He wants to know if we got seasick, and Neil reassures him we did not, speaking over a hand-held microphone as the three of us crouch awkwardly at our low window. Then the President says, ". . . This is the greatest week in the history of the world since the Creation . . ." and I lose track of the rest of his speech. As the Navy chaplain leads us in prayer, my mind wanders—greatest week! . . . Jesus Christ! . . . Jesus Christ? . . . greatest week? . . . Best pay attention to what's going on here, but nothing is. It's over, and we pull our blinds and return to our own strange little ecosystem. We have transferred from a miniature planet, *Columbia*, to a slightly larger compartment, which is intentionally isolated from earth. We have not quite returned yet, despite the President's being outside our window.

Time passes quickly and agreeably inside the MQF. We have more ceremonies to "attend" through the glass, as we trade extravagant compliments with the ship's captain and crew. Then, although it's still early afternoon in the Pacific, our Houston watches say it is past toddy time and we declare the bar officially open. A short glass of ice, a guzzle-guzzle of gin, a splash of vermouth. God, it's nice to be back! Never more am I going flying in NASA's sky. Gemini 10, Apollo 11—between the two I must have had twenty lifetimes' worth of opportunities to destroy myself; yet miraculously, here I sit, drinking my martini and pleased as hell with myself. An old fighter pilot friend of mine used to say after every flight, "Well, I cheated death again." The first couple of times I heard him, I was shocked at his sarcasm, his brashness, his cynicism, or his honesty, but what the hell, why not say it! Twice is enough; I am going to spend the rest of my days catching fish, and raising dogs and children, and sitting around on a patio drinking gin and talking to my wife. Another one? Yes, thank you, Bill, I believe I will, a big one. And John has the steaks on? Grand!

We are steaming for Pearl Harbor, where we will be transferred via a flatbed truck to a jet cargo airplane, flown to Houston, and put into the lunar receiving laboratory. The quarantine period is twenty-one days, starting at the time of our possible "infection," i.e. lunar landing day. Let's see, with a little over three days in *Columbia* and almost three in the MQF, that leaves only two weeks in the LRL. It will take us two weeks to write all our post-flight reports anyway, so as long as the mice stay healthy, our post-flight regime will not be too much different from that of previous flights.

While we are in the MQF, we could get our report writing organized, but we do not. I help John find everything he needs from *Columbia*, and we unload for our own use all the books (flight plans, check lists, etc.) we will need to refresh our memories during the debriefings and report writing. Sitting around our communal table the second afternoon with a pile of LM books, I notice that gunmetal-gray flecks from the books are dirtying the top of the table. Casually I wipe them off onto the floor with the flat of my hand. Bill Carpentier looks at me aghast: moon dust into the scuppers! I also get a gin-rummy game started with Neil, for gin rummy always shortens the hours of waiting. Waiting for what? Pearl Harbor, or LRL, or the rest of our lives? I don't know, that's something I will have to sort out later; I suppose one can sit contentedly on one's patio only after a good day's labor, and I don't have the vaguest idea what sort of labor that might be. Perhaps NASA might like to fly men to Mars, and perhaps I could help them plan it somehow. In the meantime, best focus on the business at hand, if there is any. For one thing, it doesn't seem right to abandon *Columbia* without a backward glance. Our presence in it should be marked somehow. I am not normally emotional about machines, and I consider graffiti the exclusive province of morons in train stations. Despite all that, however, I feel a powerful urge to *write* on *Columbia* somehow. Finally, on the second evening, I climb back on board its charred carcass, and on the wall of the lower equipment bay, just above the sextant mount, I write: "Spacecraft 107—alias Apollo 11—alias *Columbia*. The best ship to come down the line. God Bless Her. Michael Collins, CMP."

Pearl Harbor is wild. The sun shines brightly as we are hoisted about in our great aluminum coffin, and there are people everywhere. On our flatbed truck we are paraded slowly through the streets from dockside to Hickam Air Force Base, with stops along the way for an official welcome by the governor and unofficial delays demanded by the throngs. The whole population of Honolulu must be out there. We wave frantically, ogle the honeys, and are eternally grateful that the glass protects us from autograph seekers. Then finally, arms worn out and smiles frozen on our faces, we are swallowed into the darkened interior of a C-141 jet transport and are off for Ellington Air Force Base. Flights aboard cargo airplanes are nearly always dull, and being in a box inside the cargo compartment doesn't help matters any. Fortunately, it is non-stop and shouldn't take much over six hours, near as I can figure. A good time to sleep.

It's the middle of the night at Ellington, but that doesn't seem to have prevented half the city of Houston from turning out. They wait patiently as we are tugged out of the airplane—part way out at least. It seems there is something wrong with our truck. We have the sophistication to fly half a million miles with flawless precision, but we can't wrestle a box onto a truck. Finally it is done, and we are towed around into position next to the reviewing stand. Mayor Louis Welch welcomes us, as do a host of NASA officials, and then, blinking in the relentless glare of the TV lights, our wives are thrust forward. Pat and I speak briefly on a red telephone. "Welcome home!" she sings out. "You look great. I can't talk on these damn phones with the buttons you have to push. Can you hear me?" I nod yes, and she continues, telling me the things I want to hear. Then our time is up, and with a jerk we get underway, out through the gate and down the highway toward the LRL. Again, like Hawaii, there is bedlam in our path, except this time it is the middle of the night and not as easy to see out. We see enough to assure us we really are home, however, as familiar pizza parlors, filling stations, and old friends' faces parade past our window. Then we pull through the gates of the Manned Spacecraft Center and are backed up against the warehouse door in the side of the LRL. As soon as a germ-proof barrier is sealed between us and the LRL, the door is opened, and we are free to investigate our next home. We have returned to Houston, but not yet to the world.

Our little band is growing, however: from three to five to fifteen. We have professional cooks and housekeepers now, and even a PR man is locked in with us. Before the flight, NASA was worried about a reporter somehow crashing through the glass and joining us in quarantine, whence he would issue a stream of exclusive communiqués. Perhaps putting John Macleish in with us will dissuade such efforts, and it certainly will allow much better press coverage of our daily routine, but it pleases us not one whit. We complained before the flight to Deke about this plan, and Deke agreed with us, but apparently he has lost the battle. It's not that we don't like John, he's good company and a gentleman; it's just that it would certainly be nice to be "off duty" in here. Let's face it, there's not much trouble we can get into behind locked doors, but it sure would be nice to know that anything we might try would go unreported. Other than this minor complaint, the place is great—a palace. Instead of three of us locked in a wedge less than thirteen feet across the base, or five of us in a rectangle 9 by 35, in this LRL we must have at least twenty rooms, including bedrooms for each of us and a huge open area used as lounge, library, and dining room.

We tell the technical story of Apollo 11 over and over again. Locked behind glass panels, we conduct all-day debriefing sessions for astronauts who will man the crews that follow ours, for management, for the systems engineers, for the scientists, for the doctors, for the simulator people, for the photography analysts. When we are not talking, we are writing, preparing our pilots' report wherein, like good test pilots, we describe how we flew the flight and give our recommendations for changes. Considering the complexity of our voyage, we have amazingly little to complain about, due primarily, I think, to the spadework done by Apollos 7, 8, 9, and 10. My principal beef concerns the rendezvous procedures, in my opinion unduly complicated for the solo man. Hitting the computer buttons 850 times, not to mention being tied to the sextant, may be all well and good as long as everything aboard the command module is working properly, but it gives one precious little time to cope with problems as they arise, or even to sit back and analyze trends as the two vehicles arc around the moon toward the point of their future coupling.

As the days and briefing sessions go by, the moon out the LRL window seems to shrink, while the earth regains some of its original flatness. This is a good halfway house. It allows us one last serene opportunity to examine what we have done and to cough up each detail. Neil and Buzz, for example, explain to fascinated scientists their "flicker flashes," streaks of light they have seen inside the darkened command module. I did not notice any in flight, but now I can make my imagination work either for or against the idea and conjure up a darkness which is either absolute or penetrated occasionally by a tiny white streak crossing my retina. Unable to make an honest contribution, I stay out of the conversation. On the terrestrial side, we learn about Senator Ted Kennedy and Chappaquiddick, news which Mission Control had not seen fit to send us. We also have the excitement generated by the arrival of an attractive young woman, a lab technician "contaminated" by contact with a lunar sample, who joins our group—occupying the bedroom adjacent to mine no less.

Then there is the flood of telegrams, newspapers, and letters from all over the world, which consider Apollo 11 from a different point of view. As I pore over them in the evening, especially the newspaper editorials, I am impressed by how well they are written, but depressed by how far off the mark their essays are. Unlike Gemini 10, which ended the instant we hit the water, this flight will never truly be over in my lifetime (such is the distinct impression I get from the newspapers), but I may wish that it were. While most of the papers are extravagant in their praise (". . . an enterprise which has produced its own benefits for the human spirit . . ." as the Montreal *Star* puts it), there are disturbing countercurrents, such as that expressed by the Stockholm *Expressen:* "The moonshot . . . was imposing. But it also gives a horrible feeling to think that the U.S.A. can handle tremendous technical problems with such ease while it is considerably more difficult to cope with those of a complicated social, political, and human nature." The *Philadelphia Inquirer* asks a good question: "Will this magnificent accomplishment serve as inspiration . . . or will the inspiration be abandoned before the veiled censure of those who seem to suggest the solution of all human dilemmas lies in turning away from space to other priorities?" *The Washington Post* quotes Harvard University biochemist Dr. George Wald, who said about his students, "I am afraid that they see in this an exercise of the old and well-entrenched, an exercise in great wealth and power, heavy with military and political overtones. I am afraid that they feel a little more trapped; a little more disillusioned, a little more desperate." In many countries, the news of the landing is apparently simply being discounted as more American propaganda; and in many places the notion of man on the moon violates religious taboos, and the news causes great debate and even, such as in Mogadiscio, Somalia, fist fights in the streets. Trapped students and a violated moon? I am shocked to think that what we have done causes anyone to feel "disillusioned," much less "desperate," or that we have caused serious theological upheavals by visiting a sacred place. At least I can understand the latter, but the former defeats me completely. I hope it's not a widespread attitude; I simply cannot believe that it is.

Then there are the invitations and good-will messages that are pouring in from all directions. "Rub-a-dub-dub, three men in a tub, won't you join us?" begins one; and another, inviting me to a rodeo, says, "Anyone that can ride that whatever-it-is the way you did has gotta be a darn good cowboy." One short telegram is signed "Baudouin King of the Belgians," and Nelson Rockefeller invites us to New York for a ticker-tape parade. We also learn that Duke Ellington is playing his new composition "Moon Maiden" in the Rainbow Room and we are welcome there. Ditto the Steel Pier in Atlantic City, which offers the three of us $100,000 for a one-week stint. Another offers to name a hybrid orchid after me, and I sign a release authorizing a race horse to be called Michael Collins. May he orbit the track at unheard-of velocities, even in the mud. There are congratulatory messages from the Montgomery Police Department, the Catholic Daughters of America, the American Fighter Pilots Association, the Pope, the Peace Corps . . . on and on it goes. There are honorary memberships in a host of organizations, my favorite being the Camel Drivers Radio Club of Kabul,

Afghanistan. The most unusual telegram reads: ". . . Thirty-five years ago in our home, on July 21, 1934, we heard firsthand from our father of the adventures of Buck Rogers in his first flight to the moon. Father, Phil Nowland, originated the comic strip 'Buck Rogers' on that date. Walking on the moon was commonplace to us at Maple Avenue in Bala-Cynwyd, Pennsylvania, but it was a thrill and exciting to all of us to watch your flight into space, Buck Rogers finally coming to life. Our Best Wishes, the children of Phillip Nowland."

Most impressive of all is the letter from Charles Lindbergh, written on Pan American Flight 841 from Honolulu to Manila, and postmarked in Manila on July 28.

> *Dear Colonel Collins,*
>
> *My congratulations to you on your fascinating, extraordinary, and beautifully executed mission; and my sincere thanks for the part you took in issuing the invitation that permitted me to watch your Apollo 11 launching from the location assigned to Astronauts. (There would have been constant distractions for me in the area with VIPs, among whom I refuse to class myself—what a terrible designation!)*
>
> *I managed to intercept on television the critical portion of your mission during this orbit of my own around the world. Of course after you began orbiting the moon, television attention was concentrated on the actual landing and walk-out. I watched every minute of the walk-out, and certainly it was of indescribable interest. But it seems to me you had an experience of in some ways greater profundity—the hours you spent orbiting the moon alone, and with more time for contemplation.*
>
> *What a fantastic experience it must have been alone looking down on another celestial body, like a god of space! There is a quality of aloneness that those who have not experienced it cannot know—to be alone and then to return to one's fellow men once more. You have experienced an aloneness unknown to man before. I believe you will find that it lets you think*

*and sense with greater clarity. Sometime in the future, I would like
to listen to your own conclusions in this respect.*

*As for me, in some ways I felt closer to you in orbit than to
your fellow astronauts I watched walking on the surface of the
moon.*

*We are about to start the descent for Manila, and I must end
this letter.*

My admiration and my best wishes,

Charles A. Lindbergh

*Of course I feel sure that your sense of aloneness was regularly
broken into by Mission Control at Houston; but there must have
been intervals in between—I hope enough of them. In my flying,
years ago, I didn't have the problem of coping with radio communications.*

With mail like this, and healthy mice, the days race by.
Suddenly it is Sunday, August 10—and we are free to go. Our
cocoon has burst, for better or worse, and we are officially certified
as being fit to rejoin humanity—physically, at least. Carrying a file
of telegrams and the Lindbergh letter as my graduation present, I
emerge blinking into the Houston night, flashbulbs popping, and
get my first smell of the earth in nearly a month—warm and moist
and inviting and reassuring. I don't recall being sensitive to earth
smells before, but then perhaps my sensitivities have changed and I
will find the earth a different place from now on.

14

We shall not cease from exploration and the end of all our exploring will be to arrive where we started and know the place for the first time.

—T. S. Eliot*

Four and a half years after Neil and Buzz touched the face of another planet, I look back on the event with a mixture of pride, incredulity, and smugness. I am an optimist (I could not have flown in space were I not), and I am optimistic about the present and future state of the world and Mike Collins, but I would have to admit that things have not turned out to be the fairyland that I imagined the night I was released from the LRL. One must work and cope with the world on pretty much the same terms as before, although some alterations are possible. In my own case, I have

* From "Little Gidding" in *Four Quartets*, copyright 1943, by T. S. Eliot, copyright 1971, by Esmé Valerie Eliot. Reprinted by permission of Harcourt Brace Jovanovich Inc.

found a post-astronaut job to my liking, as has Neil. I am presently director of the Smithsonian's National Air and Space Museum in Washington, and Neil is professor of engineering at the University of Cincinnati. Buzz, on the other hand, has had a more difficult time of it, suffering bouts of depression severe enough to require hospitalization; presently retired from an abortive attempt to pick up the threads of his Air Force career, Buzz lives in a suburb of Los Angeles. Being an astronaut is a tough act to follow, as all three of us have discovered.

After six fascinating years in Houston as an astronaut, I was prompted to leave by several things. First, and most basic, I just didn't feel I could go back to the bottom of Deke Slayton's ladder and work my way up again, to be assigned as commander of one of the later lunar landing crews. It would have taken another two years, I thought (actually, as it turned out, if I had been assigned to Apollo 17, it would have taken well over three years), and I was simply not willing to spend that many days in simulators and nights in motel rooms instead of with my family. If I were leaving Deke shorthanded, or if he could have promised to get me airborne in six months (which, of course, he could not and would not), it might have been a different story. As it was, Deke had enough astronauts to fly thirty missions to the moon. Second, I wanted to leave Houston and move to Washington. Pat had never really liked Houston, and for many years I had wanted to get back to Washington, which is as close to a home town as I will ever have. My mother, one sister, and my only brother live there, as do a number of old friends from my high-school days. Furthermore, it is a vibrant, dynamic city, and I would much prefer my children to grow up there, in good schools, rather than in the hinterlands. Third, and more to the point, I was offered a job in Washington by the Secretary of State, and personally urged to take it by the President of the United States.

The first inkling I had of the job was when the three of us came to Washington in mid-September to address a joint session of Congress. Tom Paine, the NASA administrator, asked me to come see him the next day and indicated that Secretary of State William P. Rogers wanted to know if I might be interested in becoming Assistant Secretary for Public Affairs, with the mandate of increasing youth involvement in foreign affairs. I immediately started ticking off the negatives. I was an active-duty Air Force colonel, with less than three years to go to become eligible for retirement; if I took this job, I would have to dump all that and resign my commission. Youth, eh? Campus involvement meant Vietnam, with a capital V, and I could just see myself, with my short hair and buttoned-down collar, lecturing the hairies on the necessity of pacifying the villages of the Mekong delta. Paine interrupted my mumbling and sensibly suggested that I not try to decide right then and there; he merely wanted to convey to Mr. Rogers any indication that I might want to discuss the matter further. O.K., I told him, why not talk about it? The next week an appointment was set up with Rogers, but at the last minute he was not available, so I discussed the matter with Elliot Richardson, who was Undersecretary. It was a disjointed and rambling interview. Obviously Richardson had no previous warning that Rogers was considering hiring an *astronaut*, for God's sakes, and he seemed at a loss for words (which I discovered later is very, very unusual for Elliot Richardson). I didn't know what to say either. We finally agreed that one never knew where one might end up, that the vicissitudes of public service made advanced planning impossible, perhaps even undesirable. I left intrigued by the man, his paneled office, and the friendly people who had guided me in and out of it. They didn't seem one bit stuffy . . . Still, public affairs, as a new career for me? Fortunately, it wasn't something I had to decide right then, as I was scheduled to depart four days later (along with Neil and Jan, Buzz and Joan, and Pat) on a round-the-world trip lasting from the end of September to the first week in November.

The trip was altogether too swift, covering twenty-eight cities in twenty-five countries in thirty-eight days. It was tiring, and we got fed up with hotel rooms and airports ("So nice to be in your lovely city" was always safe, if unsure of just which city it was).

But despite the fatigue and the repetitive nature of the ceremonies, it was the rarest of opportunities, to cram into slightly over a month's time visits with the Queen of England, Marshal Tito, the Pope, the Emperor of Japan, the Shah of Iran, Generalissimo Franco, Baudouin King of the Belgians, King Olaf of Norway, Queen Wilhelmina of the Netherlands, the King and Queen of Thailand, and dozens of Presidents, Prime Ministers, ambassadors, and lesser lights. The trip produced some disturbing symptoms in Buzz, causing him to withdraw into stony-faced silence from time to time, but aside from this (and the obvious distress it caused Joan), we finished in good health and in good spirits, pleased with ourselves in our new diplomatic role. Along the way, I had ample opportunity to view and judge the American foreign service people at our various embassies, and I found that—with the exception of a few horses' asses—they were able and dedicated, performing useful and complex chores with a great deal of grace and precision. It wouldn't be a bad group to join.

Our trip ended on the White House lawn, with President Nixon greeting us, along with various Cabinet members. Among them was Rogers, who asked if I would stop by his office. When I did, that same afternoon, he said that he had enjoyed the speech I had given at the joint session of Congress, and he wondered whether I had written it myself. When I assured him I had, every syllable of it, he seemed relieved and pleased, and went on to say he hoped I would join his staff and that, since he knew we were spending the last night of our trip at the White House, why didn't I ask the President what he thought of the idea? I couldn't say no to that, and besides, I was as intrigued by the process as I was by what it might mean to me.

That night over cocktails, the President heard us describe our trip, talking from a huge album of pictures we gave him, and he then guided the conversation into questions of our future plans and asked if any of us would enjoy being an ambassador. Both Neil and Buzz said they didn't think so, but I used the opportunity to explain Mr. Rogers's offer, and his request that I ask the

President's opinion of it. The President didn't hesitate more than a couple of seconds; then he picked up the omnipresent telephone and spoke quietly, "The Secretary of State, please." I sat there frozen. He began talking again, telling Mr. Rogers that he thought it was an excellent idea and that he was sure any problems about retirement could be worked out (they never were). I looked past him and there sat my wife, grinning. Thus the decision was made which brought me to Washington, to be Assistant Secretary of State for Public Affairs.

After dinner, the President excused himself to do some work, and Mrs. Nixon took us on a tour of the White House and part of the Executive Office Building next door, where the President keeps a second office. To me, Mrs. Nixon had always seemed a two-dimensional figure, cut from cardboard in a frozen pose and pasted onto the pages of the Sunday supplement. Instead, we all found her charming. She was a delightful, warm hostess who really tried to make us feel at home. Her tour, which she must have conducted ad nauseum, was carried off with unexpected enthusiasm and a beautiful informality. Having just returned from our trip, Pat and I were very sensitive to such matters, and in our minds we had already awarded our own International Gold Medal to the sophisticated, gorgeous, and disarmingly friendly Empress Farah of Iran. But Pat Nixon was at least as deserving, and she made our stay at the White House the real highlight of our round-the-world trip. I only wish all Americans could meet her under similar circumstances instead of via the media, because she suffers in the translation—not that they have treated her unfairly, I don't mean that—but her warmth must be experienced firsthand.

My experiences at the State Department would fill a separate book, but suffice it to say I did go to Berkeley and other places and talk to the hairies, and I did spend long hours in Washington flying a great mahogany desk. I found that the job was not my cup of tea (I'm simply not a PR man), despite the fact that I enjoyed the people in State, especially the jovial Mr. Rogers. The foreign service officer is the most maligned person in government, invariably depicted as a "striped-pants cookie pusher" rather than the intelligent, hard-working professional he generally is. But it is true that a lot of work gets done over two-hour ceremonial luncheons, and more than once, after such an occasion, I wobbled out like a stunned ox, vowing to change jobs before I acquired gout and a faintly British accent. There were some compensations, however, such as reaffirming the fact that I could do useful, non-flying work. Although I may be wrong, I honestly believe I left the Bureau of Public Affairs, which I headed, a more effective organization than I found it. I also learned a lot about official Washington, about how the federal bureaucracy moves and shakes, and about how to operate in this strange, semi-hysterical environment. It was a superlative introduction to my present job.

It had its lighter moments too. For instance, one of my responsibilities was to sign (by machine, of course) the jillions of letters which the department prepared to answer citizen inquiries, no matter what the subject matter. In the course of this, I gained some interesting pen pals, but my favorite was a lady in New York who simply returned all correspondence with a huge BULLSHIT stamped on each page. Who knows, historians may yet record this as the Bullshit Era.

After slightly over a year in this plush purgatory, I decided it was time to move on, while I could still leave *with* my shield rather than *on* it. So I moved over to the Smithsonian Institution, where I knew there was a vacancy as director of the National Air and Space Museum. My only regret is that I left State before I had an opportunity to attend a really high-level diplomatic confrontation and write the press communiqué describing it. Instead of the canned description, "a frank and productive interchange, in a cordial atmosphere," which covered all meetings of all types, I wanted—just once—to tell it like it was. "It was a useless and futile session. His Excellency was as stubborn, pigheaded, and recalcitrant as usual, and in fact I think the surly bastard was drunk."

At the Smithsonian I find myself in much calmer waters, but it is still an interesting and challenging job. The word "museum" has a musty, dusty connotation that I don't like, and museums can be pigeonholes for discarded objects and people, but there is no reason they have to be that way. In the case of the National Air and Space Museum, it cannot be that way, for we are preparing to open a major new museum on the Mall in Washington in time for the Bicentennial Celebration. This project has had to be shepherded through the various committees of Congress, as well as the Fine Arts and National Planning bureaucracies. At this stage, the new building is fully approved and funded, and is on schedule and below budget as its 1976 opening approaches. It will contain over 200,000 square feet of exhibit space divided into twenty-five halls, plus a planetarium, auditorium, research library, cafeteria, and parking garage. Finding the right people to research and execute the exhibits and programs to fill this space properly is challenging indeed, not to mention problems of funding and administering the entire complex. The new museum will cover every aspect of man's third-dimensional progress, from balloons through the space age, and will look at some of the human implications of these developments, as well as possibilities for the future. I will be very disappointed, and surprised, if it does not turn out to be the most exciting museum in the world. After it opens, I don't know what I will do, whether I will stay on, or perhaps I have been infected by a cosmic itch which will force me to keep moving.

I am also most interested in following what happens to my space compatriots, John Young as much as Neil or Buzz. John is still hanging in there, in the astronaut business in Houston, and is working on the space shuttle, which NASA hopes to orbit before the end of this decade. Obviously John has not found ten years (and four space flights) to be as burdensome as I found six (and two), but just as obviously he is in the minority, at least in this foursome. I have not discussed the matter with the other three, nor am I apt to, as all of us tend to communicate at a shallow level about technical things, and about events rather than ideas. We are all four loners, and as a result I am not as close to any of them as the flight experiences we have shared might indicate. John is the most uncommunicative (with Neil a *distant* second—no pun intended), and I don't have any idea what flying in space has meant, or will mean, to him. I suspect that his interest in solving engineering problems is all-consuming, and that he belongs right where he is as long as he is physically able to fly; then he should move into a supporting engineering job. As for me, I envy him his flying (of T-38s as well as spacecraft), but despite that, I am glad I left Houston when I did. Being an astronaut was the most interesting job I ever expect to have, but I wanted to leave before I became stale in it, and I could tell that after Apollo 11 I could not have prevented myself from sliding downhill, in terms of enthusiasm and concentration. Hence I find myself in the weird position of saying I'm glad I no longer have one of the most fascinating jobs in the world, but I think that is an honest summation.

Now for Buzz. Buzz tells me that he doesn't at all regret flying to the moon, that it has been a positive influence in his life, and that he's glad he did it, but Buzz has had serious psychiatric difficulties since the flight. Part of his problem, it seems to me, comes from his father, who, as nearly as I can determine, used to tell him that now that he has been to the moon he has the world at his feet, if only he would assert himself. Unfortunately, that is not true, and even more unfortunately, Buzz has not the delicate touch required for self-promotion. He tends to be heavy-handed, and I expect that Air Force generals don't like to listen to Air Force colonels outline their career expectations in specific detail. Shut up and get back in line, Colonel, and I must say I agree with that point of view. On the other hand, I empathize very strongly with Buzz and his problems. If the shrinks are to be believed, passing the age of forty is a crossroads in itself, and when a confusion of future goals is superimposed on that, no wonder Buzz has had difficulties. He is a very intense, goal-oriented individual, accustomed to winning big—and not losing at all. It reminds me of ex-heavyweight champ Joe Frazier, who was hospitalized shortly after winning the title. Was he injured in the fight? No, according to his doctor, "His problem appears strictly to be the result of tension and pressure as a result of his responsibilities and plans that became a little too much for him." Joe's problem manifested itself in soaring blood pressure, while Buzz's surfaced in episodes of incapacitating depression. West Point, Air Force, M.I.T., NASA: Buzz has had at least four strict taskmasters, with Dad egging him on all the way. Suddenly it is over: Buzz the pilot fish has been thrown clear of the shark Apollo and is swimming around, desperately looking for another streamlined creature of speed and danger to attach himself to. There aren't any, Buzz, but I earnestly pray you will find some placid whale an adequate substitute for the shark.

If Buzz is a case study in postpartum depression, then why isn't Neil similarly afflicted? Well, for openers, Neil is altogether different. He *is* number one, not two, and he could put together a lifetime (and not a bad life style) of simply being that. Fortunately, Neil is not a PR man and seeks a life of altogether different fabric than that of a huckster for NASA or for anyone else, even for Neil. He has great balance and perspective, he is an intuitive historian, and he has had an abiding interest in teaching. "I know I could make a million dollars in personal appearances on the outside," Neil was reported to have said not long ago, "but I just want to be a university professor and be permitted to do my research." I believe that, and I think Neil was wise in pointing his life in that direction. I have friends in high-pressure Washington who criticize Neil for dropping out, for failing to get out and "sell the program," but they don't understand Neil or his problem or perhaps even "the program." As the first human being to step onto another planet, Neil will be a unique person for all his life, and I think he must ration himself in a tasteful and sensible fashion. He has done that at Cincinnati, where he lives in a figurative castle surrounded by a moat full of dragons. When he chooses, which is not often, he can lower the drawbridge and sally forth; but more important, when he chooses, he can retreat with honor and dignity and direct his attention to teaching courses in dynamics and flight testing, about which he knows a great deal. Neil knows what he is doing, and he is doing it well.

Which brings us to Mike Collins. Fortunately, I have been a poor student all my life, and my parents, concealing their disappointment, seldom pushed me. Consequently, the pressures in my various jobs have been mostly self-induced. This fact, plus a native laziness, has saved me from experiencing the full flavor of the Buzz syndrome, although I share with him a mild melancholy about future possibilities, for it seems to me that the list of exciting things to do here on earth has diminished greatly in the wake of the lunar landings. I just can't get excited about things the way I could before Apollo 11; I seem gripped by an earthly ennui which I don't relish, but which I seem powerless to prevent. I am more impervious to minor problems now; when two of my people come to me red-faced and huffing over some petty dispute, I feel like telling them, "Well, the earth continues to turn on its axis, undisturbed by your problem; take your cue from it, and work it out by yourselves; it really doesn't amount to much anyway." Of course, I don't say that, because it obviously means a great deal to them, but not many things seem quite as vital to me any more. My threshold of measuring what is important has been raised; it takes a lot more to make me nervous or to make me blow my cool. There are fewer good jobs around. Part of this stems from having received a number of terrestrial honors, and part from having been privileged to see the earth from a great distance. The three of us have been entertained by kings and queens; we have received the Collier Trophy, the Harmon Trophy, the Hubbard Medal, the Presidential Medal of Freedom, etc., etc.; we have addressed a joint session of Congress and a hundred lesser audiences. Through it all, the earth continues to turn on its axis; I can see it doing so, and I am less impressed by my own disturbance to that serene motion, or by that of my fellow man.

That doesn't mean I have acquired a complete guru-like detachment—far from it. I still get irritated, and I still express irrational annoyance. For example, if one more fat cigar smoker blows smoke in my face and yells at me, "What was it really like up there?" I think I may bury my fist in his flabby gut; I have *had it*, with the same question over and over again. It is the curse of flying in space, this business of answering the same question one million times. There should be a statute of limitations on it. A close second is autographing things, especially "To Cousin Esmeralda, and Baby Jane, and all the boys at the fire station, and put down the date, and sign your name so we can read it." Jesus, lady, I don't do that well by my banker. Kids collecting autographs I can understand, and I don't think I have ever barked at one of them, but pushy adults are something else. A perceptive PR man told me one time, shortly after Apollo 11 that there was a special place in hell reserved for autograph seekers; I didn't know what he meant then, but I do now.

On the other hand, I would have to admit that notoriety is not an unmixed catastrophe. While I want desperately to be able to take my kids to the zoo unnoticed, I don't mind one bit crawling on board the jetliner, clutching my tourist ticket, and having the little honey recognize me and whisk me up into first class, there to feed and nurture me and protect me from the nasty autograph seekers. I don't mind *that* breach of anonymity one little bit. Nor am I above telling the police officer when I am stopped for speeding, "Gosh, Officer, I guess I just haven't learned to slow down since I got back from the moon." So I suppose I am a phony who wants it both ways. It's not unlike the starlet who wants to be remembered for her acting ability rather than her great body, except that in my case it's not a question of "for what" but "by whom."

Howard Muson in *The New York Times Magazine* describes the returning astronaut as "the wandering hero back among his tribe, after stealing the sacred fire and grappling with terrifying demons, condemned to ask tough questions." It is an apt description, although I never felt I stole anybody's fire (I merely carried it through the sky), and the demons with whom I have grappled have generally been dressed by Lord and Taylor's. But tough questions, that is right on target. The toughest, of course, is, Was it worth it all? It certainly was to me personally, which obviously makes me suspect as an objective witness to the expenditure of $25 billion of the taxpayers' money. Besides, I frankly gave little thought to the financial end of the space program, just as I never considered what percentage of the GNP Flash Gordon might reasonably twit away exploring the caverns of Mongo. Furthermore, I think it is premature to make a judgment on the manned space program and its possible value to mankind. We simply don't know yet what it may mean to us. Fred Hoyle, the British astronomer, suggested as early as 1948 that the first picture of the whole earth would unleash a flood of new ideas. Supporters of the space program say that the current interest in ecology is one such byproduct. Opponents say baloney, and besides you don't have to fly men in space to obtain such photographs. Supporters say that the moon rocks will ultimately allow scientists to determine the origin of the solar system, and that this is a fundamental bit of knowledge. Opponents say the moon, a sterile and lifeless rock pile, cannot possibly be of much interest. Opponents say that it is immoral to support exploration while our cities rot. Supporters say it cannot be examined on an either/or basis, that our national economic and budgetary process prevents the simple transfer of funds from one project to another, that without a space program our cities would still rot. Opponents say we must focus our priorities on our planet and its ills; supporters say the solution to these problems depends on studying the earth from orbit. That kind of technology got us into this mess, say the detractors. Only advanced technology can get us out of it, say the space buffs.

As the argument ebbs and flows, I think a couple of points are worth making. First, Apollo 11 was perceived by most Americans as being an end, rather than a beginning, and I think that is a dreadful mistake. Frequently, NASA's PR department is blamed for this, but I don't think NASA could have prevented it. It's simply the American way, to view a televised spectacular and think of it as the Super Bowl. Then followed confusion and a trace of irritation. Why was the Super Bowl being played over and over again? When Apollo 13 blew its oxygen tanks and the networks interrupted their regular programs to report it, they received angry phone calls for their efforts. By Apollo 16, *The Washington Post's* inimitable Nicholas von Hoffman headlined it TWO KLUTZES ON THE MOON. The magic was gone, despite all the talk of more sophisticated scientific instruments and more extensive explorations using the lunar rover. The only thing that could have titillated the public and gotten the momentum back was a manned expedition to Mars, and that seemed impractical even to the program's most ardent supporters. So the focus returned from moon to earth, and the orbiting Skylab was loaded with cameras to record in as much detail as possible the ravages to our planet, as a first step in repairing the damage. So that's where we are.

The second point which I think is worth examining is the American tendency to be faddish, supporting "in" things blindly and just as blindly rejecting the passé. This line of thought tends to force evaluation on an either/or basis. Either cure cancer or fly in space. Either clean up the environment or fly in space. There doesn't seem to be a willingness to do all these things on a balanced basis. Furthermore, I think Americans are grossly misinformed about how much of their tax dollar goes into space. Not long ago I initiated an informal polling of the first one hundred adults who walked into one of the Smithsonian buildings. Among the dozen questions was: Which agency consumes more of the tax dollar, NASA or HEW? I expected a few suckers to say NASA, but I was flabbergasted to find that fifty said HEW, forty NASA, and ten declined to guess. The last time I looked up the figures, the HEW budget was $75 billion ($93 billion if you count social security), while NASA's was $3 billion. Now that's either a very quiet $75 billion, or a very noisy $3 billion. My personal opinion is that canceling the space program would have little effect on HEW's projects even if the money could be directly transferred. Recently I noticed an advertisement by my old friend North American Aviation, maker of the command module, under its new corporate name of Rockwell International. The ad doesn't even mention the fact that Rockwell has the contract for the space shuttle, but talks only about business jets, trucks, and looms for manufacturing women's clothing. They know which side their bread is buttered on.

Perhaps my own involvement in Apollo 11 causes me to put too much emphasis on it, but it seems to me that the space program was too popular before the first lunar landing and not popular enough after it. The pendulum swings back and forth across the truth, and it will take some time for future historians to put Apollo in its proper perspective. How long, I don't know, but technology moves with ever-increasing swiftness and compresses all our lives whether we like it or not. Consider Effie Corum Pelton, who watched Halley's comet from her California desert window in 1910, when the Wrights were just getting started. The great Muroc Dry Lake, at what is now Edwards Air Force Base, was given the Corum name, although spelled backward, and became the kindergarten for astronauts, as the machines assigned there flew ever faster and higher. A fellow optimist, Mrs. Pelton summed it up very nicely, I think: "Halley's comet truly was an omen of the future of Muroc. We made the transition from horse and buggy to the moon!" That's a long way to go in a lifetime, and makes one want to sit back and assimilate the past, rather than pressing on into the future. Yet our nation's strength has always derived from our youthful pioneers, from the first colonization of the East Coast. NASA was born in the Space Act of 1958; yet space exploration in this country really began when Columbus landed. Some people were never content to huddle in protective little clumps along the East Coast but pushed westward as boldly as circumstances permitted. When horizontal exploration met its limits, it was time to try the vertical, and thus has it been since, ever higher and faster.

Now we have the capability to leave the planet, and I think we should give careful consideration to taking that option. Man has always gone where he has been able to go, it is a basic satisfaction of his inquisitive nature, and I think we all lose a little bit if we choose to turn our backs on further exploration. Exploration produces a mood in people, a widening of interest, a stimulation of the thought process, and I hate to see it wither. Our universe should be explored by microscope and by telescope, but I don't believe the argument that less emphasis on one will cause a more powerful focus on the other. When man fails to push himself to the possible limits of his universe in a physical sense, I think it causes a mental slackening as well, and we are all the poorer for it. Space is the only physical frontier we have left, and I believe its continued exploration will produce real, if unpredictable, benefits to all of us who remain behind on this planet. That one cannot spell out in any detail what these benefits will be does not contradict or deny their existence. We all know examples of unexpected by-products of research (such as penicillin) and of man's serendipity in a new environment, but my favorite story goes back to 1783, when Ben Franklin witnessed the first public launching, in Paris, of a hydrogen balloon. Of what possible use was this new invention, a skeptic asked Franklin, and he replied, "Of what use is a newborn babe?" I suppose one might ask why, if I am so strongly in favor of continued exploration, I didn't practice what I now preach and stay in the space program. I must admit I do feel a twinge of guilt occasionally, especially when someone says to me, as one intelligent lady did, "Oh, you sold out at the top of the market, didn't you?" I guess what I am saying is that it is important that mankind explores, but it doesn't matter a damn whether Mike Collins or any other one individual does. Mike Collins has had his turn and now will gladly watch from the sidelines.

So far I have emphasized the negative aftereffects of having flown in space, but they really are minor compared to the deep sense of satisfaction I feel. My eyes have been privileged to see extraordinary scenes; their recollection and meaning far outweigh the aggravation of answering the same old questions and signing the same old envelopes. A more serious problem is how to prevent the rest of my life from being an anticlimax, but I am confident that I am working that out. Although I certainly don't expect ever again to do anything as dramatic as carrying the fire of space flight,

I do expect always to have interesting projects in work, so that I can devote my energies to planning the future rather than ruminating over the past.

Meanwhile, my life style has not changed as much as I would have guessed. One would think that doing something as bizarre as flying to the moon would result in equally outlandish behavior here, such as spending one week on safari in Kenya and the next diving for abalone off Catalina Island. However, it just doesn't work that way. By and large, one brings back from the moon the same limitations of pocketbook, imagination, and taste that one took on the trip, and one is stuck with them. I spend more time at school board meetings than I do at bullfights, and more in supermarkets than in nightclubs. There is money hanging around, but it is tainted PR money, trading great piles of greenbacks for tiny bits of soul, in an undetermined but unsatisfactory ratio. For example, I have been offered $50,000 to do beer commercials, and I love beer, but somehow it seems a grubby thing to do. It would involve bringing TV crews to my place of work and into my home, and would flash my family across the screen in a cheap context. It would be the same invasion of privacy that I abhorred during the *Life* contract days, but for the greater glory of a particular brand of beer rather than for the space program. However, when called upon, I have felt an obligation to do TV commercials for free, for savings bonds, despite the fact that I cringe at my friends' descriptions of them (mercifully, I have never seen one). So I remain flat broke, and I rationalize it by saying that it is a good thing, that it forces me to focus on the future, and that it keeps me lean and hungry in my outlook.

I am also planning to leave a lot of things undone. Part of life's mystery depends on future possibilities, and mystery is an elusive quality which evaporates when sampled frequently, to be followed by boredom. For example, catching various types of fish is on my list of good things to do, but I would be reluctant to rush into it, even if I had the time. I want no part of destroying fishing as a mysterious sport. In similar fashion, I want to walk over the recently constructed bridge connecting Europe and Asia, but I don't think I will. I think having it to walk over is better than having walked over it. On the other hand, I'll have to admit I don't feel at all that way about the moon. I very much appreciate having the good fortune to have been one of the very few men privileged to fly that far away from home, and I feel more contented since the trip than I did before.

I also find that my two space flights have changed my perception of the earth. Of course, Apollo 11 also changed my perception of the moon, but I don't regard that as being nearly as important. There seem to be two moons now, the one I see in my back yard and the one I remember from up close. Intellectually, I know they are one and the same, but emotionally they are separate entities. The small moon, the one I have known all my life, remains unchanged, except that I now know it is three days away. The new one, the big one, I remember primarily for its vivid contrast with the earth. I really didn't appreciate the first planet until I saw the second one. The moon is so scarred, so desolate, so monotonous, that I cannot recall its tortured surface without thinking of the infinite variety the delightful planet earth offers: misty waterfalls, pine forests, rose gardens, blues and greens and reds and whites that are missing entirely on the gray-tan moon. When I was about ten years old, I had a lot of dental work done, and I found it painful and traumatic. Desperate for some way to relieve my anxiety, I discovered that I really could remove my mind from my body (for short periods, and not during the time of deepest drilling) by imagining that I had flown up near the ceiling and was looking down upon the dentist and his prostrate victim, who was my age but not me. It was him that was hurting, not me. In similar fashion, I can now lift my mind out into space and look back at a midget earth. I can see it hanging there, surrounded by blackness, turning slowly in the relentless sunlight. When things are not going well here on earth, when a toothache of one sort or another begins, I can gain a bit of solace and perspective by making this mental trip. After I learned to fly airplanes, but before I flew in space, I kept a secret place in the cumulus clouds to which I could retreat, but the view from a greater distance is even more supportive.

I really believe that if the political leaders of the world could see their planet from a distance of, let's say, 100,000 miles, their outlook could be fundamentally changed. That all-important border would be invisible, that noisy argument suddenly silenced. The tiny globe would continue to turn, serenely ignoring its subdivisions, presenting a unified façade that would cry out for unified understanding, for homogeneous treatment. The earth *must* become as it appears: blue and white, not capitalist or Communist; blue and white, not rich or poor; blue and white, not envious or envied. I am not a naïve man. I don't believe that a glance from 100,000 miles out would cause a Prime Minister to scurry back to his parliament with a disarmament plan, but I do think it would plant a seed that ultimately could grow into such concrete action. Just because borders are invisible from space doesn't mean that they are not real—they are, and I like them. I feel just as thankful today that I live in the United States of America as I did before flying in space, and I have no desire for this country to merge into a United States of the World. What I am saying, however, is that all countries must begin thinking of solutions to their problems which benefit the entire globe, not simply their own national interests. The smoke from the Saar Valley may pollute half a dozen other countries, depending on the direction of the wind. We all *know* that, but it must be *seen* to make an indelible impression, to produce an emotional impact that makes one argue for long-term virtues at the expense of short-term gains. I think the view from 100,000 miles could be invaluable in getting people together to work out joint solutions, by causing them to realize that the planet we share unites us in a way far more basic and far more important than differences in skin color or religion or economic system. The pity of it is that so far the view from 100,000 miles has been the exclusive property of a handful of test pilots, rather than the world leaders who need this new perspective, or the poets who might communicate it to them. Of course, we could always pass out whole-earth photographs and have everyone study them, and if there is any truth in my 100,000-mile premise, the results should be the same. Unfortunately, it doesn't work that way. Seeing the earth on an 8-by-10-inch piece of paper, or ringed by the plastic border of a television screen, is not only not the same as the real view but even worse—it is a pseudo-sight that denies the reality of the matter. ("Oh, I've seen everything those astronauts have seen.") To actually be 100,000 miles out, to look out four windows and find nothing but black infinity, to finally locate the blue-and-white golf ball in the fifth window, to know how fortunate we are to be able to return to it—all these things are required, in addition to merely gauging its size and color. While the proliferation of photos constantly reminds us of the earth's dimensions, the photos deceive us as well, for they transfer the emphasis from the *one* earth to the multiplicity of reproduced images. There is but one earth, tiny and fragile, and one must get 100,000 miles away from it to appreciate fully one's good fortune in living on it.

If I could use only one word to describe the earth as seen from the moon, I would ignore both its size and color and search for a more elemental quality, that of fragility. The earth appears "fragile," above all else. I don't know why, but it does. As we walk its surface, it seems solid and substantial enough, almost infinite as it extends flatly in all directions. But from space there is no hint of ruggedness to it; smooth as a billiard ball, it seems delicately poised in its circular journey around the sun, and above all it seems fragile. Once this concept of apparent earthly fragility is introduced, one questions whether it is real or imagined, and that leads inexorably to an examination of its surface. There we find things are very fragile indeed. Is the sea water clean enough to pour over your head, or is there a glaze of oil on its surface? Is the sky blue and the cloud white, or are both obscured by yellow-brown airborne filth? Is the riverbank a delight or an obscenity? The difference between a blue-and-white planet and a black-and-brown one is delicate indeed.

We rush about like busy ants, bringing immense quantities of subsurface solids, liquids, and gases up from their hiding places, and converting them into quickly discarded solids, liquids, and waste gases which lie on or just above the surface as unholy evidence of our collective insanity. The entropy of the planet, its unavailable energy, is increasing at an alarming rate; the burning of fossil fuels is an irreversible process and can only be slowed down. At the same time, the sun shines on us whether we like it or not; yet we are making but feeble efforts to focus this energy for our use. For that matter, the sun's energy, which is produced by converting hydrogen into helium, can probably be duplicated by creating our own little thermonuclear reactors here on earth, if we put every effort into the attempt. These problems and their solutions are becoming increasingly well known and I'm sure would have been recognized had there been no space program. Anyone who has viewed our planet from afar can only cry out in pain at the knowledge that the pristine blue and whiteness he can still close his eyes and see is an illusion masking an ever more senseless ugliness below. The beauty of the planet from 100,000 miles should be a goal for all of us, to help in our struggle to make it as it appears to be.

Seeing the earth from a distance has changed my perception of the solar system as well. Ever since Copernicus' theory (that the earth was a satellite of the sun, instead of vice versa) gained wide acceptance, men have considered it an irrefutable truth; yet I submit that we still cling emotionally to the pre-Copernican, or Ptolemaic, notion that the earth is the center of everything. The sun comes up at dawn and goes down at dusk, right? Or as the radio commercial describes sunset: "When the sun just goes away from the sky . . ." Baloney. The sun doesn't rise or fall: it doesn't move, it just sits there, and we rotate in front of it. Dawn means that we are rotating around into sight of it, while dusk means we have turned another 180 degrees and are being carried into the shadow zone. The sun never "goes away from the sky." It's still there sharing the same sky with us; it's simply that there is a chunk of opaque earth between us and the sun which prevents our seeing it. Everyone knows that, but I really *see* it now. No longer do I drive down a highway and wish the blinding sun would set; instead I wish we could speed up our rotation a bit and swing around into the shadows more quickly. I do not have to force myself to call this image to mind; it is there, and occasionally, I use it for other things, although admittedly I have to stretch a bit. "What a pretty day" makes me think that it's always a pretty day somewhere; if not here, then we just happen to be standing in the wrong place. "My watch is fast" translates into no, it's not, it's just that you should be standing farther to the east.

I'm not completely cured, though; I still say "in" this world, instead of "on" it, and I still think of the North Pole as being "up" and the South Pole as "down," which is absurd. Give a hundred people a picture of the earth, identify the North Pole for them, and a hundred will hold the photo with the North Pole toward their head and the South Pole toward their feet. Of course, what they are really doing, if they are standing up, is pointing the South Pole at the center of the earth and, if they are standing at the equator, pointing the North Pole at some spot in the sky, which, as the earth turns, traces a circle intersecting the plane of the ecliptic at 23½ degrees. Now why people persist in this foolishness I don't know. In my living room I have a small framed photograph showing a thin crescent against a black background. Even though the colors are wrong, people always say, "Oh, the moon!"; but it is the earth. The earth isn't ever supposed to be a crescent, I suppose.

Finally, flying in space has changed my perception of myself. Outwardly, I seem to be the same person, and my habits are about the same. Oh, I seem to be spending money a bit more freely now, and I am inclined to put more energy into my family and less into my job, but basically I am the same guy. My wife confirms the fact. I didn't find God on the moon, nor has my life changed dramatically in any other basic way. But although I may feel I am the same person, I also feel that I am different from other people. I have been places and done things you simply would not believe, I feel like saying; I have dangled from a cord a hundred miles up; I have seen the earth eclipsed by the moon, and enjoyed it. I have seen the sun's true light, unfiltered by any planet's atmosphere. I have seen the ultimate black of infinity in a stillness undisturbed by any living thing. I have been pierced by cosmic rays on their endless journey from God's place to the limits of the universe, perhaps there to circle back on themselves and on my descendants. If Einstein's special theory is true, my travels have made me younger by a fraction of a second than if I had stayed always on the earth's surface. The molecules in my body are different, and will remain so until the seven-year biological cycle causes them to be replaced, one by one. Although I have no intention of spending the rest of my life looking backward, I do have this secret, this precious thing, that I will always carry with me. I have not been able to do these things because of any great talent I possess; rather, it has all been the roll of the dice, the same dice that cause the growth of cancer cells, or an aircraft ejection seat to work or not. In my life so far I have been very, very lucky. Even the bad things, like the surgeon's knife, have turned out to have fortunate consequences. These events confirm my native optimism, although I have seen too many promising young lives snuffed out not to know that it can happen to me. Any death seems premature, but I really believe my own will seem *less* premature, because of what I have been able to do. At what I hope is the midway point in my life (I am forty-three), my eyes have already been privileged to see more than most men see in all their years. It is perhaps a pity that my eyes have seen more than my brain has been able to assimilate or evaluate, but like the Druids at Stonehenge, I have attempted to bring order out of what I have observed, even if I have not understood it fully.

Unfortunately, my feelings cannot be conveyed by the clever arrangement of stone pillars. I am condemned to the use of words. I know, because after the flight of Gemini 10, I tried to use paint, and it was a total flop. My oceans filled the canvas with a muddy monotony; my clouds were anemic instead of ethereal; my spacecraft were lumpy and misshapen. The feeling was all wrong. After that, I didn't even attempt to paint, but I did try to convey in words the magic of carrying the fire, just as the god Apollo carried the sun across the sky in his chariot. For magic is assuredly there: changing urine particles into angels is magic; tumbling end over end with velvet smoothness is magic. Even some of the things which happen on the ground during a flight are magic, such as watching Apollo 8 carrying men away from the earth for the first time in history, an event in many ways more awe-inspiring than landing on the moon. For me, at that moment, Apollo 8 was an electronic beam in Mission Control. Surely that was magic, but how to express it?

> The moving line skims, sure and swift,
> Green as a snake across the wall.
> A linear lie of circular progress, it tells us nothing,
> Except that man must keep his sensors saturated.

Or wondering about the moon, before the flight of Apollo 11. The geologists did their best to take the magic out of it, but would the moon really be proof of the Law of Least Astonishment?

> Rocks without fossils,
> Wells without water.
> Will there be no surprises,
> When I am unpinned by gravity?

Once in flight, how to convey the awesome feeling of looking out four windows and finding nothing? Should I try the fifth window?

> The Cosmos is a social club,
> The Zenith a radio set.
> Cosmology is an unfamiliar word,
> But, Lord—out my window is real.

I could even write an "Ode to the Moon's Far Side," if I thought it would help explain things.

Cold stones jumbled in a heap.
Lifeless plains, sharing only the sun
With a verdant recollection I must keep,
Till I next see one:
One penny, one peony, one misty waterfall.
For me a choice—to hear a voice,
Or slip on by it all.

I don't know what magic the future may hold. I hope that in my lifetime man will visit Mars, for that would be magic; but it may not happen. In the meantime, there is magic aplenty for me here on earth. I find it in such things as the Guzman Medal, awarded to Armstrong, Aldrin, and Collins by the French Academy of Sciences in 1969. This medal had been waiting to be awarded since 1889. It went to the first persons "to find the means of communicating with a heavenly body—Mars excluded." The exception of Mars from the award was made by the founder, Mrs. Anna Emile Guzman, "because that planet appears to be sufficiently well known." Now that is magic.

But I don't want to end this book with poetry or magic. Tom Paine has called Apollo "a triumph of the squares," and I think he is right. Accordingly, I should like to end it with the language of one square, Mike Collins, addressing a group of squares. The date was September 16, 1969, and the place a joint session of Congress in Washington, my home town. For me, that date and these words marked the end of Apollo 11, and closed an extraordinary chapter in my life.

MR. SPEAKER, MR. PRESIDENT, MEMBERS OF CONGRESS, AND DISTIN-GUISHED GUESTS:

One of the many things I have very much enjoyed about working for the Space Agency and the Air Force is that they have always given me free rein—even to the extent of addressing this most august assemblage without coaching, without putting any words into my mouth. Therefore, my brief remarks are simply those of a free citizen, living in a free country, and expressing free thoughts which are purely my own.

Many years before there was a space program, my father had a favorite quotation: "He who would bring back the wealth of the Indies, must take the wealth of the Indies with him." This we have done. We have taken to the moon the wealth of this nation, the vision of its political leaders, the intelligence of its scientists, the dedication of its engineers, the careful craftsmanship of its workers, and the enthusiastic support of its people.

We have brought back rocks, and I think it's a fair trade. For just as the Rosetta Stone revealed the language of ancient Egypt, so may these rocks unlock the mystery of the origin of the moon, and indeed even of our earth and solar system.

During the flight of Apollo 11, in the constant sunlight between the earth and the moon, it was necessary for us to control the temperature of our spacecraft by a slow rotation, not unlike that of a chicken on a barbecue spit. As we turned, the earth and the moon alternately appeared in our windows. We had our choice. We could look toward the moon, toward Mars, toward our future in space, toward the New Indies, or we could look back toward the earth, our home, with the problems spawned over more than a millennium of human occupancy.

We looked both ways. We saw both, and I think that is what our nation must do. We can ignore neither the wealth of the Indies nor the realities of the immediate needs of our cities, our citizens, our civics.

We cannot launch our planetary probes from a springboard of poverty, discrimination, or unrest; but neither can we wait until each and every terrestrial problem has been solved. Such logic two hundred years ago would have prevented expansion westward past the Appalachian Mountains, for assuredly, the Eastern seaboard was beset by problems of great urgency then, as it is today.

Man has always gone where he has been able to go. It's that

simple. He will continue pushing back his frontier, no matter how far it may carry him from his homeland.

Someday in the not-too-distant future, when I listen to an earthling step out onto the surface of Mars or some other planet, as I heard Neil step out onto the surface of the moon, I hope I hear him say: "I come from the United States of America."

APPENDIX

PROJECT MERCURY

Flight	Date	Crew	Remarks
FREEDOM 7	May 5, 1961	Al Shepard	Suborbital; first American in space
LIBERTY BELL 7	July 21, 1961	Gus Grissom	Same type of flight as Shepard's
FRIENDSHIP 7	February 20, 1962	John Glenn	First American in orbit
AURORA 7	May 24, 1962	Scott Carpenter	Repeat of Glenn's flight
SIGMA 7	October 3, 1962	Wally Schirra	Six revolutions (double Glenn's three)
FAITH 7	May 15, 1963	Gordon Cooper	Long-duration flight (thirty-four hours)

PROJECT GEMINI

Flight	Date	Crew	Remarks
GEMINI 3	March 23, 1965	Gus Grissom (2) * John Young	Cautious first flight test, limited to three revolutions
GEMINI 4	June 3–7, 1965	Jim McDivitt Ed White	First American space walk
GEMINI 5	August 21–9, 1965	Gordon Cooper (2) Pete Conrad	Long duration flight (eight days)
GEMINI 6	December 15–16, 1965	Wally Schirra (2) Tom Stafford	First space rendezvous, using 7 as target
GEMINI 7	December 4–18, 1965	Frank Borman Jim Lovell	Long long-duration flight (fourteen days)

* Numbers (2, 3, 4) in parentheses after names indicate second, third, or fourth flight for that individual.

Flight	Date	Crew	Remarks
GEMINI 8	March 16, 1966	Neil Armstrong Dave Scott	First docking in space; cut short by thruster malfunction
GEMINI 9	June 3–6, 1966	Tom Stafford (2) Gene Cernan	Three different rendezvous; space walk by Cernan
GEMINI 10	July 18–21, 1966	John Young (2) Mike Collins	First use of Agena's power; rendezvous with second Agena; space walk
GEMINI 11	September 12–15, 1966	Pete Conrad (2) Dick Gordon	Rendezvous and docking; space walk; altitude record (850 miles)
GEMINI 12	November 11–15, 1966	Jim Lovell (2) Buzz Aldrin	Rendezvous and docking; space walk

PROJECT APOLLO

Flight	Date	Crew	Remarks
APOLLO 1	January 27, 1967	Gus Grissom (3) Ed White (2) Roger Chaffee	Pad 34, Cape Kennedy. Fire in the spacecraft killed all three
APOLLO 7	October 11–22, 1968	Wally Schirra (3) Donn Eisele Walt Cunningham	First test flight, but eleven-day duration instead of Gemini 3's three revs (increased confidence)
APOLLO 8	December 21–7, 1968	Frank Borman (2) Jim Lovell (3) Bill Anders	Bold second manned flight, taking CSM all the way to the moon on Christmas Eve
APOLLO 9 (CM–Gumdrop) (LM–Spider)	March 3–13, 1969	Jim McDivitt (2) Dave Scott (2) Rusty Schweickart	Back in earth orbit again, for the first flight of the LM/CSM combination

Flight	Date	Crew	Remarks
APOLLO 10 (CM–Charlie Brown) (LM–Snoopy)	May 18–26, 1969	Tom Stafford (3) John Young (3) Gene Cernan (2)	Dress rehearsal for the lunar landing, taking LM to within fifty thousand feet of landing site
APOLLO 11 (CM–Columbia) (LM–Eagle)	July 16–24, 1969	Neil Armstrong (2) Mike Collins (2) Buzz Aldrin (2)	First lunar landing at Tranquility Base, July 20, 1969
APOLLO 12 (CM–Yankee Clipper) (LM–Intrepid)	November 14–24, 1969	Pete Conrad (3) Dick Gordon (2) Al Bean	Pinpoint landing near Surveyor III site in Oceanus Procellarum
APOLLO 13 (CM–Odyssey) (LM–Aquarius)	April 11–17, 1970	Jim Lovell (4) Jack Swigert Fred Haise	CSM oxygen tanks blew, causing Apollo 13 to limp home (just barely) using the LM's oxygen and electricity
APOLLO 14 (CM–Kitty Hawk) (LM–Antares)	January 31–February 9, 1971	Al Shepard (2) Stu Roosa Ed Mitchell	Third lunar landing, in the Fra Mauro region. Shepard hits a golf ball, Mitchell tries ESP
APOLLO 15 (CM–Endeavor) (LM–Falcon)	July 26–August 7, 1971	Dave Scott (3) Al Worden Jim Irwin	First landing in the lunar mountains (Hadley-Apennine region). First use of Rover vehicle
APOLLO 16 (CM–Casper) (LM–Orion)	April 16–27, 1972	John Young (4) Ken Mattingly Charlie Duke	Landed in the Descartes region, thought to be an area of extensive volcanism
APOLLO 17 (CM–America) (LM–Challenger)	December 7–19, 1972	Gene Cernan (3) Ron Evans Jack Schmitt	Landed in the Taurus–Littrow region; Jack Schmitt first scientist (geologist) to fly

PROJECT SKYLAB

Flight	Date	Crew	Remarks
SKYLAB 2	May 25–June 22, 1973	Pete Conrad (4) Joe Kerwin Paul Weitz	Twenty-eight days in the earth orbiting laboratory. Crew installed improvised sun shield and released stuck solar panel
SKYLAB 3	July 28–September 25, 1973	Al Bean (2) Owen Garriott Jack Lousma	Fifty-nine days, continuing important earth and sun studies
SKYLAB 4	November 16, 1973–February 8, 1974	Jerry Carr Ed Gibson Bill Pogue	Record duration (eighty-four days) proves man can go to Mars

APOLLO-SOYUZ TEST PROJECT

Flight	Date	Crew	Remarks
APOLLO CSM	Scheduled for July 1975	Tom Stafford (4) Deke Slayton Vance Brand	Rendezvous, dock, and socialize with a Russian crew in earth orbit